INDEX TO REVIEWS,
SYMPOSIA VOLUMES AND MONOGRAPHS
IN ORGANIC CHEMISTRY

INDEX TO REVIEWS, SYMPOSIA VOLUMES AND MONOGRAPHS IN ORGANIC CHEMISTRY

For the Period 1963-1964

Compiled and Edited by

NORMAN KHARASCH

and

WALTER WOLF

University of Southern California
Los Angeles, California

SYMPOSIUM PUBLICATIONS DIVISION

PERGAMON PRESS

OXFORD · LONDON · EDINBURGH · NEW YORK
TORONTO · SYDNEY · PARIS · BRAUNSCHWEIG

Pergamon Press Ltd., Headington Hill Hall, Oxford
4 & 5 Fitzroy Square, London, W.1
Pergamon Press (Scotland) Ltd., 2 & 3 Teviot Place, Edinburgh 1
Pergamon Press Inc., 44–01 21st Street, Long Island City, New York 11101
Pergamon of Canada, Ltd., 6 Adelaide Street East, Toronto, Ontario
Pergamon Press (Aust.) Pty. Ltd., 20-22 Margaret Street, Sydney, New South Wales
Pergamon Press S.A.R.L., 24 rue des Écoles, Paris 5e
Vieweg & Sohn GmbH, Burgplatz 1, Braunschweig

Copyright © 1966
Pergamon Press Ltd.

First edition 1966

Library of Congress Catalog Card No. 61–18796

Printed in Great Britain by Ipswich Litho Press Limited
(3131/66).

FOREWORD

This third volume of AN INDEX TO REVIEWS, SYMPOSIA VOLUMES AND MONOGRAPHS IN ORGANIC CHEMISTRY is for the period 1963-1964. It covers works in the English, French and German languages, including also English translations of Russian studies.

The excellent response which has been given to the two preceding volumes of the INDEX has been encouraging. These earlier volumes covered the periods 1940-1960 and 1961-1962, and we trust that the present volume for 1963-1964 will also be warmly accepted by scientists all over the world.

The three volumes of the INDEX now cover a span of twenty-five years during which organic chemistry has had its most rapid development. Used together, they now represent a basic library tool with which all organic chemists, students and many other scientists will find it useful to be well acquainted.

As with the previous volumes, the 1963-1964 edition will allow research workers, teachers and students to locate quickly those current reviews which may be pertinent to their work. This volume is one that can be used daily, in conjunction with library research, report and manuscript preparation, and for selecting references for study and bibliographical use.

The 1963-1964 Index contains about twenty percent more entries than did the 1961-1962 issue, again showing the rapid growth of the literature of organic chemistry. The great majority of the articles and books cited in this issue are fairly complete, authoritative reviews and monographs which are particularly useful as a first step in assessing the total original literature.

We cannot claim that the listings are fully exhaustive, but the major articles have been included, and these will quickly lead to the key original sources in the various subjects.

The Appendix again includes some references which have been omitted in the course of compiling the INDEX and there also is given a current list of addresses of publishers. This latter list was found to be useful by many persons and we are pleased to include it once more. The Appendix also lists other sources in which reviews are published, but which do so only occasionally, or may be less accessible to users of this INDEX.

We wish to express our appreciation for assistance to the following persons:

Mrs. Hazel Lord, science librarian at the University of Southern California, and her staff, for kindly helping us in many ways when obtaining items for the text. Frank A. Billig and Philip Lewis, for their work in completing the critical subject index to this volume. Mr. Gary Alexander, Book editor of Pergamon Press, New York, for his encouragement to carry on with this project and assistance with the publication of it. And, particularly, to Elenora R. Kharasch, whose basic assistance in the library work, typing and compilation of the INDEX made it possible for us to undertake this project.

We would be pleased to have suggestions from readers for any improvements which could be considered for the 1965-1966 edition, which is already in preparation.

Los Angeles, California

Norman Kharasch
Walter Wolf

CONTENTS

PART I. REVIEWS IN JOURNALS AND PERIODIC PUBLICATIONS

ADVANCES IN ANALYTICAL CHEMISTRY AND INSTRUMENTATION	1
ADVANCES IN CANCER RESEARCH	2
ADVANCES IN CARBOHYDRATE CHEMISTRY	2
ADVANCES IN CATALYSIS AND RELATED SUBJECTS	4
ADVANCES IN CHEMICAL PHYSICS	5
ADVANCES IN CHEMOTHERAPY	7
ADVANCES IN CLINICAL CHEMISTRY	8
ADVANCES IN DRUG RESEARCH	9
ADVANCES IN ENZYMOLOGY AND RELATED SUBJECTS OF BIOCHEMISTRY	10
ADVANCES IN FLUORINE CHEMISTRY	11
ADVANCES IN FOOD RESEARCH	11
ADVANCES IN HETEROCYCLIC CHEMISTRY	12
ADVANCES IN INORGANIC AND RADIOCHEMISTRY	14
ADVANCES IN LIPID RESEARCH	15
ADVANCES IN ORGANIC CHEMISTRY: METHODS AND RESULTS	16
ADVANCES IN ORGANOMETALLIC CHEMISTRY	16
ADVANCES IN PHOTOCHEMISTRY	17
ADVANCES IN PHYSICAL ORGANIC CHEMISTRY	19
ADVANCES IN PROTEIN CHEMISTRY	20
ADVANCES IN SPECTROSCOPY	20
ADVANCES IN TRACER METHODOLOGY	21
AMERICAN SCIENTIST	23
THE ANALYST	23
ANALYTICAL CHEMISTRY	24
ANGEWANDTE CHEMIE	27
ANNALES DE CHIMIE	34
ANNALES PHARMACEUTIQUES FRANÇAISES	35
ANNUAL REVIEW OF BIOCHEMISTRY	36
ANNUAL REVIEW OF NUCLEAR SCIENCE	37
ANNUAL REVIEW OF PHYSICAL CHEMISTRY	38
ANNUAL REVIEW OF PHYTOPATHOLOGY	39
ANTIBIOTICA ET CHEMOTHERAPIA	40
THE AUSTRALIAN JOURNAL OF SCIENCE	40
BULLETIN DE LA SOCIETÉ DE CHIMIE BIOLOGIQUE	41
BULLETIN DE LA SOCIETÉ CHIMIQUE DE FRANCE	42
CHEMICAL REVIEWS	47
CHEMIKER-ZEITUNG	50
CHEMISTRY AND INDUSTRY	52
CHIMIA	56
CHIMIE ET INDUSTRIE	59

CHROMATOGRAPHIC REVIEWS	61
DEVELOPMENTS IN APPLIED SPECTROSCOPY	63
ENDEAVOUR	65
EXPERIENTIA	66
FORTSCHRITTE DER ARZNEIMITTELFORSCHUNG	67
FORTSCHRITTE DER CHEMIE ORGANISCHER NATURSTOFFE	68
FORTSCHRITTE DER CHEMISCHEN FORSCHUNG	69
FORTSCHRITTE DER HOCHPOLYMEREN-FORSCHUNG	70
JOURNAL OF CHEMICAL EDUCATION	71
JOURNAL DE CHIMIE PHYSIQUE ET DE PHYSICO-CHIMIE BIOLOGIQUE	75
JOURNAL OF PHARMACEUTICAL SCIENCES	78
JOURNAL OF PHARMACY AND PHARMACOLOGY	80
JOURNAL OF POLYMER SCIENCE	81
MEDICINAL CHEMISTRY	95
DIE NATURWISSENSCHAFTEN	96
NUCLEONICS	97
NUCLEUS	98
ÖSTERREICHISCHE CHEMIKER-ZEITUNG	99
ORGANIC REACTIONS	100
PROCEEDINGS OF THE CHEMICAL SOCIETY	101
PROGRESS IN INORGANIC CHEMISTRY	102
PROGRESS IN MEDICINAL CHEMISTRY	102
PROGRESS IN NUCLEIC ACID RESEARCH	103
PROGRESS IN ORGANIC CHEMISTRY	104
PROGRESS IN PHYSICAL ORGANIC CHEMISTRY	105
PROGRESS IN REACTION KINETICS	106
PROGRESS IN THE CHEMISTRY OF FATS AND OTHER LIPIDS	107
PURE AND APPLIED CHEMISTRY	108
QUARTERLY REVIEWS	117
RADIATION RESEARCH	119
RECENT PROGRESS IN HORMONE RESEARCH	121
RECORD OF CHEMICAL PROGRESS	123
REVIEWS OF PURE AND APPLIED CHEMISTRY	125
THE ROYAL INSTITUTE OF CHEMISTRY JOURNAL	126
RUSSIAN CHEMICAL REVIEWS	127
SCIENCE	132
SURVEY OF PROGRESS IN CHEMISTRY	135
SVENSK KEMISK TIDSKRIFT	136
TALANTA	139
TETRAHEDRON	140
VITAMINS AND HORMONES	142
ZEITSCHRIFT FÜR ANALYTISCHE CHEMIE	143
ZEITSCHRIFT FÜR VITAMIN-, HORMON-, UND FERMENTFORSCHUNG	146

PART II. REVIEWS IN SYMPOSIA, COLLECTIVE VOLUMES AND NON-PERIODICAL PUBLICATIONS

ADVANCES IN CHEMISTRY	149
ADVANCES IN ENZYMIC HYDROLYSIS OF CELLULOSE AND RELATED MATERIALS	153
AUSGEWAHLTE PHYSIKALISCHE METHODEN DER ORGANISCHEN CHEMIE	154
BIOGENESIS OF NATURAL COMPOUNDS	155
BIOMEDICAL APPLICATIONS OF GAS CHROMATOGRAPHY	156
BIOMEDICAL FRONTIERS IN MEDICINE	157
BORON, METALLO-BORON COMPOUNDS AND BORANES	157
CATALYSIS AND CHEMICAL KINETICS	158
CHELATING AGENTS AND METAL CHELATES	159
THE CHEMISTRY OF CATIONIC POLYMERIZATION	160
THE CHEMISTRY OF HETEROCYCLIC COMPOUNDS	161
CHEMISTRY OF ORGANIC SULFUR COMPOUNDS IN PETROLEUM AND PETROLEUM PRODUCTS	162
CHEMISTRY OF WOOD	163
ELECTROCHEMISTRY	164
ELECTRONIC ASPECTS OF BIOCHEMISTRY	165
ESSAYS IN CO-ORDINATION CHEMISTRY	167
FARADAY SOCIETY — DISCUSSIONS	168
FATTY ACIDS	170
FREE-ELECTRON THEORY OF CONJUGATED MOLECULES	171
FRIEDEL-CRAFTS AND RELATED REACTIONS	173
GAS CHROMATOGRAPHY	176
HIGH POLYMERS	177
HIGH PRESSURE PHYSICS AND CHEMISTRY	179
HORMONAL STEROIDS	180
INFRA-RED SPECTROSCOPY AND MOLECULAR STRUCTURE	181
INTERNATIONAL UNION OF PURE AND APPLIED CHEMISTRY	182
LESS COMMON MEANS OF SEPARATION	182
MAGNETIC AND ELECTRIC RESONANCE AND RELAXATION	183
MASS SPECTROMETRY	184
MASS SPECTROMETRY OF ORGANIC IONS	185
METHODEN DER ORGANISCHEN CHEMIE	186
MOLECULAR ORBITALS IN CHEMISTRY, PHYSICS AND BIOLOGY	187
NEW BIOCHEMICAL SEPARATIONS	188
NITRO COMPOUNDS	189
NON-STOICHIOMETRIC COMPOUNDS	190
PARAMAGNETIC RESONANCE	190
PEPTIDES	191
PERSPECTIVES IN BIOLOGY	194
PHYSICAL METHODS IN HETEROCYCLIC CHEMISTRY	195
PHYSICAL METHODS IN ORGANIC CHEMISTRY	196
PHYSICAL PROPERTIES OF THE STEROID HORMONES	197

PROCEEDINGS OF THE FIRST INTERNATIONAL PHARMACOLOGICAL MEETING 198
PROGRESS IN BIOPHYSICS AND MOLECULAR BIOLOGY 203
THE PROTEINS 204
PTERIDINE CHEMISTRY 205
RAPID MIXING AND SAMPLING TECHNIQUES IN BIOCHEMISTRY 208
RODD'S CHEMISTRY OF CARBON COMPOUNDS 209
THE ROYAL INSTITUTE OF CHEMISTRY OF GREAT BRITAIN AND IRELAND 210
THEORY AND STRUCTURE OF COMPLEX COMPOUNDS 211
THERMAL STABILITY OF POLYMERS 212
THIN-LAYER CHROMATOGRAPHY 214
VACUUM MICROBALANCE TECHNIQUES 216
VITAMINE 217

PART III

MONOGRAPHS ON ORGANIC CHEMISTRY: 1963–1964 221
    Author Index 239
    Subject Index 269
    Addresses of Publishers 323

PART I
REVIEWS IN JOURNALS AND PERIODIC PUBLICATIONS
Pages 1-146

ADVANCES IN ANALYTICAL CHEMISTRY AND INSTRUMENTATION

This is a collection of review articles, in English. Edited by C.N. Reilley and published by Interscience Publishers, Inc., New York.

Title	Authors	Reference
Progress in qualitative organic analysis	F. Feigl, R. Belcher and W.I. Stephen	2, 1-34 (1963)
Laboratory pH measurements	G. Mattock	2, 35-121 (1963)
Application of hanging mercury drop electrodes in analytical chemistry	W. Kemula and Z. Kublik	2, 123-177 (1963)
Mass spectra of organic molecules	S. Meyerson and J.D. McCollum	2, 179-218 (1963)
Some techniques in organic polarography	P. Zuman	2, 219-253 (1963)
Reaction rate methods	H.B. Mark, Jr. L.J. Papa and C.N. Reilley	2, 255-385 (1963)
Atomic absorption spectroscopy	R. Lockyer	3, 1-29 (1964)
Photometric titrations	A.L. Underwood	3, 31-104 (1964)
Analytical applications of enzyme-catalyzed reactions	W.J. Blaedel and G.P. Hicks	3, 105-142 (1964)
Ion sources and detectors for the mass spectroscopic study of solids	L.F. Herzog, D.J. Marshall, B.R.F. Kendall and L.A. Cambey	3, 143-181 (1964)
Galvanic analysis	P. Hersch	3, 183-249 (1964)
Linear elution adsorption chromatography	L.R. Snyder	3, 251-313 (1964)
Concepts and column parameters in gas chromatography	J.C. Giddings	3, 315-367 (1964)
Thin-layer chromatography	R. Maier and H.K. Mangold	3, 369-477 (1964)

ADVANCES IN CANCER RESEARCH

This is a collection of review articles, in English. Edited by A. Haddow and S. Weinhouse and published by Academic Press, New York-London.

Title	Authors	Reference
Mechanisms of resistance to anticancer agents	R.W. Brockman	7, 129-234 (1963)
Nuclear proteins of neoplastic cells	H. Busch and W.J. Steele	8, 41-120 (1964)

ADVANCES IN CARBOHYDRATE CHEMISTRY

This is an annual collection of review articles, in English. The present editors are M.L. Wolfrom and R.S. Tipson. Published by Academic Press, New York-London.

Title	Authors	Reference
Photochemistry of carbohydrates	G.O. Phillips	18, 9-59 (1963)
Paper electrophoresis of carbohydrates	H. Weigel	18, 61-97 (1963)
Chemistry of osotriazoles	H. El Khadem	18, 99-121 (1963)
Developments in the chemistry of thio sugars	D. Horton and D.H. Hutson	18, 123-199 (1963)
Trehaloses	G.G. Birch	18, 201-225 (1963)
Naturally occurring C-glycosyl compounds	L.J. Haynes	18, 227-258 (1963)
Chemistry of the amino sugars derived from antibiotic substances	J.D. Dutcher	18, 259-308 (1963)
Biosynthesis of saccharides from glycopyranosyl esters of nucleotides ("Sugar nucleotides")	E.F. Neufeld and W.Z. Hassid	18, 309-356 (1963)
Physical properties of solutions of polysaccharides	W. Banks and C.T. Greenwood	18, 357-398 (1963)
Crystal-structure analysis in carbohydrates chemistry	G.A. Jeffrey and R.D. Rosenstein	19, 7-22 (1964)
Infrared spectroscopy and carbohydrate chemistry	H. Spedding	19, 23-49 (1964)

Nuclear magnetic resonance	L.D. Hall	19,	51-93 (1963)
Gas-liquid chromatography of carbohydrate derivatives	C.T. Bishop	19,	95-147 (1964)
The action of hydrogen peroxide on carbohydrates and related compounds	G.J. Moody	19,	149-179 (1964)
3-Deoxyglycosuloses (3-deoxyglycosones) and the degradation of carbohydrates	E.F.L.J. Anet	19,	181-218 (1964)
Structure and some reactions of cellulose	D.M. Jones	19,	219-246 (1964)
Wood hemicelluloses: Part I.	T.E. Timell	19,	247-302 (1964)
The pneumococcal polysaccharides	M.J. How, J.S. Brimacombe and M. Stacey	19,	303-358 (1964)

This is an annual collection of review articles, in English. Edited by D.D. Eley, H. Pines and P.B. Weisz and published by Academic Press, New York-London.

Title	Authors	Reference
Quantum conversion in chloroplasts	M. Calvin	14, 1-34 (1963)
The catalytic decomposition of formic acid	P. Mars, J.J.F. Scholten and P. Zwietering	14, 35-113 (1963)
Application of spectrophotometry to the study of catalytic systems	H.P. Leftin and M.C. Hobson, Jr.	14, 115-201 (1963)
Hydrogenation of pyridines and quinolines	M. Freifelder	14, 203-253 (1963)
Modern methods in surface kinetics (Flash desorption, field emission microscopy and ultrahigh vacuum techniques)	G. Ehrlich	14, 255-427 (1963)
Catalytic oxidation of hydrocarbons	L. Ya. Margolis	14, 429-501 (1963)
The mechanism of the hydrogenation of unsaturated hydrocarbons on transition metal catalysts	G.C. Bond and P.B. Wells	15, 91-226 (1964)
Electronic spectroscopy of adsorbed gas molecules	A. Terenin	15, 227-284 (1964)
The catalysis of isotopic exchange in molecular oxygen	G.K. Boreskov	15, 285-339 (1964)

ADVANCES IN CHEMICAL PHYSICS

This is a collection of review articles, in English. Edited by I. Prigogine and published by Interscience Publishers (a division of John Wiley and Sons) New York - London.

Title	Authors	Reference
New developments in the one-electron theory of π-electron systems	H. Hartmann	5, 1-31 (1963)
Spectroscopy of transition-group complexes	Chr. K. Jorgensen	5, 33-146 (1963)
Convex molecules in gaseous and crystalline states	T. Kihara	5, 147-188 (1963)
Theories on the magnetic properties of compounds	S. Koide and T. Oguchi	5, 189-240 (1963)
Forbidden transitions in organic and inorganic systems	A.D. Liehr	5, 241-260 (1963)
The formal statistical theory of transport processes	J.A. McLennan, Jr.	5, 261-317 (1963)
Quantum mechanical interpretation of nuclear quadruple coupling data	E. Scrocco	5, 319-352 (1963)
Collision theory of chemical reaction rates	B. Widom	5, 353-386 (1963)
Toward an analytic theory of chemical reactions	H. Aroeste	6, 1-83 (1964)
Many-electron theory of atoms, molecules and their interactions	O. Sinanoğlu	6, 315-412 (1964)
Ionic solvation	J. Stecki	6, 413-458 (1964)
Melting mechanisms of crystals	A.R.J.P. Ubbelohde	6, 459-480 (1964)
The Structure and Properties of Biomolecules and Biological Systems. Part I.	J. Duchesne (Editor)	
Electronic structures in quantum biochemistry	J.I. Fernandez-Alonso	7, 3-83 (1964)
Quantum mechanical considerations on some properties of DNA	T.A. Hoffmann and J. Ladik	7, 84-158 (1964)
Electronic structure and magnetic properties of hemo-proteins, particularly of hemoglobins	M. Kotani	7, 159-181 (1964)

Magnetic susceptibilities and the chemical bond in hemoproteins	G. Schoffa	7, 182-198 (1964)
Thermal effects on protein, nucleic acid viruses	E.C. Pollard	7, 201-237 (1964)
Adsorption of water on solid proteins with special reference to haemoglobin	D.D. Eley and R.B. Leslie	7, 238-258 (1964)
The effects of ionizing radiations on some fibrous proteins	R. Braams and G. Van Herpen	7, 259-281 (1964)
Electronic conduction in organic molecular solids	D.R. Kearns	7, 282-338 (1964)
The electronic properties of deoxyribonucleic acid	P. Douzou and C. Sadron	7, 339-358 (1964)
The properties of metalporphyrin and similar complexes	P.S. Braterman, R.C. Davies and R.J.P. Williams	7, 359-407 (1964)
Prospects for the use of infrared spectroscopy in biology	J. Lecomte	7, 408-434 (1964)
Infrared spectra of nucleic acids and related compounds	T. Shimanouchi, M. Tsuboi and Y. Kyogoku	7, 435-498 (1964)
The study of specific molecular interactions by nuclear magnetic relaxation measurements	O. Jardetzky	7, 499-531 (1964)
Recent advances in EPR spectroscopy	B. Smaller	7, 532-583 (1964)
Photoprotection from far ultraviolet effects in cells	J. Jagger	7, 584-601 (1964)
Some recent developments in the study of paramagnetic species of biological interest	A. Ehrenberg	7, 602-627 (1964)
ESR Investigations on different plant systems	C.S. Nicolau	7, 628-644 (1964)
The use of product inhibition and other kinetic methods in the determination of mechanisms of enzyme action	C. Walter	7, 645-724 (1964)

ADVANCES IN CHEMOTHERAPY

This is a collection of review articles, in English. Edited by A. Goldin and F. Hawking and published by Academic Press, New York - London.

Title	Authors	Reference
Historical perspectives in chemotherapy	E.K. Marshall, Jr.	1, 1-8 (1964)
Quantitative concepts in the clinical study of drugs	C.G. Zubrod	1, 9-34 (1964)
Chemoprophylaxis and chemotherapy of viral diseases	R.L. Thompson	1, 85-131 (1964)
The *Vinca* alkaloids	N. Neuss, I.S. Johnson, J.G. Armstrong and C.J. Jansen	1, 133-174 (1964)
Drug synergism in antineoplastic chemotherapy	J.M. Venditti and A. Goldin	1, 397-498 (1964)
New concepts of the use of inhibitors in chemotherapy	N.D. Kaplan and M. Friedkin	1, 499-522 (1964)

ADVANCES IN CLINICAL CHEMISTRY

This is an annual collection of review articles, in English. Edited by H. Sobotka and C.P. Stewart and published by Academic Press, New York - London.

Title	Authors	Reference		
Enzymatic determinations of glucose	A.H. Free	6,	67-96	(1963)
Inherited metabolic disorders: Errors of phenylalanine and tyrosine metabolism	L.I. Woolf	6,	97-230	(1963)
Normal and abnormal human hemoglobins	T.H.J. Huisman	6,	231-361	(1963)
Principles and applications of atomic absorption spectroscopy	A. Zettner	7,	1-62	(1964)
Aspects of disorders of the kynurenine pathway of tryptophan metabolism in man	L. Musajo and C.A. Benassi	7,	63-135	(1964)
The clinical biochemistry of the muscular dystrophies	W.H.S. Thomson	7,	137-197	(1964)
Mucopolysaccharides in disease	J.S. Brimacombe and M. Stacey	7,	199-234	(1964)
Proteins, mucosubstances, and biologically active components of gastric secretion	G.B. Jerzy Glass	7,	235-372	(1964)
Fractionation of macromolecular components of human gastric juice by electrophoresis, chromatography, and other physicochemical methods	G.B. Jerzy Glass	7,	373-479	(1964)

ADVANCES IN DRUG RESEARCH

This is a collection of review articles, in English. Subject and author indexes are included. Edited by N.J. Harper and A.B. Simmonds and published by Academic Press, Paris-London-New York.

Title	Authors	Reference
Penicillins and related subjects	F.P. Doyle and J.H.C. Nayler	1, 1-69 (1964)
Physiological transport of drugs	L.S. Schanker	1, 71-106 (1964)
Antitussives	F.P. Doyle and M.D. Mehta	1, 107-159 (1964)
Adrenergic neurone blocking agents	F.C. Copp	1, 161-189 (1964)

This is an annual collection of review articles, usually in English, edited by F.F. Nord and published by Interscience Publishers (a division of John Wiley and Sons) New York - London.

Title	Authors	Reference
Photosynthesis; energetics and related topics	J.A. Bassham	25, 39-117 (1963)
The chemistry of light emission	W.D. McElroy and H.H. Seliger	25, 119-166 (1963)
Coenzyme Q (Ubiquinone)	Y. Hatefi	25, 275-328 (1963)
Biological methylation	D.M. Greenberg	25, 395-431 (1963)
Recent developments in the biochemistry of amino sugars	R.W. Jeanloz	25, 433-456 (1963)
Phytochrome and its control of plant and growth development	H.W. Siegelman and S.B. Hendricks	26, 1-33 (1964)
Sugar nucleotides and the synthesis of carbohydrates	V. Ginsburg	26, 35-88 (1964)
Formation of the secondary and tertiary structure of enzymes	F.B. Straub	26, 89-114 (1964)
Hydrogen transfers with pyridine nucleotides (in German)	H. Sund, H. Diekmann and K. Wallenfels	26, 115-191 (1964)
Advances in the vitamin B_{12} field (in German)	K. Bernhauer, O. Muller and F. Wagner	26, 233-281 (1964)

ADVANCES IN FLUORINE CHEMISTRY

This is a collection of review articles, in English. Edited by M. Stacey, J.C. Tatlow and A.G. Sharpe and published by Butterworths Inc., Washington, D.C.

Title	Authors	Reference
Effects of adjacent perfluoroalkyl groups on carbonyl reactivity	H.P. Braendlin and E.T. McBee	3, 1-18 (1963)
Perfluoroalkyl derivatives of the elements	H.C. Clark	3, 19-58 (1963)
Mechanisms of fluorine displacement	R.E. Parker	3, 63-91 (1963)
Nitrogen fluorides and their inorganic derivatives	C.B. Colburn	3, 92-116 (1963)
The organic fluorochemicals industry	J.M. Hamilton, Jr.	3, 117-180 (1963)
The preparation of organic fluorine compounds by halogen exchange	A.K. Barbour, L.J. Belf and M.W. Buxton	3, 181-270 (1963)

ADVANCES IN FOOD RESEARCH

This is a collection of review articles, in English, publishing occasionally articles of a chemical nature. Edited by C.O. Chichester, E.M. Mrak and G.F. Stewart and published by Academic Press, New York - London.

Title	Authors	Reference
Chemistry of nonenzymic browning. 1. The reaction between aldoses and amines	T.M. Reynolds	12, 1-52 (1963)
Physiology, chemistry, and technology of passion fruit	J.S. Pruthi	12, 203-282 (1963)
Utilization of synthetic gums in the food industry	M. Glicksman	12, 283-366 (1963)
Recent advances in the freeze-drying of food products	R.F. Burke and R.V. Decareau	13, 1-88 (1964)
Fundamentals of low-temperature food preservation	O. Fennema and W.D. Powrie	13, 219-347 (1964)

ADVANCES IN HETEROCYCLIC CHEMISTRY

This is a collection of review articles, in English. Edited by A.R. Katritzky and published by Academic Press, New York and London.

Title	Authors	Reference
Recent advances in the chemistry of thiophenes	S. Gronowitz	<u>1</u>, 1-124 (1963)
Reactions of acetylenecarboxylic acids and their esters with nitrogen-containing heterocyclic compounds	R.M. Acheson	<u>1</u>, 125-165 (1963)
Heterocyclic pseudo bases	D. Beke	<u>1</u>, 167-188 (1963)
Aza analogs of pyrimidine and purine bases of nucleic acids	J. Gut	<u>1</u>, 189-251 (1963)
Quinazolines	W.L.F. Armarego	<u>1</u>, 253-309 (1963)
Prototropic tautomerism of heteroaromatic compounds:		
I. General discussion and methods of study	A.R. Katritzky and J.M. Lagowski	<u>1</u>, 311-338 (1963)
II. Six-membered rings	A.R. Katritzky and J.M. Lagowski	<u>1</u>, 339-437 (1963)
III. Five-membered rings and one hetero atom	A.R. Katritzky and J.M. Lagowski	<u>2</u>, 1-26 (1963)
IV. Five-membered rings with two or more hetero atoms	A.R. Katritzky and J.M. Lagowski	<u>2</u>, 27-81 (1963)
Three-membered rings with two hetero atoms	E. Schmitz	<u>2</u>, 83-130 (1963)
Free-radical substitutions of heteroaromatic compounds	R.O.C. Norman and G.K. Radda	<u>2</u>, 131-177 (1963)
The action of metal catalysts on pyridines	G.M. Badger and W.H.F. Sasse	<u>2</u>, 179-202 (1963)
Recent advances in quinoxaline chemistry	G.W.H. Cheeseman	<u>2</u>, 203-244 (1963)
The reactions of diazomethane with heterocyclic compounds	R. Gompper	<u>2</u>, 245-286 (1963)
The acid-catalyzed polymerization of pyrroles and indoles	G.F. Smith	<u>2</u>, 287-309 (1963)
1,3-Oxazine derivatives	Z. Eckstein and T. Urbanski	<u>2</u>, 311-342 (1963)

The present state of selenazole chemistry	E. Bulka	2, 343-364	(1963)
Recent developments in isoxazole chemistry	N.K. Kochetkov and S.D. Sokolov	2, 365-422	(1963)
The quaternization of heterocyclic compounds	G.F. Duffin	3, 1-56	(1964)
The reactions of heterocyclic compounds with carbenes	C.W. Rees and E.E. Smithen	3, 57-78	(1964)
The carbolines	R.A. Abramovitch and I.D. Spenser	3, 79-207	(1964)
Applications of the Hammett equation to heterocyclic compounds	H.H. Jaffé and H. Lloyd Jones	3, 209-261	(1964)
1,2,3,4-Thiatriazoles	K.A. Jensen and C. Pedersen	3, 263-284	(1964)
Nucleophilic heteroaromatic substitution	G. Illuminati	3, 285-371	(1964)
Pentazoles	I. Ugi	3, 373-383	(1964)

ADVANCES IN INORGANIC AND RADIOCHEMISTRY

This is an annual collection of review articles, in English. Edited by H.J. Eméleus and A.G. Sharpe and published by Academic Press, Inc., New York.

Title	Authors	Reference
The chemistry of gallium	N.N. Greenwood	5, 91-134 (1963)
Chemical effects of nuclear activation in gases and liquids	I.G. Campbell	5, 135-214 (1963)
The borazines	E.K. Mellon, Jr. and J.J. Lagowski	5, 259-305 (1963)
Decaborane-14 and its derivatives	M.F. Hawthorne	5, 307-345 (1963)
The structure and reactivity of organophosphorus compounds	R.F. Hudson	5, 347-398 (1963)
Complexes of the transition metals with phosphines, arsines, and stibines	G. Booth	6, 1-69 (1964)
Anhydrous metal nitrates	C.C. Addison and N. Logan	6, 71-142 (1964)
Chemical reactions in electric discharges	A.S. Kana'an and J.L. Margrave	6, 143-206 (1964)
The chemistry of astatine	A.H.W. Aten, Jr.	6, 207-223 (1964)
The chemistry of silicon-nitrogen compounds	U. Wannagat	6, 225-278 (1964)
Peroxy compounds of transition metals	J.A. Connor and E.A.V. Ebsworth	6, 279-381 (1964)
The direct synthesis of organosilicon compounds	J.J. Zuckerman	6, 383-432 (1964)
The Mossbauer effect and its application in chemistry	E. Fluck	6, 433-489 (1964)

This is an annual collection of review articles, in English. Edited by R. Paoletti and D. Kritchevsky and published by Academic Press, New York-London.

Title	Authors	Reference		
The structural investigation of natural fats	M.H. Coleman	1,	1-64	(1963)
Physical structure and behavior of lipids and lipid enzymes	A.D. Bangham	1,	65-104	(1963)
Chromatographic investigations in fatty acid biosynthesis	M. Pascaud	1,	253-284	(1963)
The plant sulfolipid	A.A. Benson	1,	387-394	(1963)
Triglyceride structure	R.J. Vander Wal	2,	1-16	(1964)
Bacterial lipids	M. Kates	2,	17-90	(1964)
Phosphatidylglycerols and lipoamino acids	M.G. Macfarlane	2,	91-125	(1964)
The brain phosphoinositides	J.N. Hawthorne and P. Kemp	2,	127-166	(1964)
The synthesis of phosphoglycerides and some biochemical applications	L.L.M. Van Deenen and G.H. De Haas	2,	167-234	(1964)

ADVANCES IN ORGANIC CHEMISTRY: METHODS AND RESULTS

This is a collection of review articles, in English. Edited by R.A. Raphael, E.C. Taylor and H. Wynberg and published by Interscience Publishers, Inc. (a division of John Wiley and Sons) New York-London.

Title	Authors	Reference
Mass spectrometry as a structural tool	R.I. Reed	3, 1-73 (1963)
Phosphorylation	D.M. Brown	3, 75-157 (1963)
Selectively removable amino protective groups used in the synthesis of peptides	R.A. Boissonnas	3, 159-190 (1963)
Protective groups	J.F.W. McOmie	3, 191-294 (1963)
Enamines	J. Szmuszkovicz	4, 1-113 (1963)
Synthetic methods in the carotenoid and vitamin A fields	O. Isler and P. Schudel	4, 115-224 (1963)
The coupling of acetylenic compounds	G. Eglinton and W. McCrae	4, 225-328 (1963)

ADVANCES IN ORGANOMETALLIC CHEMISTRY

This is a collection of review articles, in English. Edited by F.G.A. Stone and R. West and published by Academic Press, New York-London.

Title	Authors	Reference
Diene iron carbonyl complexes and related species	R. Pettit and G.F. Emerson	1, 1-46 (1964)
Reactions of organotin hydrides with organic compounds	H.G. Kuivila	1, 47-87 (1964)
Organic substituted cyclosilanes	H. Gilman and G.L. Schwebke	1, 89-141 (1964)
Fluorocarbon derivatives of metals	P.M. Treichel and F.G.A. Stone	1, 143-220 (1964)
Conjugate addition of Grignard reagents to aromatic systems	R.C. Fuson	1, 221-238 (1964)
Infrared and Raman spectral studies of π complexes formed between metals and C_nH_n rings	H.P. Fritz	1, 239-316 (1964)

ADVANCES IN PHOTOCHEMISTRY

This is a collection of review articles, in English. Edited by W.A. Noyes, Jr., G.S. Hammond and J.N. Pitts, Jr. and published by Interscience Publishers, Inc. (a division of John Wiley and Sons), New York-London.

Title	Authors	Reference
The "vocabulary" of photochemistry	J.N. Pitts, Jr. F. Wilkinson and G.S. Hammond	1, 1-21 (1963)
The photochemistry of aromatic hydrocarbon solutions	E.J. Bowen	1, 23-42 (1963)
Photochemical gas phase reactions in the hydrogen-oxygen system	D.H. Volman	1, 43-82 (1963)
Photochemistry of the cyclic ketones	R. Srinivasan	1, 83-113 (1963)
Addition of atoms to olefins in the gas phase	R.J. Cvetanovic	1, 115-182 (1963)
A new approach to mechanistic organic photochemistry	H.E. Zimmerman	1, 183-208 (1963)
Isotopic effects and the mechanism of energy transfer in mercury photosensitization	H.E. Gunning and O.P. Strausz	1, 209-274 (1963)
Photochromism	R. Dessauer and J.P. Paris	1, 275-321 (1963)
Photochemical rearrangements of organic molecules	O.L. Chapman	1, 323-420 (1963)
Some problems of structure and reactivity in free radical and molecule reactions in the gas phase	S.W. Benson	2, 1-23 (1964)
Mechanisms and rate constants of elementary gas phase reactions involving hydroxyl and oxygen atoms	L.I. Avramenko and R.V. Kolesnikova	2, 25-62 (1964)
Photochemical reactions of sulfur and nitrogen heteroatomic organic compounds	A. Mustafa	2, 63-136 (1964)
Photochemical processes in halogenated compounds	J.R. Majer and J.P. Simons	2, 137-181 (1964)
The chemistry of ionic states in solid saturated hydrocarbons	L. Kevan and W.F. Libby	2, 183-218 (1964)
Preparation, properties and reactivity of methylene	W.B. DeMore and S.W. Benson	2, 219-261 (1964)

Some recent developments in the photochemistry of organic nitrites and hypohalites	M. Akhtar	*2*,	263-303 (1964)
Phosphorescence and delayed fluorescence from solutions	C.A. Parker	*2*,	305-383 (1964)
Photoionization and photodissociation of aromatic molecules by vacuum ultraviolet radiation	A. Terenin and F. Vilessov	*2*,	385-421 (1964)
Unimolecular decomposition and some isotope effects of simple alkanes and alkyl radicals	B.S. Rabinovitch and D.W. Setser	*3*,	1-82 (1964)
Gaseous photooxidation reactions	D.E. Hoare and G.S. Pearson	*3*,	83-156 (1964)
Vacuum ultraviolet photochemistry	J.R. McNesby and H. Okabe	*3*,	157-240 (1964)
Electronic energy transfer between organic molecules in solution	F. Wilkinson	*3*,	241-268 (1964)

ADVANCES IN PHYSICAL ORGANIC CHEMISTRY

This is an annual collection of review articles, in English. Edited by V. Gold and published by Academic Press, Inc., New York and London.

Title	Authors	Reference
Entropies of activation and mechanisms of reactions in solution	L.L. Schaleger and F.A. Long	1, 1-33 (1963)
A quantitative treatment of directive effects in aromatic substitution	L.M. Stock and H.C. Brown	1, 35-154 (1963)
Hydrogen isotope exchange reactions of organic compounds in liquid ammonia	A.I. Shatenshtein	1, 155-201 (1963)
Planar and non-planar aromatic systems	G. Ferguson and J.M. Robertson	1, 203-281 (1963)
The identification of organic free radicals by electron spin resonance	M.C.R. Symons	1, 283-363 (1963)
The structure of electronically excited organic molecules	J.C.D. Brand and D.G. Williamson	1, 365-423 (1963)
Isotopes and organic reaction mechanisms	C.J. Collins	2, 1-91 (1964)
Use of volumes of activation for determining reaction mechanisms	E. Whalley	2, 93-162 (1964)
Hydrogen isotope effects in aromatic substitution reactions	H. Zollinger	2, 163-200 (1964)
The reactions of energetic tritium and carbon atoms with organic compounds	A.P. Wolf	2, 201-277 (1964)

ADVANCES IN PROTEIN CHEMISTRY

This is an annual collection of review articles, in English. Edited by C.B. Anfinsen, Jr., M.L. Anson and J.T. Edsall in 1963; and by C.B. Anfinsen, Jr., M.L. Anson, J.T. Edsall and F.M. Richards in 1964. Published by Academic Press, New York-London.

Title	Authors	Reference		
Recent studies on the structure of tobacco virus	F.A. Anderer	18,	1-35	(1963)
Assembly and stability of the tobacco mosaic virus particle	D.L.D. Caspar	18,	37-121	(1963)
The dissociation and association of protein structures	F.J. Reithel	18,	123-226	(1963)
The amino acid composition of some purified proteins	G.R. Tristram and R.H. Smith	18,	227-318	(1963)
The hemoglobins	G. Braunitzer, K. Hilse, V. Rudloff and N. Hilschmann	19,	1-72	(1964)
Hemoglobin and myoglobin	A. Fanelli, E. Antonini and A. Caputo	19,	73-222	(1964)
Linked functions and reciprocal effects in hemoglobin: A second look	J. Wyman, Jr.	19,	223-286	(1964)
Thermodynamic analysis of multicomponent solutions	E.F. Cassa and H. Eisenberg	19,	287-395	(1964)

ADVANCES IN SPECTROSCOPY

This is a collection of review articles, in English. Edited by H.W. Thompson and published by Interscience Publishers (a division of John Wiley and Sons) New York. No volumes appeared in 1963 or 1964.

ADVANCES IN TRACER METHODOLOGY

This is a collection of review articles and symposia on Tracer Methodology. Edited by S. Rothchild and published by Plenum Press, New York. Volume I includes the proceedings of the 5th Annual Symposium on Tracer Methodology, held in Washington, D.C., October 20, 1961, and selected papers from the first four Annual Symposia and from published issues of "Atomlight". This is a publication of the New England Nuclear Corporation.

Title	Authors	Reference
The tritium gas exposure method. Individual parts of the symposium were presented by each author, covering specific areas.	K.E. Wilzbach, C. Rosenblum and H.T. Meriwether	$\underline{1}$, 1-49 (1963)
	H.J. Dutton and R.F. Nystrom	
	K.E. Wilzbach	
	R.K. Crane, G.R. Drysdale and K.H. Hawkins	
	A. Markovitz, J.A. Cifonelli and J.I. Gross	
	S. Rothchild	
	R.F. Nystrom	
Tritium labeling by other methods	S. Rothchild, J.W. Harlan	$\underline{1}$, 50-63 (1963)
	R.E. Ober, S.A. Fusari, G.L. Coffey, G.W. Gwynn and A.J. Glazki	
	M. Gut and M. Hayano	
Radiation decomposition	B.M. Tolbert	$\underline{1}$, 64-68 (1963)
Liquid scintillation counting	J.R. Arnold, J.J. Kabara, N. Spafford, M.A. McKendry and N.L. Freeman	$\underline{1}$, 69-166 (1963)
	L.A. Baillie, D. Steinberg, J.F. Snell, H. Jeffay	
	J.W. Harlan, R.B. Loftifield	
	G. Popjak, A.E. Lows and D. Moore	
	H.J. Dutton	
	B.D. Rupe, W.F. Bousquet and J.E. Christian	

ADVANCES IN TRACER METHODOLOGY (Contd.)

	V.P. Guinn and C.D. Wagner	
Gas counting	B.M. Tolbert C.V. Robinson R.L. Wolfgang	1, 167-184 (1963)
Combustion methods	D.A. Buyske R. Kelly J. Florini S. Gordon and E. Peets	
	H. Sheppard and W. Rodegker	
Instrumentation	W.A. Higinbotham	1, 195-200 (1963)
Health physics	J.S. Handloser	1, 201-202 (1963)
Measuring turnover rates	D.B. Zilversmit M.D. Kamen J.M. Reiner H. Jeffay	1, 205-221 (1963)
Tritium as a tracer isotope	M.L. Eidinoff V.P. Guinn J.F. Snell E.H. La Brosse	1, 222-264 (1963)
	W.E. Baumgartner L. Lazer and A. Dalziel D.R. Carr	
Tracer methodology	R.E. Peterson C.H. Wang	1, 265-294 (1963)
Autoradiography	W.L. Hughes J.H. Taylor	1, 291-313 (1963)
	W.E. Kisieleski R. Baserga and J. Vaupotic	
	V. Nair and L.J. Roth	
Clinical methodology	C.C. Lushbaugh and R.L. Schich	1, 314-325 (1963)

AMERICAN SCIENTIST

This is the quarterly journal of the Sigma Xi-Resa Societies. Edited by H.S. Taylor and published by the Society of Sigma Xi.

Title	Authors	Reference
Lasers	A.K. Levine	51, 14-31 (1963)
Investigations of crystalline imperfections and properties of materials	S. Weissmann	51, 193-226 (1963)
The genetic code	T.H. Jukes	51, 227-245 (1963)
People and pesticides	T.H. Jukes	51, 355-361 (1963)
Atom to Adam	M. Calvin and G.J. Calvin	52, 163-186 (1964)
Polymeric sulfur and other polysulfide polymers	A.V. Tobolsky	52, 358-364 (1964)
DNA and the chemistry of inheritance	B. Commoner	52, 365-388 (1964)
Valency and the chemical bond	J.W. Linnett	52, 459-475 (1964)

THE ANALYST

This is a monthly journal, in English, and has included review articles since 1955. It has been published since 1876 under the auspices of the Society of Public Analysts and other Analytical Chemists. Edited by J.B. Attrill and published for the Society by W. Heffer and Sons, Ltd., Cambridge, England.

Title	Authors	Reference
Classification of methods for determining particle size. Analytical Methods Committee (Particle Size Analysis Sub-Committee)	E.Q. Laws (Chairman)	88, 156-187 (1963)
Methods of separation of long-chain unsaturated fatty acids	A.T. James	88, 572-582 (1963)
Circular dichroism	R.D. Gillard	88, 825-828 (1963)
Thermogravimetric analysis	A.W. Coats and J.P. Redfern	88, 906-924 (1963)
Some analytical problems involved in determining the structure of proteins and peptides	D.G. Smyth and D.F. Elliot	89, 81-94 (1964)
The Faraday effect, magnetic rotatory dispersion and magnetic circular dichroism	J.G. Dawber	89, 755-762 (1964)

ANALYTICAL CHEMISTRY

This is a monthly journal, which occasionally publishes review articles and symposia. A publication of the American Chemical Society, Washington, D.C.

Title	Authors	Reference
Advances in Gas Chromatography. The 2nd International Symposium, held in Houston, Texas, March 23-26, 1964.		
Improved ionization cross-section detectors	J.E. Lovelock, G.R. Shoemake and A. Zlatkis	36, 1410-1415 (1964)
Specific detection of halogens and phosphorus by flame ionization	A. Karmen	36, 1416-1421 (1964)
Performance and characteristics of ultrasonic gas chromatograph effluent detector	F.W. Noble, K. Abel and P.W. Cook	36, 1421-1427 (1964)
Concentration and mass flow rate sensitive detectors in gas chromatography	I. Halá'sz	36, 1428-1430 (1964)
Performance of argon detectors in the field-intensified current region	J.Z. Knapp and A.S. Meyer	36, 1430-1441 (1964)
Specific retentions of monofunctional organic solutes in monofunctional hexadecyl derivatives	A.B. Littlewood	36, 1441-1451 (1964)
Gas chromatographic determination of the carbonyl forms of sugars	E. Bayer and R. Widder	36, 1452-1455 (1964)
Pressure changes during passage of a solute through a theoretical plate in gas liquid chromatography	R.P.W. Scott	36, 1455-1461 (1964)
Determination of carbohydrates in glycolipids and gangliosides by gas chromatography	C.C. Sweeley and B. Walker	36, 1461-1466 (1964)
Determination of estrogen in low and high titer urines using thin layer and gas liquid chromatography	H.H. Wotiz and S.C. Chattoraj	36, 1466-1472 (1964)
Temperature limitations of stationary phases in gas chromatography	S.J. Hawkes and E.F. Mooney	36, 1473-1477 (1964)
A new gas chromatographic method for measuring gaseous diffusion coefficients and obstructive factors	J.H. Knox and L. McLaren	36, 1477-1482 (1964)
Measurement and interpretation of the C terms of gas chromatography	J.C. Giddings and P.D. Schettler	36, 1483-1489 (1964)

ANALYTICAL CHEMISTRY (Contd.)

Title	Authors	Reference
Large-bore capillary and low-pressure drop packed columns	R. Teranishi and T.R. Mon	36, 1490-1492 (1964)
Particle-to-column diameter ratio effect on band spreading	J.C. Sternberg and R.E. Poulson	36, 1492-1502 (1964)
Wide-range programmed temperature gas chromatography in the separation of very complex mixtures	C. Merritt, Jr. J.T. Walsh D.A. Forss P. Angelini and S.M. Swift	36, 1502-1508 (1964)
Capillary column efficiencies in gas chromatography – Mass spectral analyses	R. Teranishi R.G. Buttery W.H. McFadden T.R. Mon and J. Wasserman	36, 1509-1512 (1964)
Simulated distillation by gas chromatography	L.E. Green L.J. Schmauch and J.C. Worman	36, 1512-1516 (1964)
Gas chromatographic determination of energy of hydrogen bonds in the adsorbed layers of alcohols on graphitized carbon black	L.D. Belyakova A.V. Kiselev and N.V. Kovaleva	36, 1517-1519 (1964)
Automatic capillary gas chromatography and sampling of distillation products	D.W. Grant	36, 1519-1522 (1964)
Use of gas capillary columns with modified internal area in gas chromatography	F.A. Bruner and G.P. Carton	36, 1522-1526 (1964)
Effect of pore size of silica gels in the separation of hydrocarbons	A.V. Kiselev Yu. S. Nikitin R.S. Petrova K.D. Shcherbakova and Ya. I. Yashin	36, 1526-1533 (1964)
Gas chromatographic separation of hydrocarbons (C_1 to C_8) by carbon number using packed capillary columns	W. Schneider and H. Bruderreck	36, 1533-1540 (1964)
Gas chromatographic studies of catecholamines, tryptamines, and other biological amines. I. Catecholamines and related compounds	C.J.W. Brooks and E.C. Horning	36, 1540-1545 (1964)
II. Tryptamine-related amines and catecholamines	E.C. Horning M.G. Horning W.J.A. Vanden-Heuvel K.L. Knox B. Holmstedt and C.J.W. Brooks	36, 1546-1549 (1964)

ANALYTICAL CHEMISTRY (Contd.)

Characterization and separation of amines by gas chromatography	W.J.A. Vanden-Heuvel, W.L. Gardner and E.C. Horning	36,	1550-1560 (1964)
Analysis of sulfur compounds with electron capture / hydrogen flame dual channel gas chromatography	D.M. Oaks, H. Hartmann and K.P. Dimick	36,	1560-1565 (1964)
Applications of isotopic exchange in gas chromatography	J. Tadmor		

Fisher Award Symposium. (Honoring John Mitchell, Jr.) Division of Analytical Chemistry, 147th National Meeting, American Chemical Society, Philadelphia, Pa., April, 1964.

Aquametric methods in organic analysis (Fisher Award Address)	J. Mitchell, Jr.	36,	2050-2052 (1964)
Mild, rapid, hydroxyl determination for primary alcohols	J.A. Floria, I.W. Dobratz and J.H. McClure	36,	2053-2055 (1964)
A low-level carbon-14 counting technique	L.T. Freeland	36,	2055-2057 (1964)
New differential thermal analysis technique for measuring isothermal crystallization rates of high polymers	J. Chiu	36,	2058-2061 (1964)
Problems of direct determination of trace nickel in oil by X-ray emission spectrography	E.L. Gunn	36,	2086-2090 (1964)

ANGEWANDTE CHEMIE
(International Edition)

This is a monthly publication, printed both in German and in English (International) edition. Only references to the latter one, a complete translation version, are given below. Edited by W. Foerst and published by Verlag Chemie, GMBH, Weinheim and Academic Press, New York-London.

Title	Authors	Reference
Valence isomerization in compounds with strained rings	E. Vogel	2, 1-11 (1963)
Stereochemistry of displacement reactions at phosphorus atoms	R.F. Hudson and M. Green	2, 11-20 (1963)
Reactive dyes containing vinylsulfonyl groups	J. Heyna	2, 20-23 (1963)
Syntheses in the 3-aminoquinazol-4-one and 3-aminobenzotriazin-4-one series	S. Petersen H. Herlinger E. Tietze and W. Siefken	2, 24-29 (1963)
Polycondensation dyes (A new principle for the preparation of wash-fast cotton dyeings)	K. Schimmelschmidt H. Hoffmann and E. Baier	2, 30-31 (1963)
Low-temperature polymerization of ethylene	H. Bestian K. Clauss H. Jensen and E. Prinz	2, 32-41 (1963)
Some illustrative applications of NMR spectroscopy to organic chemistry	J.D. Roberts	2, 53-59 (1963)
Metal π-complexes with di- and oligo-olefinic ligands	E.O. Fischer and H. Werner	2, 80-93 (1963)
Cyclooligomerization of butadiene and transition metal π-complexes	G. Wilke	2, 105-115 (1963)
Thermal rearrangements	W. Von E. Doering and W.R. Roth	2, 115-122 (1963)
Fulvenes as isomers of benzenoid compounds	K. Hafner K.H. Häfner C. König et al.	2, 123-134 (1963)
(Halogenomethyl)pyridines and (halogenomethyl)quinolines	W. Mathes and H. Schuly	2, 144-149 (1963)
Recent developments in organotin chemistry	W.P. Neumann	2, 165-175 (1963)

Chemistry and biochemistry of insect hormones	P. Karlson	_2_,	175-182 (1963)
Gas chromatographic investigation of racemization during peptide synthesis	F. Weygand, A. Prox, L. Schmidhammer and W. König	_2_,	183-188 (1963)
New Methods of Preparative Organic Chemistry Syntheses using pyridinium salts	F. Kröhnke	_2_,	225-238 (1963)
The biogenesis of ergot alkaloids	F. Weygand and H.-G. Floss	_2_,	243-247 (1963)
Macromolecules with non-uniform solubilities Heterogels and heteropolymers	Ch. Sadron	_2_,	248-250 (1963)
The Mossbauer effect and its significance in chemistry	E. Fluck, W. Kerler and W. Neuwirth	_2_,	277-287 (1963)
Emulsion polymerization of vinyl monomers under the influence of ionizing radiation	D. Hummel	_2_,	295-308 (1963)
New Methods in Preparative Organic Chemistry Syntheses with S-triazine	Ch. Grundmann	_2_,	309-323 (1963)
Principles and techniques of diffuse-reflectance spectroscopy	G. Kortum, W. Braun and G. Herzog	_2_,	333-341 (1963)
The biosynthesis of alkaloids. I.	K. Mothes and H.R. Schütte	_2_,	341-357 (1963)
Methods of peptide synthesis	T. Wieland and H. Determann	_2_,	358-370 (1963)
Base-catalyzed reactions of ketones with hydrogen sulfide	R. Mayer, G. Hiller, M. Nitzschke and J. Jentzsch	_2_,	370-373 (1963)
Formation and cleavage of dihydroxydiarylmethane derivatives	H. Schnell and H. Krimm	_2_,	373-379 (1963)
Production of polyhydric alcohols from wood polysaccharides	V.I. Sharkov	_2_,	405-409 (1963)
Oligomers and pleionomers of synthetic fiber-forming polymers	H. Zahn and G.B. Gleitsman	_2_,	410-420 (1963)

Recent studies on the centrally active endogenous amines	J. W. Daly and B. Witkop	2,	421-440 (1963)
The biosynthesis of alkaloids. II.	K. Mothes and H. R. Schutte	2,	441-458 (1963)
Syntheses of heterocyclic compounds from amino guanidine	F. Kurzer and L. E. A. Godfrey	2,	459-476 (1963)
Some developments in the chemistry of psycho-therapeutic agents	E. Jucker	2,	493-507 (1963)
Recent developments in the chemistry of metal silyls of the type $M(SiR_3)_n$	E. Wiberg, O. Stecher, H. J. Andrascheck, L. Kreuzbichler and E. Staude	2,	507-515 (1963)
Permethylation of oligomeric and polymeric carbohydrates and quantitative analysis of the cleavage products	K. Wallenfels and G. Bechtler	2,	515-523 (1963)
Further contributions to the knowledge of semimetals	W. Klemm and H. Niermann	2,	523-530 (1963)
Recent investigations of sulfur-nitrogen compounds	O. Glemser et al.	2,	530-540 (1963)
Modern methods for the separation and continuous measurement in the gas phase of compounds labelled with radiocarbon (Radio reaction gas chromatography)	F. Drawert and O. Bachmann	2,	540-545 (1963)
1,3-Dipolar cycloadditions. Past and future	R. Huisgen	2,	565-598 (1963)
Nitrenes	L. Horner and A. Christmann	2,	599-608 (1963)
An acylol of dimethylketone	H. Bestian and D. Gunther	2,	608-613 (1963)
Preparation of carbon tetrachloride from phosgene	O. Glemser et al.	2.	613-615 (1963)
Kinetics and mechanism of 1,3-dipolar cyclo-additions	R. Huisgen	2,	633-645 (1963)
New developments in the chemistry of polyphosphanes	E. Wiberg, M. Vanghemen and G. Muller-Schiedmayer	2,	646-654 (1963)

Syntheses of s-triazine and substituted s-triazines	H. Bredereck F. Effenberger A. Hofmann and M. Hajek	2,	655-659 (1963)
Some aspects of the chemistry of organoselenium compounds	J. Gosselck	2,	660-669 (1963)
Preparation of N-alkoxyureas and their use as selective herbicides	O. Scherer G. Hörlein and K. Härtel	2,	670-673 (1963)
Neighboring group and substituent effects in organosulfur compounds	K.D. Gundermann	2,	674-683 (1963)
Reactions of olefins with the titanium-carbon bond	H. Bestian and K. Clauss	2,	704-714 (1963)
Isothiazoles	F. Hübenett F.H. Flock W. Hansel et al.	2,	714-719 (1963)
Type and extent of isomerization of straight-chain monoolefins during the Koch carboxylic acid synthesis	K.E. Moller	2,	719-722 (1963)
Phenol oxidation reactions	H. Musso	2,	723-735 (1963)
Proton transfer, acid-base catalysis, and enzymatic hydrolysis. Part I: Elementary processes	M. Eigen	3,	1-19 (1964)
The course of the Willgerodt-Kindler reaction of alkyl aryl ketones	F. Asinger W. Schäfer K. Halcour et al.	3,	19-28 (1964)
Recent developments in the study of lysozymes	P. Jolles	3,	28-36 (1964)
Defensive substances of the arthropods, their isolation and identification	H. Schildknecht K. Holoubek K.H. Weis and H. Krämer	3,	73-82 (1964)
The valence tautomerism of cyclooctatetraene	R. Huisgen and F. Mietzsch	3,	83-85 (1964)
Recent developments in the chemistry of aminoboranes	K. Niedenzu	3,	86-92 (1964)
The mechanism of the action of organometallic catalysts	H. Sinn and F. Patat	3,	93-101 (1964)
Complex formation with organoaluminum compounds	H. Lehmkuhl	3,	107-114 (1964)

ANGEWANDTE CHEMIE (International Edition) (Contd.)

Title	Authors	Citation
Recent developments in the chemistry of pteridines	W. Pfleiderer	3, 114-132 (1964)
Structure and aromatic character of non-benzenoid cyclically conjugated systems	K. Hafner	3, 165-173 (1964)
Transformations of organoboranes at elevated temperatures	R. Köster et al.	3, 174-185 (1964)
Coordination chemistry and catalysis (Investigations on the synthesis of cyclooctatetraene by the method of W. Reppe)	G.N. Schrauzer with P. Glockner and S. Eichler	3, 185-191 (1964)
Oxidative condensation of quaternary phenolic bases	B. Franck, G. Blaschke and G. Schlingloff	3, 192-200 (1964)
New chemical and biochemical developments in the vitamin B_{12} field	K. Bernhauer, O. Müller and F. Wagner	3, 200-211 (1964)
New methods of preparative organic chemistry. IV. Cyclization of dialdehydes with nitromethane	F.W. Lichtenthaler	3, 211-224 (1964)
Polar addition of hydrogen halides onto olefins	J.S. Dewar and R.C. Fahey	3, 245-249 (1964)
New methods in preparative organic chemistry. IV. Synthesis of naturally occurring unsaturated fatty acids by sterically controlled carbonyl olefination	L.D. Bergelson and M.M. Shemyakin	3, 250-260 (1964)
Tetramethylurea as a solvent and reagent	A. Lüttringhaus and H.W. Dirksen	3, 260-269 (1964)
The peroxyester reaction	G. Sosnovsky and S.-O. Lawesson	3, 269-276 (1964)
Aliphatic thioketones	R. Mayer, J. Morgenstern and J. Fabian	3, 277-286 (1964)
Organosodium and organopotassium compounds Part I: Properties and reactions	M. Schlosser	3, 287-306 (1964)
Structure and specificity of organic chelating agents	E. Bayer et al.	3, 325-332 (1964)
Three-membered rings containing two hetero-atoms	E. Schmitz	3, 333-341 (1964)

Reactions of sodium hydrazide with organic compounds	Th. Kauffmann et al.	3,	342-353 (1964)
Recent advances in the synthesis of 19-norsteroids	T.B. Windholz and M. Windholz	3,	353-361 (1964)
Organosodium and organopotassium compounds Part II: Preparation and synthetic applications	M. Schlosser	3,	362-373 (1964)
Biosynthesis, structure, and biological action of the lipids of the tubercle bacillus	E. Lederer	3,	393-400 (1964)
New methods of preparative organic chemistry. IV. Syntheses with nascent quinones	H.W. Wanzlick et al.	3,	401-408 (1964)
New reactive dyes	K.G. Kleb E. Siegel and K. Sasse	3,	408-416 (1964)
Acrolein polymers	R.C. Schulz	3,	416-423 (1964)
Complexes of tetradentate ligands containing phosphorus and arsenic	L.M. Venanzi	3,	453-460 (1964)
The chemistry of new heat-resistant films and fibers	F.T. Wallenberger	3,	460-470 (1964)
A synthetic route to the Carrin system	E. Bertele H. Boos J.D. Dunitz et al.	3,	490-496 (1964)
Intramolecular free-radical reactions	K. Heusler and J. Kalvoda	3,	525-538 (1964)
The preparation of catena compounds by directed synthesis	G. Schill and A. Lüttringhaus	3,	546-547 (1964)
The modes of reaction of ambident cations	S. Hünig	3,	548-560 (1964)
Relations between structure and reactivity of ambifunctional nucleophilic compounds	R. Gompper	3,	560-570 (1964)
The phenomenon of phototrophy	W. Luck and H. Sand	3,	570-580 (1964)
Free radicals and free-radical reactions of monovalent and divalent sulfur	U. Schmidt	3,	602-608 (1964)
Electron affinity of organic molecules	G. Briegleb	3,	617-632 (1964)
Structure and acidity of organic compounds	W. Simon	3,	661-668 (1964)

Surface oxides of carbon	H.-P. Boehm, E. Diehl, W. Heck and R. Sappok	3,	669-677 (1964)
The design of phosphorylating agents	V.M. Clark, D.W. Hutchinson, A.J. Kirby and S.G. Warren	3,	678-685 (1964)
Stereospecific homopolymerization of cyclopentene	G. Natta, G. Dall'asta and G. Mazzanti	3,	723-729 (1964)
Photochemical substitutions on metal carbonyls and their derivatives	W. Strohmeier	3,	730-737 (1964)
Phosphonium salt formation of the second kind	H. Hoffmann and H.J. Diehr	3,	737-746 (1964)
Recoil reactions of tritium with organic compounds	F. Schmidt-Bleek and F.J. Rowland	3,	769-776 (1964)
Proton addition complexes of aromatic hydrocarbons	H.H. Perkampus and E. Baumgarten	3,	776-783 (1964)
Gradient and low temperature thin layer chromatography	E. Stahl	3,	784-791 (1964)
Old and new processes of multipliative distribution (Liquid-liquid extraction)	W. Fischer	3,	791-799 (1964)

This is a bi-monthly journal, in French, published since 1790, which since 1962 publishes reviews, articles and theses in all areas of chemistry. Edited by A. Willemart and published by Masson et Cie, Paris.

Title	Authors	Reference
Research on mobile hydrogen compounds (in French)	J.A. Gautier	8, 5-25 (1963)
Chemical transformation of steroids by the action of micro-organisms (in French)	J.R. Pasqualini	8, 27-43 (1963)
Contribution to the magneto-optical study of π bonds between aliphatic carbon atoms (in French)	J.-F. Labarre	8, 45-83 (1963)
Hydration of acetylenic compounds (in French)	M. Miocque, N.N. Hung and V.Q. Yen	8, 157-174 (1963)
Research on cycloheptanes. (Photochemical reactions of diphenyl-1,4-cycloheptadiene-1,3) (in French)	P. Courtot	8, 197-227 (1963)
Contribution to the study of the properties of the benzoyl-1-isoquinolines and of their transformation into phenyl-1-benzazepines-3 (in French)	J. Chazerain	8, 255-284 (1963)
Studies on phellonic acid (in French)	L. Duhamel	8, 315-346 (1963)
Contribution to the study of the organic amides of phosphoric acid (in French)	R. Burgada	8, 347-381 (1963)
Influence of intra-molecular hydrogen bonds on the conformations measured in solution of β-diol and polyvinylic alcohol molecules (in French)	H. Buc	8, 431-456 (1963)
Action of organomagnesium compounds on α-amino-tetrahydropyrans (in French)	A. Gaumeton	8, 457-486 (1963)
The problems of the structural analogues of emetine (in French)	C. Viel	8, 503-513 (1963)
Research on the preparation of homoveratrylamine and on its uses for the synthesis of bis-isoquinolic compounds (in French)	C. Viel	8, 515-544 (1963)
Contribution to the kinetic study of the mechanism of nucleophilic substitution (in French)	J. Landais	8, 575-601 (1963)

ANNALES DE CHIMIE (Contd.)

Title	Authors	Reference
On the molecular configuration of some derivatives of pyridine and of N-phenylsuccinimide (in French)	J. Barassin	8, 637-666 (1963)
Study of N-amine derivatives of methyl-1 piperazine, piperidine and morpholine (in French)	F. Erb-Debruyne	9, 73-93 (1964)
Recent studies on alkaloids of pseudo-cinchona and yohimbe (in French)	J. Poisson	9, 99-121 (1964)
Alcoholation reactions of carbon in liquid ammonia (in French)	M. Miocque and M. Duchon-d'Engenieres	9, 123-134 (1964)
New nucleophilic and electrophilic substitutions in allylic and propargylic series (in French)	M. Andrac	9, 287-315 (1964)
Effect of structural factors on melting and transition points of polyolefins (in French)	A. Le Bris and G. Lefebvre	9, 333-348 (1964)
Physico-chemical properties of organo-phosphor compounds (in French)	G. Mavel	9, 349-357 (1964)
Contribution to the study of phenyl-2 indanedione-1,3 derivatives (in French)	P. Aubrun	
Infrared spectrography of ethylenic alcohols (in French)	Y. Armand and A. Arnaud	9, 433-460 (1964)
Use of some inclusion compounds in organic chemistry (in French)	C. Asselineau and J. Asselineau	9, 461-479 (1964)

ANNALES PHARMACEUTIQUES FRANCAISES

This is a monthly journal, in French. It is the official bulletin of the Societies of Pharmacy of Paris and the Provinces. Published by Masson et Cie, Paris.

Title	Authors	Reference
Biology of amino acids	M. Vigneron	21, 425 (1963)
Some aspects of the metabolism of amino acids in plants	A. Santos-Ruiz	21, 529-541 (1963)
Some aspects of the metabolism of amino acids in plants	A. Santos-Ruiz	21, 619-638 (1963)
Importance and reactivity of α-ketonic acids	P. Cordier	22, 225-241 (1964)

This is an annual collection of review articles, in English, published since 1932. Edited in 1963 by E.E. Snell, J.M. Luck, F.W. Allen and G. Mackinney and by E.E. Snell, J.M. Luck, P.D. Boyer and G. Mackinney in 1964. Published by Annual Reviews, Inc., Palo Alto, California

Title	Authors	Reference
Lost in the twentieth century	A. Szent-Gyorgyi	32, 1-14 (1963)
Ribonucleic acids — Biosynthesis and degradation	A. Stevens	32, 15-42 (1963)
Chemistry of the carbohydrates	R.U. Lemieux and D.R. Lineback	32, 155-184 (1963)
Complex lipids	D.J. Hanahan and G.A. Thompson, Jr.	32, 215-240 (1963)
Structure of proteins	F.M. Richards	32, 269-300 (1963)
The hemoglobins	W.A. Schroeder	32, 301-320 (1963)
Gas chromatography	E.C. Horning and W.J.A. Vanden-Heuvel	32, 709-754 (1963)
Mass spectrometry	K. Biemann	32, 755-780 (1963)
Mechanisms of certain phosphotransferase reactions: Correlation of structure and catalysis in some selected enzymes	J.P. Hummel and G. Kalnitsky	33, 15-50 (1964)
Polysaccharides	D.H. Northcote	33, 51-74 (1964)
The chemistry of lipids	R.O. Brady and E.G. Trams	33, 75-100 (1964)
The chemistry and metabolism of recently isolated amino acids	L. Fowden	33, 173-204 (1964)
Genetic coding for protein structure	J.C. Bennett and W.J. Dreyer	33, 205-234 (1964)
Chemistry and metabolic action of non-steroid hormones	R. Schwyzer	33, 259-286 (1964)
Vitamin and coenzyme function: Vitamin B_{12} and folic acid	L. Jaenicke	33, 287-312 (1964)

ANNUAL REVIEW OF BIOCHEMISTRY (Contd.)

Title	Authors	Reference
Application of nuclear magnetic resonance in biochemistry	A. Kowalsky and M. Cohn	33, 481-518 (1964)
Biochemistry of cancer	H. Busch and W.C. Starbuck	33, 519-570 (1964)
Metabolism of drugs and toxic substances	L. Shuster	33, 571-596 (1964)
The chemistry of peptides and proteins	C.H.W. Hirs	33, 597-632 (1964)

ANNUAL REVIEW OF NUCLEAR SCIENCE

This is an annual collection of review articles, in English. Edited by E. Segre, L.I. Schiff, G. Friedlander and W.E. Mayerhof and published by Annual Reviews, Inc., in cooperation with the National Research Council of the National Academy of Sciences.

Title	Authors	Reference
Methods and applications of activation analysis	E.V. Sayre	13, 145-162 (1963)
Physics, chemistry, and meteorology of fallout	R. Bjornerstedt and K. Edvarson	13, 505-534 (1963)
Radiation effects on macromolecules of biological importance	F. Hutchinson	13, 535-564 (1963)
The chemistry of the actinide elements	B.B. Cunningham	14, 323-346 (1964)
Quantitation of cellular radiobiological responses	G.F. Whitmore and J.E. Till	14, 347-374 (1964)

This is an annual collection of review articles, in English. Edited by H. Eyring and published by Annual Reviews, Inc., Palo Alto, California.

Title	Authors	Reference		
Fast reactions in solution	L. DeMaeyer and K. Kustin	14,	5-34	(1963)
Electronic spectra of molecular crystals	O. Schnepp	14,	35-60	(1963)
Heterogeneous catalysis	G.-M. Schwab and K. Gossner	14,	177-204	(1963)
Solutions of nonelectrolytes	J.A. Barker	14,	229-250	(1963)
Thermodynamics of solutions at low temperatures	D. White and C.M. Knobler	14,	251-270	(1963)
Nucleophilic reactivity	J.F. Bunnett	14,	271-290	(1963)
High polymers	R.E. Hughes and C.A. von Frankenberg	14,	291-312	(1963)
Radiation chemistry	P.J. Dyne, D.R. Smith and J.A. Stone	14,	313-334	(1963)
Nuclear magnetic resonance	S. Meiboom	14,	335-358	(1963)
Infrared vibrational-rotational spectroscopy	A.H. Nielsen	14,	359-386	(1963)
Fast reactions in solution	G.G. Hammes	15,	13-30	(1964)
Solutions of nonelectrolytes	A.G. Williamson	15,	63-82	(1964)
Surface chemistry	R.L. Burwell, Jr. and J.B. Peri	15,	131-154	(1964)
Chemical and electrochemical electron-transfer theory	R.A. Marcus	15,	155-196	(1964)
Electron spin resonance	D. Kivelson and C. Thomson	15,	197-224	(1964)
Microwave spectroscopy	D.R. Lide, Jr.	15,	225-250	(1964)
Quantum theory of atoms and molecules	O. Sinanoglu and D. Fu-tai Tuan	15,	251-280	(1964)

ANNUAL REVIEW OF PHYSICAL CHEMISTRY (Contd.)

Title	Authors	Reference
Optical rotation	D.J. Caldwell and H. Eyring	15, 281-310 (1964)
Quantum processes in photosynthesis	G.W. Robinson	15, 311-348 (1964)
High pressure chemistry	S.D. Hamann	15, 349-370 (1964)
Nucleic acids	I. Tinoco, Jr. and D.N. Holcomb	15, 371-394 (1964)
Contractile processes in fibrous macromolecules	L. Mandelkern	15, 421-448 (1964)
Thermochemistry and bond dissociation energies	H.A. Skinner	15, 449-468 (1964)
Solutions of electrolytes	J.N. Agar	15, 469-488 (1964)
High resolution nuclear magnetic resonance	D.M. Grant	15, 489-528 (1964)

ANNUAL REVIEW OF PHYTOPATHOLOGY

These are the first two volumes of a new series in Annual Reviews, intended to evaluate and summarize the current advances in given areas within the field of plant pathology. Articles are in English and emphasize the biological approach, including also some chemical inter-relations. Edited by J.G. Horsfall and published by Annual Reviews, Palo Alto, California.

Title	Authors	Reference
Growth regulators in plant disease	L. Sequeira	1, 5-30 (1963)
Chemistry and physiology of fungicidal action	R.G. Owens	1, 77-100 (1963)
Selective toxicity of chemicals to soil microorganisms	W.A. Kreutzer	1, 101-126 (1963)
Degradation of cellulose	B. Norkrans	1, 325-350 (1963)
Phytoalexins	I.A.M. Cruickshank	1, 351-374 (1963)
Ozone damage to plants	S. Rich	2, 253-266 (1964)
Soil fungicides	K.H. Domsch	2, 293-320 (1964)
Microbial degradation of synthetic compounds	D. Woodcock	2, 321-340 (1964)

ANTIBIOTICA ET CHEMOTHERAPIA

This is a collection of review articles, in French, German or English, mainly in the area of clinical chemistry. Published by S. Karger, Basel - New York.

Title	Authors	Reference
2nd International Symposium of Chemotherapy. New Antibiotics Side Effects of Modern Chemotherapy	H.P. Kuemmerle, P. Preziosi and R. Reutchnick (Editors)	
New antifungal antibiotics (in French)	E. Drouhet	11, 21-56 (1963)
The rifamycins (in English)	G. Ginnchi	11, 61-83 (1963)
New tetracyclin derivatives (in German)	H. Knothe	11, 97-117 (1963)
The new synthetic penicillins (in English)	E.T. Knudsen	11, 118-135 (1963)
New tetracycline derivatives (in English)	V. Capraio	11, 177-179 (1963)
Antibiotic actinoidin : its isolation and chemical properties (in English)	H.G. Brazhnikova, N.M. Lomakina, M.J. Yurina and M.F. Lavrova	11, 253-260 (1963)
Pharmacokinetics and Drug Dosage	E. Freerksen, L. Dettli, E. Kruger-Thiemer and E. Nelson (Editors)	
Kinetics of the acetylation and excretion of sulfonamides and a comparison of two models (in English)	E. Nelson	12, 29-40 (1964)
Relation of physical and chemical properties of chemotherapeutic agents to their biological activity (in German)	J. Seydel	12, 135-147 (1964)

THE AUSTRALIAN JOURNAL OF SCIENCE

This is a monthly journal, in English. Edited by K.W. Knox and published for the Australian and New Zealand Association for the Advancement of Science.

Title	Authors	Reference
Atomic emission spectroscopy	S.C. Baker	26, 171-177 (1963)

BULLETIN DE LA SOCIÉTÉ DE CHIMIE BIOLOGIQUE

This is a monthly journal, usually in French, of the Biochemical Society of France. Most articles are of an experimental nature but occasionally lectures are published. Edited by Y. Raoul and published by Masson et Cie, Paris.

Title	Authors	Reference
Chemical factors that control ionic movements during nervous system activity (in French)	D. Nachmansohn	45, 29-54 (1963)
Hyaluronic acid and permeability factors in tissues (in French)	W. Pigman	45, 185-202 (1963)
Biosynthesis of ATP (in French)	E. Racker	45, 321-337 (1963)
Some experimental evidence concerning the function of soluble RNA in the synthesis of proteins (in French)	F. Chapeville	45, 835-855 (1963)
Action of adenilic nucleosides and nucleotides on the biosynthesis of porphyrins (in French)	A. Gajdos and M. Gajdos-Török	45, 1227-1246 (1963)
Chemistry of photosynthesis (in French)	O. Warburg, G. Krippahl, K. Jetschmann and A. Lehmann	46, 9-23 (1964)
New nucleotides (in French)	W.E. Cohn	46, 239-253 (1964)
Glycolipids (in French)	J. Polonovski	46, 833-857 (1964)
The synthesis of proteins upon ribosomes (in English)	J.D. Watson	46, 1399-1425 (1964)
On the transcription of DNA sequences by RNA polymerase (in English)	P. Berg and M. Chamberlin	46, 1427-1440 (1964)
A contribution to the understanding of the primary quantum conversion in photosynthesis (in English)	R.H. Ruby, I.D. Kuntz, Jr. and M. Calvin	46, 1595-1605 (1964)
Primary reactions in photosynthesis (in French)	P. Joliot and J. Lavorel	46, 1607-1626 (1964)
Structure and activity of pancreatic ribonuclease (in French)	S. Moore	46, 1739-1744 (1964)
The active center in enzyme action (in English)	D.E. Koshland, Jr.	46, 1745-1755 (1964)

BULLETIN DE LA SOCIÉTÉ CHIMIQUE DE FRANCE

This is a monthly journal, usually in French, of the French Chemical Society. Most articles are of an experimental nature but reviews and symposia are included. Edited by G. Kersaint and published for the French Chemical Society by Masson et Cie, Paris.

Title	Authors	Reference	
Conformational analysis (in French)	W. Hückel	1-7	(1963)
Five-membered rings (in French)	W. Hückel	8-10	(1963)
Studies of certain organic derivatives of pyrophosphoric acids (in French)	J. Michalski	11-16	(1963)
Principles of x-ray spectroscopy (in French)	G. Pomey	185-191	(1963)
Inorganic heterocyclic chemistry. II. (in French)	H. Garcia-Fernandez	416-424	(1963)
Aldolic reactions in the gaseous phase. XV. Reactions between benzoic aldehyde and aliphatic aldehydes (in French)	S. Malinowski, W. Kiewlicz and E. Sołtys	439-444	(1963)
Present biochemical trends (in French)	E. Lederer	942-947	(1963)

International Colloquium on Organometallic Compounds, held in Paris, 24-28 September, 1962.

Title	Authors	Reference	
Structure of the carbon-metal bond in certain groups of organometallic compounds (in French)	R. Daudel	1345-1352	(1963)
Alkaline metallo-organics: A chapter of organic anionic chemistry (in German)	G. Wittig	1352-1356	(1963)
Some recent and current studies on organolithium compounds and pseudo-organolithium compounds (in English)	H. Gilman	1356-1359	(1963)
The use of lithium in the synthesis of silicon-nitrogen compounds (in English)	E.G. Rochow	1360-1364	(1963)
The transmetalation reaction as a source of new organolithium reagents. A review of recent work at M.I.T. (in English)	D. Seyferth, M.A. Weiner, L.G. Vaughan et al.	1364-1367	(1963)
The influence of the nature of the alkaline metal in the selective alcoylation of a tertiary acetylenic alcohol (in French)	J.A. Gautier, M. Miocque and M. Duchon D'Engenières	1368-1372	(1963)

Multi-center and assisted mechanistic pathways in the reactions of organometallic compounds (in English)	R.E. Dessy and F. Paulik	1373-1383 (1963)
On the synthesis of potically active magnesiumorganic and lithiumorganic compounds by means of mercury organic compounds (in English)	O.A. Reutov	1383-1385 (1963)
On the genesis, structure and reactivity of solvated organometallics (in French)	M. Andrac, F. Gaudemar, M. Gaudemar et al.	1385-1395 (1963)
Study of the composition of Grignard solutions (in French)	A. Kirrmann, R. Hamelin and S. Hayes	1395-1403 (1963)
Tonometric study of organomagnesium solutions (in French)	S. Hayes	1404-1407 (1963)
Studies on diethylmagnesium solutions (in French)	A. Kirrmann, M. Vallino and J.F. Fauvarque	1408-1410 (1963)
The mechanism of the action of organomagnesium solutions on ketones (in French)	R. Hamelin	1411-1417 (1963)
Unsolvated organomagnesium complexes (in English)	D. Bryce Smith	1418-1422 (1963)
On the preparations of polyfunctional organometallic reagents from the combinations of the type $R.C \equiv C.CH_2.COOH$, $R.CH = C = CH.CH_2.COOH$, $N \equiv C—CH_2—COOH$ and their use in synthesis (in French)	D. Ivanoff and B. Blagoev	1422-1428 (1963)
Study of the behavior of α-aminonitrils as a function of the nature of organometallic reagents (in French)	G. Chauvière, B. Tchoubar and Z. Welvart	1428-1433 (1963)
Recent findings in the field of organomagnesium compounds (in French)	H. Normant	1434-1437 (1963)
Eliminations in the Grignard reaction (in French)	J. Ficini and H. Normant	1438-1440
Formation of organomagnesium compounds from halogenated ketones (in French)	C. Feugeas and H. Normant	1441-1443 (1963)
Contribution to the study of vinylic magnesiums and related compounds (in French)	G. Fontaine, C. André, C. Jolivet and P. Maitte	1444-1448 (1963)

The influence of the solvent in the preparation and reactions of organometallics (of magnesium and aluminum), particularly in the case of α-acetylenic and α-ethylenic derivatives (in French)	L. Miginiac-Groizeleau	1449-1452 (1963)
Action of ethyl γ-bromocrotonate on aliphatic ketones in the presence of zinc (in French)	J. Colonge and J. Varagnat	1453-1455 (1963)
Progress of organometallic syntheses using aluminum derivatives (in French)	K. Ziegler	1456-1461 (1963)
Aluminum-alkenes and -alkynes (in German)	G. Wilke and W. Schneider	1462-1467 (1963)
Alkyl group exchange and association in aluminum alkyl compounds (in English)	E.G. Hoffmann	1467-1471 (1963)
Studies in the field of organometallic compounds of zinc and cadmium (in French)	K.A. Kotchechkov N.I. Cheverdina and I.E. Paleeva	1472-1474 (1963)
Research on organoaluminum and α-unsaturated organozinc derivatives (in French)	M. Gaudemar	1475-1482 (1963)
Steric orientation and the Reformatsky reaction (in French)	M. Mousseron J. Neyrolles and Y. Beziat	1483-1490 (1963)
Reaction of organomagnesium in hydrocarbons. Existence of a magnesium complex of tert-butyl acetate (in French)	J.E. Dubois S. Molnarfi F. Hennequin et al.	1491-1496 (1963)
On the reactivity of organocadmium derivatives (in French)	F. Tatibouët and P. Fréon	1496-1500 (1963)
Coordination synthesis on metal centers (in English)	H.H. Zeiss	1500-1506 (1963)
Recent studies on the synthesis and on certain chemical properties of glycerophosphatidic acids, of glycerocephalines and of glycerolecithines (in French)	P.E. Verkade	1993-2005 (1963)
Organic compounds of trivalent phosphorus with a P–N bond (in French)	R. Burgada	2334-2343 (1963)
Recent progress in the chemistry of lignines (in French)	M.S. Adjangba	2344-2358 (1963)
Polarography of organic substances in non-aqueous media (in French)	G. LeGuillanton	2359-2375 (1963)

Mechanisms of stereospecific polymerization by transition metals (in French)	F. Dawans and P. Teyssie	2376-2392 (1963)
On the structure and pelletierine, an alkaloid of pomegranate (in French)	G. Drillien and C. Viel	2393-2400 (1963)
Desoxymercuration (in French)	M.M. Kreevoy	2431-2432 (1963)
Polyphosphoric acid in organic chemistry (in French)	J.-P. Marthe and S. Munavalli	2679-2692 (1963)
Reactive species and the intermediary complexes in heterolytic halogenation of organic compounds (in French)	E. Chilov	2902-2909 (1963)
Recent findings in the chromatography of amino acids (in French)	M. Jutisz and P. de la Llosa	2913-2923 (1963)
Chromatography of peptides on resin columns and on ion exchange columns (in French)	P. Jollès	2923-2932 (1963)
Qualitative and quantitative chromatography of dinitrophenyl-amino acids (in French)	G. Biserte and Y. Moschetto	2932-2938 (1963)
New methods in paper chromatography of dinitrophenyl-amino acids. Improvements and attempts at systematization of the methods of paper chromatography of amino acids (in French)	R.L. Munier and G. Sarrazin	2939-2953 (1963)
Chromato-electrophoresis of amino acids and peptides (in French)	G. Biserte M. Dautrevaux and P. Boulanger	2954-2959 (1963)
The use of volatile buffers in ion exchange column chromatography (in French)	P. Padieu N. Maleknia and A.-M. Thireau	2960-2964 (1963)
The original family, the sydnones (in French)	Y. Noel	173-177 (1964)
Anionic polymerizations without termination reactions (in French)	P. Sigwalt	423-435 (1964)
Automatic potentiometric titrations (in French)	J. Tacussel	1155-1166 (1964)
Apocynacea alkaloids, as intermediaries in steroid syntheses (in French)	R. Goutarel	1665-1672 (1964)
Modern aspects of lipochemistry (in French)	Ch. Paquot	1673-1679 (1964)

Coupling constants between nuclei in nuclear magnetic resonance (in French)	J.-Cl. Muller	2027-2032 (1964)
Coupling between nuclei in nuclear magnetic resonance and double magnetic resonance (in French)	J. Parello	2033-2062 (1964)
Applications of the diene synthesis to the preparation of 17-keto steroids in the estrone and D-homo-estrone series (in French)	I.V. Torgov, T.I. Sorkina and I.I. Zaretskaya	2063-2068 (1964)
The use of solvents in organic chemistry (in French)	B. Tchoubar	2069-2079 (1964)
The use of diedral angles in conformational analysis. Geometrical and energetic correlations in relation to single rings and the structure of steroids (in French)	R. Bucourt	2080-2092 (1964)
Use of mass spectrometry in structural chemistry (in French)	M. Plat	2361-2372 (1964)
Use of electron spin resonance for the study of dynamic systems (in French)	J.M. Sommer	2372-2380 (1964)
Use of ultrasound in organic chemistry (in French)	P. Renaud and H. Gillier	2381-2391 (1964)
Coulometry with compensated intensity (in French)	Cl. Schoedler	2401-2407 (1964)
Nuclear magnetic resonance (in French)	A. Kirrmann	2695 (1964)
Kinetic studies using NMR techniques (in French)	J.J. Delpuech	2697-2710 (1964)
NMR studies of fluorine (in French)	Cl. Beguin	2711-2714 (1964)
NMR spectra of organometallic compounds (in French)	S. Hayes	2715-2718 (1964)

CHEMICAL REVIEWS

This is a quarterly journal, in English. Published since 1924 by the <u>American Chemical Society</u> and dedicated exclusively to reviews in all areas of chemistry. Edited by R.L. Shriner.

Title	Authors	Reference
Active nitrogen	G.G. Mannella	63, 1-20 (1963)
Oxidations and dehydrogenations with N-bromo-succinimide and related N-haloimides	R. Filler	63, 21-43 (1963)
<u>Meta</u>-Bridged aromatic compounds	R.W. Griffin, Jr.	63, 45-54 (1963)
Cyclization of N-halogenated amines (The Hofmann-Löffler Reaction)	M.E. Wolff	63, 55-64 (1963)
Thermochromism	J.H. Day	63, 65-80 (1963)
Chemical applications of nuclear magnetic double resonance	J.D. Baldeschwieler and E.W. Randall	63, 81-121 (1963)
The catalytic hydrogenolysis of small carbon rings	J. Newham	63, 123-137 (1963)
Solvent extraction of inorganic species	Y. Marcus	63, 139-170 (1963)
Linear free energy relationships	P.R. Wells	63, 171-219 (1963)
Properties of organic solutions of heavy metal soaps	N. Pilpel	63, 221-234 (1963)
Carbenes. Reactive intermediates containing divalent carbon	E. Chinoporos	63, 235-268 (1963)
Synthetic reversible oxygen-carrying chelates	L.H. Vogt, Jr. H.M. Faigenbaum and S.E. Wiberley	63, 269-277 (1963)
Heterocyclic quinones from 2,3-dichloro-1,4-naphthoquinone	M.F. Sartori	63, 279-296 (1963)
Resolution of optical isomers by crystallization procedures	R.M. Secor	63, 297-309 (1963)
The reactions of mesityl oxide	M. Hauser	63, 311-324 (1963)
Effects of the environment on the fluorescence of aromatic compounds in solution	B.L. van Duuren	63, 325-354 (1963)

Atomic chlorination of simple hydrocarbon derivatives in the gas phase	G. Chiltz P. Goldfinger G. Huybrechts et al.	63, 355-372 (1963)
The Fischer indole synthesis	B. Robinson	63, 373-401 (1963)
Organogermanium chemistry	D. Quane and R.S. Bottei	63, 403-442 (1963)
Structure and reactivity of the oxyanions of transition metals	A. Carrington	63, 443-460 (1963)
Ring-chain tautomerism	P.R. Jones	63, 461-487 (1963)
The chemistry of imines	R.W. Layer	63, 489-510 (1963)
Recent advances in the chemistry of pyrrole	E. Baltazzi and L.I. Krimen	63, 511-556 (1963)
Organic sulfites	H.F. van Woerden	63, 557-571 (1963)
The chemistry of the vitamin B_{12} group	R. Bonnett	63, 573-605 (1963)
The oxygen fluorides	A.G. Streng	63, 607-624 (1963)
Glutamic acid: Chemical syntheses and resolutions	C.W. Huffman and W.G. Skelly	63, 625-644 (1963)
Uses of isotopes in addition polymerization	G. Ayrey	63, 645-667 (1963)
Articles for chemical reviews Suggestions to authors	R.L. Shriner	64, 1-6 (1964)
Hindered and chelated 1,2-enediols	C.A. Buehler	64, 7-18 (1964)
Chemistry of the aliphatic polynitro compounds and their derivatives	P. Noble, Jr. F.G. Borgardt and W.L. Reed	64, 19-57 (1964)
The aporphine alkaloids	M. Shamma and W.A. Slusarchyk	64, 59-79 (1964)
The rearrangement of ketoxime O-sulfonates to amino ketones (The Neber rearrangement)	C. O'Brien	64, 81-89 (1964)
Organic and inorganic hypofluorites	C.J. Hoffman	64, 91-98 (1964)

Copper (II) complexes with subnormal magnetic moments	M. Kato H.B. Jonassen and J.C. Fanning	64, 99-128 (1964)
The chemistry of the sydnones	F.H.C. Stewart	64, 129-147 (1964)
Preparation and properties of imido intermediates (Imidogens)	R.A. Abramovitch and B.A. Davis	64, 149-185 (1964)
Zone melting of organic compounds	W.R. Wilcox R. Friedenberg and N. Back	64, 187-220 (1964)
Isocoumarins. Developments since 1950	R.D. Barry	64, 229-260 (1964)
Separation and analysis of various forms of hydrogen by adsorption and gas chromatography	S. Akhtar and H.A. Smith	64, 261-276 (1964)
Adamantane: Consequences of the diamondoid structure	R.C. Fort, Jr. and P. von R. Schleyer	64, 277-300 (1964)
Anionic free radicals	B.J. McClelland	64, 301-315 (1964)
Mechanisms of nucleophilic subsitution in phosphate esters	J.R. Cox, Jr. and O.B. Ramsay	64, 317-352 (1964)
Chemistry of butenolides	Y.S. Rao	64, 353-388 (1964)
Solid allotropes of sulfur	B. Meyer	64, 429-451 (1964)
Electron spin resonance spectra of oriented radicals	J.R. Morton	64, 453-471 (1964)
Nitrones	J. Hamer and A. Macaluso	64, 473-495 (1964)
Review of the chemistry of cyclopropene compounds	F.L. Carter and V.L. Frampton	64, 497-525 (1964)
Diffusion of dissolved gases in liquids	D.M. Himmelblau	64, 527-550 (1964)
Lactonic alkaloids	A.R. Pinder	64, 551-572 (1964)
The acyloin condensation as a cyclization method	K.T. Finley	64, 573-589 (1964)

CHEMIKER-ZEITUNG

This is a bi-weekly journal, in German. It publishes articles and information geared to industrial chemists and chemical engineers, as well as selected review articles. Edited by E. Baum and published by Dr. Alfred Hüthig Verlag, GMBH, Heidelberg.

Title	Authors	Reference
Processes for the synthesis of styrene	F. Klema	87, 115-117 (1963)
Organic compounds with mobile hydrogen. II. Synthetic part: Acetylenic synthesis of plurifunctional side chains	J.A. Gautier	87, 189-196 (1963)
The modern concepts of the nature of the chemical bond	E. Fluck	87, 499-508 (1963)
The modern concepts of the nature of the chemical bond	E. Fluck	87, 539-548 (1963)
Thin-layer chromatographic separation of 2,4-dinitrophenylhydrazones	A. Mehlitz, K. Gierschner and Th. Minas	87, 573-576 (1963)
The basis of nuclear resonance and electronic resonance	K. Scheffer	87, 619-627 (1963)
Stereospecific polymerization in homogeneous media	W. Kern and D. Braun	87, 799-807 (1963)
Are we using the information explosion efficiently	G.H. Hiller	88, 155-159 (1964)
Electrochemical reduction of polycyclic carbonic acids	W. Fuchs and A. Schiffers	88, 187-190 (1964)
Identification of polycyclic aromatic hydrocarbons with the aid of chromatographic methods	H.-J. Petrowitz	88, 235-241 (1964)
Gas chromatographic analysis of complex mixtures (alcohols, esters, ketones, terpenes)	F. Drawert and A. Rapp	88, 267-270 (1964)
Thin layer chromatography of aromatic nitrogen compounds	G. Pastuska and H.-J. Petrowitz	88, 311-314 (1964)
The chemical basis of bioluminescence	E. Korten	88, 385-388 (1964)
To the memory of Gerhardt Domagk		88, 395-397 (1964)
Preparative gas chromatography	K. Derge	88, 511-515 (1964)
Dielectric properties of gas chromatographic absorption agents	F. Oehme	88, 657-661 (1964)

Stereoregular and optically active polymers	R. Hatz	<u>88</u>,	855-866 (1964)
Stereoregular and optically active polymers	R. Hatz	<u>88</u>,	887-891 (1964)
Binding properties of phosphorous	E. Fluck	<u>88</u>,	951-962 (1964)

CHEMISTRY AND INDUSTRY

This is a weekly journal, in English. This journal publishes review articles and short communications. Edited by T.F. West and published by the Society of Chemical Industry, London.

Title	Authors	Reference
The chemicals from acetylene	S.A. Miller	4-15 (1963)
Recent applications of nuclear magnetic resonance spectroscopy to problems of organic chemistry. II.	R.A.Y. Jones and A.R. Katritzky	99-104 (1963)
Flavonoid compounds	R. Robinson	108-110 (1963)
Mechanical properties of polymers	A.W. Birley	154-155 (1963)
The production of acetylene by the partial oxidation of methane	E.M. Hughes, E.H. Howland, P. Grootenhuis and N.P.W. Moore	189-196 (1963)
Alkaloids containing the Ph - N - C - N system	B. Robinson	218-227 (1963)
Chemistry and mental diseases	G. Spencer	311-318 (1963)
Modern very high pressure techniques	C.H.L. Goodman	326-328 (1963)
Recent developments in the chemistry of carbohydrates	W.G. Overend	342-354 (1963)
Examination of unsaturated glyceride oils by gas chromatography	L.A. O'Neill and S.M. Rybicka	390-392 (1963)
Industrial applications of the metal organic compounds	J.H. Harwood	430-436 (1963)
The chemistry of wood	R.H. Farmer	438-440 (1963)
Criteria for beta-elimination in organic phosphates	A. Lapidot, D. Samuel and B. Silver	468-471 (1963)
The science of surface coatings	L. Valentine	523-525 (1963)
Steroid progress 1962	V. Petrow	548-554 (1963)
Electrophilic addition to carboxylic acids and their derivatives	M.H. Palmer	589-594 (1963)

CHEMISTRY AND INDUSTRY (Contd.)

Title	Author(s)	Pages
Oxidation processes in industry	C.A. Finch	602-603 (1963)
An introduction to the principles and technique of gas chromatography	G.E. Howard	622-628 (1963)
The modern approach to inorganic chemistry	N.N. Greenwood	666-668 (1963)
Carbohydrate chemistry	M. Stacey	669-670 (1963)
Some applications of radioisotopes in the chemical industry	P. Johnson, R.M. Bullock and J. Whiston	750-756 (1963)
Horizons in disinfection and antisepsis (Jubilee Memorial Lecture)	G. Sykes	885-893 (1963)
Cellulose derivatives	G.B. Turner	920-928 (1963)
Recent progress in autoxidation of monoethenoid fatty acids and their esters	J.H. Skellon and D.M. Wharry	929-932 (1963)
Theoretical aspects of phthalic anhydride manufacture	J.A. Allen	1225-1229 (1963)
Ampholytic surface active agents (A criticism of the term ampholytic as at present applied to surface active agents)	C.D. Moore and R.B. Hardwick	1268-1269 (1963)
Some aspects of recent flavour research	C.H. Lea	1406-1413 (1963)
Developments in microwave spectroscopy	D.J. Millen	1472-1482 (1963)
Zone electrophoresis	P.N. Campbell	1543-1544 (1963)
Biosynthesis of cellulose	J.A. Gascoigne	1580-1584 (1963)
Physical chemistry of detergents	H.E. Garrett	1606-1616 (1963)
Twenty-five years of nutrition	A.E. Bender	1668-1675 (1963)
Spectrometric identification of organic compounds	J.C. Roberts	1676-1677 (1963)
Chemical control of plant parasitic nematodes in the United Kingdom	J.E. Peachey	1736-1740 (1963)
International Symposium on Nitro Compounds, Warsaw, 18-20, September, 1963	M.J. Kamlet	1741-1748 (1963)

The examination of catalysts by physical and physico-chemical methods	F.S. Stone	1810-1916 (1963)
The impact of polymers in the paper industry	F.L. Hudson	13-15 (1964)
Dr. L.H. Baekeland	D.R. Houwink	50-51 (1964)
The long spacings of polymorphic crystalline forms of long-chain compounds with special reference to triglycerides	F.D. Gunstone	84-89 (1964)
Cycloaliphatic epoxides: Their synthesis and properties	H. Batzer	179-189 (1964)
About perfumery	J. Pickthall and D.E. Butterfield	259-262 (1964)
Perfumery chemicals from pinene	B.D. Sully	263-267 (1964)
Propylene oxide and its derivatives	A.C. Fyvie	384-388 (1964)
The handling and storage of process materials	T.R. Helcke	400-402 (1964)
Industrial developments in the detergent field	H. Stupel	470-472 (1964)
The discovery of the anaesthetic halothane -- an example of industrial research (25th Hurter Memorial Lecture)	J. Ferguson	818-824 (1964)
Medicines debt to the sulphonamide group (The Jubilee Memorial Lecture of the Society of Chemical Industry)	F.L. Rose	858-865 (1964)
Lipids revisited	G.S. Hartley	1012-1013 (1964)
The paint industry as a user of heavy organic chemicals	H.G. Rains	1047-1058 (1964)
Optical rotatory power: Some recent developments	S.F. Mason	1286-1290 (1964)
Additives in plastics	R.N. Haward	1442-1455 (1964)
Insect cross resistance phenomena; Their practical and fundamental implications	F.P.W. Winteringham and P.S. Hewlett	1512-1518 (1964)
Utilisation of hydrocarbons by micro-organisms	M.J. Johnson	1532-1537 (1964)
Oxidative coupling	J.R. Lewis	1672-1677 (1964)

CHEMISTRY AND INDUSTRY (Contd.)

Applications of instrumental techniques in structural monosaccharide chemistry	R.J. Ferrier and N.R. Williams	1696-1707 (1964)
Microbial synthesis in industry and its relation to microbial physiology	S.J. Pirt	1772-1777 (1964)
Kinetic studies in the oxidation of organic compounds with the halogens	I.R.L. Barker	1936-1942 (1964)
Laboratory hazards	L. Bretherick	1978 (1964)
Phosphine explosion	H.A.C. McKay	1978 (1964)
Some aspects of gas drying by adsorption	R. Worthington	2110-2120 (1964)

CHIMIA

This is a monthly journal, in English, French or German. It contains review articles, symposia and short communications. Edited by H. Nitschmann and published for the Swiss Chemical Society by H.R. Sauerlander & Company, Zurich.

Title	Authors	Reference		
Action of radiation on cellulose and plant supportive tissue (in German)	H. Lück and F. Dell	17,	1-8	(1963)
Solid state polymerization reactions (in English)	S. Okamura	17,	9-16	(1963)
Application of high resolution mass spectrometry in organic chemistry (in English)	R.D. Craig B.N. Green and J.D. Waldron	17,	33-42	(1963)
Recent applications of electron spin resonance in chemistry (in English)	A. Horsfield	17,	42-51	(1963)
The determination of the form and size of macromolecules by rheological methods (in German)	A. Peterlin	17,	65-76	(1963)
Thermochemical titration (in English)	J. Jordan	17,	101-109	(1963)
The gibberelins (in German)	H. Kaufmann	17,	120-128	(1963)
Quantitative aspects of gas chromatography (in English)	R.S. Evans and R.P.W. Scott	17,	137-144	(1963)
Zone melting and column crystallization, a new separation and purification process for substances that crystallize (in German)	H. Schildknecht	17,	145-159	(1963)
The determination of molecular structure by x-ray methods (in English)	J.D. Dunitz	17,	169-175	(1963)
New applications of nuclear magnetic resonance spectroscopy (in German)	R.F. Zürcher	17,	175-184	(1963)
Spectropolarimetry. II. (in German)	H.G. Leemann and K. Stich	17,	184-198	(1963)
Preparative gas chromatography (in German)	E. Bayer	17,	199-206	(1963)
Disulfur difluoride (in German)	F. Seel and D. Gölitz	17,	207-208	(1963)
Chemical problems of food technology (in German)	H. Neukom	17,	225-233	(1963)

Forensic chemistry today (in German)	H. Brandenberger	<u>17</u>,	249-257 (1963)
Psycho-pharmaceutics and their determination on misuses (in German)	J. Baumler	<u>17</u>,	257-262 (1963)
Documentation of peptides (in German)	G. Langer	<u>17</u>,	292-298 (1963)
The plant plastids in light of their chemistry and biochemistry (in German)	E.C. Grob	<u>17</u>,	341-348 (1963)
O-Substituted hydroxylamines and their derivatives (in German)	A.O. Ilvespää and A. Marxer	<u>18</u>,	1-16 (1964)
New mechanistic interpretation of E_2 reactions (in English)	J. Csapilla	<u>18</u>,	37-48 (1964)

High Pressure Technology in Chemistry. I.
A symposium held in Zurich, October 25-26, 1963.

High pressure physics as basic research (in German)	E. Kuss	<u>18</u>,	75-89 (1964)
Equipment and research methods in chemistry of high pressures (in French)	L. Deffet	<u>18</u>,	89-93 (1964)
The significance of the solvation of groups with ionic charge in organic compounds on the substitution effect on chemical equilibria and reaction velocity (in German)	G. Schwarzenbach	<u>18</u>,	107-110 (1964)
The high pressure polymerization of ethylene (in German)	H. Hopff	<u>18</u>,	129-131 (1964)
The oxy-synthesis of propionaldehyde (in German)	P. Guyer and E. Bosshard	<u>18</u>,	131-134 (1964)
Growth of polymers by insertion of monomers (in German)	F. Patat	<u>18</u>,	233-243 (1964)
The steric build-up of 1-2-metallocomplex dyes (in German)	G. Schetty	<u>18</u>,	244-251 (1964)
Polar and steric contributions of saturated substituents to the acid-base equilibria of aqueous amines (in English)	G.N. Chremos and H.K. Zimmerman	<u>18</u>,	265-270 (1964)
Synthesis of high temperature resistant polymers (in English)	H.F. Mark	<u>18</u>,	270-275 (1964)
Research-problems in the field of fibre-active dyes (in English)	I.D. Rattee	<u>18</u>,	293-300 (1964)

The significance of zone-melting preparative chemistry (in German)	A. Gäumann	<u>18</u>,	300-305 (1964)
Methods for the isolation, purification and analysis of glycoproteins −−A brief review (in English)	K. Schmid	<u>18</u>,	321-327 (1964)
Some modern developments in polarography (in French)	E.T. Verdier	<u>18</u>,	328-336 (1964)
Radiolysis of liquid hexane (in German)	T. Gäumann	<u>18</u>,	345-348 (1964)
Configuration and confirmation analysis with the aid of NMR using steroids as an example (in German)	R.F. Zürcher	<u>18</u>,	349-354 (1964)
The principles of build-up of aromatic hydrocarbons (in German)	E. Clar	<u>18</u>,	375-386 (1964)

CHIMIE ET INDUSTRIE

This is a monthly journal, in French. It contains review articles and short communications. Edited by L. Jacque and published by Societé De Productions Documentaires, Paris 7.

Title	Authors	Reference
Polyacetals and their industrial applications	M. Letort	89, 155-167 (1963)
Characterization by nuclear magnetic resonance of the double bonds in organic chemistry	G. Martin	89, 168-184 (1963)
The present state of macromolecule science	G. Champetier	89, 269-278 (1963)
Contribution to the study of the Knoevenagel-Claisen condensation reactions	R. Pallaud T. Chaudron B. Marcot and P. Delest	89, 283-292 (1963)
Attempts to mechanize chemical documentation	M. Fournier	89, 374-378 (1963)
A comparison between solid and liquid propellants	C. Napoly	89, 417-427 (1963)
Late investigations on the substituted derivatives of metal carbonyls	M. Bigorgne R. Poilblanc G. Bouquet and D. Benlian	89, 428-432 (1963)
Processing, without plasticizers, of polyvinylchloride and copolymers of vinyl chloride	A. Achterberg	89, 433-447 (1963)
Research development on high polymers at the Milan Polytecnico	G. Natta	89, 545-557 (1963)
New methods in organic synthesis	M.G.J. Beets	89, 667-682 (1963)
Isoprene from propylene: the most eceonomic way of producing a natural synthetic rubber	R. Landau R.W. Simon and G.S. Schaffel	90, 37-43 (1964)
The new synthetic processes in organic chemistry	M.G.J. Beets	90, 44-57 (1964)
Some research work in the field of organic tin compounds	G.J.M. Van der Kerk	90, 251-258 (1964)
Synthesis of mercaptans	P. Bapseres	90, 358-369 (1964)
Characteristic infrared absorption of heterocyclic compounds, basic components of photographic sensitizers, variation of the heteroatomicity and the substituents	P. Bassignana C. Cogrossi and M. Gandino	90, 370-378 (1964)

Title	Authors	Vol.	Pages	Year
Contribution to the study of hydrotropy	G. Noisillier, J.L. Pierre and P. Arnaud	90,	505-510	(1964)
Organic synthesis in the photographic industry	A. Van Dormael	90,	619-628	(1964)
Propylene oxidation by oxygen in the liquid phase. Propylene oxide synthesis	F. Lanos, G.-M. Clément and F. Pouliguen	91,	47-56	(1964)
By-products in the manufacture of caprolactam. The nitrosation of cyclo-hexane	C. Matasa	91,	57-64	(1964)
The synthesis of acrylic acid as a technical example for acetylene chemistry	T.H. Toepel	91,	139-145	(1964)
Contribution to the study of hydrotropy	G. Noisillier, J.-L. Pierre and P. Arnaud	91,	263-268	(1964)
The oxidation of n-paraffins into alcohols	J.-E. Germain and J.-M. Cognion	91,	519-528	(1964)
Contribution to the study on the oxidation of fats	R. François and M. Loury	91,	650-653	(1964)
Recent views on the synthesis of chlorohydrins, their dehydrochlorination and hydrolysis	J. Myszkowski and A.Z. Zielinski	91,	654-666	(1964)
Use of peracids in organic synthesis	V.J. Karnojitzki	92,	381-394	(1964)
Enzymatic catalysis	J. Yon	92,	509-518	(1964)
Measure of the wetting power of surface agents	Commission for Methods of Analysis	92,	519-532	(1964)
Study of the formation of phosgene by electric-arc discharge in chlorinated hydrocarbons	P. Jay	92,	533-537	(1964)
Alkoxy-metals, their success and future in industrial chemistry	K. Ziegler	92,	631-644	(1964)
Molecular structure and physiological activity	C. Viel	92,	645-657	(1964)

This is an annual series of review articles, in English, dealing with progress in chromatography, electrophoresis, and related methods. Up to 1963 (volume 5) reviews were usually published in the Journal of Chromatography and in Chromatographic Reviews. Since 1963, publication of reviews was discontinued in the Journal of Chromatography. Edited by M. Lederer and published by Elsevier Publishing Co., Amsterdam-London-New York.

Title	Authors	Reference
Protein mobilities and iron binding constants evaluated by zone electrophoresis (In memory of Prof. Dr. Med. Carlos Lobo Onell, Santiago de Chile, 1885-1962)	H. Waldmann-Meyer	5, 1-45 (1963)
Furnishing a laboratory for paper and thin-layer chromatography	E. von Arx and R. Neher	5, 46-57 (1963)
Paper chromatography and chemical structure. I. Tankless or flat-bed chromatography. A method for the accurate determination of R_m values	J. Green and S. Marcinkiewicz	5, 58-64 (1963)
Paper chromatography and chemical structure. II. The chromatography of phenols, alkoxyphenols, coumaranols and chromanols. The use of group and atomic ΔR_m values. Steric and electronic effects in chromatography	S. Marcinkiewicz, J. Green and D. McHale	5, 65-90 (1963)
Paper chromatography and chemical structure. III. The correlation of complex and simple molecules. The calculation of R_m values for tocopherols, vitamins K, ubiquinones and ubichromenols from R_m.(phenol). Effects of unsaturation and chain branching	J. Green, S. Marcinkiewicz and D. McHale	5, 91-116 (1963)
Paper chromatography and chemical structure. IV. Intramolecular hydrogen bonding	S. Marcinkiewicz and J. Green	5, 117-122 (1963)
Paper chromatography and chemical structure. V. Tautomerism. The determination of tautomeric equilibrium by paper chromatography. Thienol and p-nitrosophenols	J. Green and S. Marcinkiewicz	5, 123-134 (1963)
Paper chromatography and chemical structure. VI. Tautomerism and intramolecular hydrogen bonding in the same molecule. O-Nitrosophenols	S. Marcinkiewicz and J. Green	5, 135-140 (1963)
Paper chromatography and chemical structure. VII. The separation of *meta*- and *para*-derivatives of benzene	S. Marcinkiewicz and J. Green	5, 141-157 (1963)
Paper chromatography and chemical structure. VIII. Hyperconjugation	J. Green and S. Marcinkiewicz	5, 158-160 (1963)

CHROMATOGRAPHIC REVIEWS (Contd.)

A comprehensive bibliography of separations of organic substances by counter current distribution	C.G. Casinovi	5,	161-207 (1963)
Paper chromatography of oestrogens	R.E. Oakey	5,	208-222 (1963)
Gas chromatography in inorganic chemistry	J. Tadmor	5,	223-235 (1963)
Commercial equipment for gas chromatography	G.S. Learmonth	6,	1-18 (1964)
Centrifugal chromatography	Z. Deyl, J. Rosmus and M. Pavlicek	6,	19-52 (1964)
Chromatography of free nucleotides	J.J. Saukkonen	6,	53-80 (1964)
Paper chromatography of iodoamino acids and related compounds	L.G. Plaskett	6,	91-109 (1964)
Chromatographic techniques for pesticide residue analysis	G. Zweig	6,	110-128 (1964)
Liquid exchangers: Separations on inert supports impregnated with liquid ion exchangers	E. Cerrai	6,	129-153 (1964)
Polymeric coordination compounds. The synthesis and applications of selective ion-exchangers and polymeric chelate compounds	G. Nickless and G.R. Marshall	6,	154-190 (1964)
Application of paper ionophoresis and electrochromatography to the study of metal complexes in solution	E. Blasius and W. Preetz	6,	191-213 (1964)

DEVELOPMENTS IN APPLIED SPECTROSCOPY

These volumes contain the proceedings of the Annual Symposia on Applied Spectroscopy. Although the symposia, which are sponsored by the Chicago Section of the American Chemical Society, have been held since 1959, publication was not begun until the 1961 symposium was presented as Volume 1 in the series. Volume 1 was edited by W.D. Ashby, volume 2 was edited by J.R. Ferraro and J.S. Ziomek and volume 3 was edited by J.E. Forrette and E. Lanterman. Published by Plenum Press, New York.

Title	Authors	Reference
These are the proceedings of the 12th Annual Symposium on Spectroscopy, held in Chicago, Illinois, May 15-18, 1961.	W.D. Ashby (Editor)	
Fluorescent x-ray spectroscopy: Determination of trace elements	W.J. Campbell and J.W. Thatcher	1, 31-62 (1962)
Advances in high-resolution x-ray spectrometry	E.L. Jossem	1, 63-75 (1962)
Establishing and controlling analytical curves	T.P. Schreiber	1, 129-136 (1962)
Sample problem in Raman spectroscopy	M.C. Tobin	1, 205-214 (1962)
Advances in ionization detectors The electron affinity detector	S.J. Clark	1, 215-225 (1962)
These are the proceedings of the 13th Annual Symposium on Spectroscopy, held in Chicago, Illinois, April 3 - May 3, 1962.	J.R. Ferraro and J.S. Ziomek (Editors)	
World-wide communication of spectroscopic information	F.F. Cleveland	2, 2-5 (1963)
Methods of storing, retrieving, and correlating infrared spectral data	F.H. Dyke, Jr.	2, 6-14 (1963)
Raman and infrared spectra of the dicyanamide ion	A.J. Perkins	2, 43-51 (1963)
Raman spectral data, assignments, potential energy constants, and calculated thermodynamic properties for CH_2Cl_2, $CHDCl_2$, and CD_2Cl_2	F.E. Palma and K. Sathianandan	2, 58-64 (1963)
Ortho-substituent effects and the fundamental NH_2 stretching vibrations in anilines	P.J. Krueger	2, 65-88 (1963)
Solvent effects on the infrared spectra of organophosphorus compounds	J.R. Ferraro	2, 89-95 (1963)
Anomalous chemical shifts in the proton magnetic resonance spectra of the dimethylcyclohexanes and related hydrocarbons	N. Muller and W.C. Tosch	2, 98-114 (1963)

DEVELOPMENTS IN APPLIED SPECTROSCOPY (Contd.)

Title	Authors	Vol.	Pages	Year
Determination of fat in corn and corn germ by wide-line nuclear magnetic resonance techniques	T.F. Conway and R.J. Smith	2,	115-127	(1963)
Far ultraviolet analysis of steroids and other biologically active substances	W.F. Ulrich	2,	130-141	(1963)
Time resolution spectroscopy	F.D. Harrington	2,	162-197	(1963)
Carbon internal standard in the spectrochemical analysis of lubricating oil	J.H. LaSell	2,	239-252	(1963)
X-ray spectroscopy in biology and medicine. Preliminary report on the microanalysis of human tissues for iron, zinc, and potassium	J.C. Mathies and P.K. Lund	2,	326-334	(1963)
Analysis of nonmetallics by x-ray fluorescence techniques	B.R. Boyd and H.T. Dryer	2,	335-349	(1963)
A review of preparative gas chromatography	J.A. Perry	2,	368-382	(1963)
A completely automatic preparative scale gas chromatograph	J.M. Kauss, J. Peters and C.B. Euston	2,	383-393	(1963)
Programmed-temperature preparative chromatography	N. Brenner and D.R. Bresky	2,	394-408	(1963)
A remote injection and fraction collection apparatus for preparative chromatography	W.H. Penney and J.P. Windey	2,	409-414	(1963)
Temperature-programmed capillary columns	F.L. Boys	2,	415-425	(1963)
A technique for the identification of volatile flavor components in foods	R. Jensen and S.W. Leslie	2,	426-430	(1963)
Some applications of gas chromatography to forensic chemistry	D.T. Dragel, E. Beck and A.H. Principe	2,	431-438	(1963)
These are the proceedings of the 14th Annual Mid-America Spectroscopy Symposium, held in Chicago, Illinois, May 20-23, 1963.	J.E. Forrette and E. Lanterman (Editors)			
The use of thin-layer chromatography with infrared spectroscopy	M.K. Snavely and J.G. Grasselli	3,	119-141	(1964)
The characterization of saturated aliphatic esters in the 15 - 40 micron region	J.J. Lucier and F.F. Bentley	3,	142-157	(1964)

DEVELOPMENTS IN APPLIED SPECTROSCOPY (Contd.)

Title	Authors	Reference
Flame photometric investigations of the factors affecting the emissivity of metal chelates	H.C. Eshelman and J. Armentor	3, 190-195 (1964)
Application of electron spin resonance techniques to biochemical systems: Current information on photosynthetic systems	J.J. Heise and R.W. Treharne	3, 340-360 (1964)
Electron spin resonance spectra of gamma radiolysis products of solid acetonitrile	D. Dunbar, D. Hale, L. Harrah et al.	3, 361-373 (1964)

ENDEAVOUR

This is a quarterly journal, in English. It publishes review articles designed to record the progress of the sciences "in the service of mankind". Edited by T.I. Williams and published by Imperial Chemical Industries, Limited, London.

Title	Authors	Reference
Luminescence in animals	J.A.C. Nicol	22, 37-41 (1963)
Ligand-field theory	L.E. Orgel	22, 42-47 (1963)
The aromatic fluorocarbons and their derivatives	J.C. Tatlow	22, 89-95 (1963)
Ion exchangers	D. Reichenberg	22, 123-128 (1963)
Microwave spectroscopy	T.M. Sugden	22, 129-133 (1963)
The formal duplication of DNA	J. Cairns	22, 141-145 (1963)
The communication of information	C. Cherry	23, 13-17 (1964)
Teichoic acids and the bacterial cell wall	J. Baddiley	23, 33-37 (1964)
The relief of pain —— The search for the ideal analgesic	K.W. Bentley	23, 97-101 (1964)
Molecular sieves	R.M. Barrer	23, 122-130 (1964)
Physical measurement by gas chromatography	J.H. Purnell	23, 142-147 (1964)
Structural plant polysaccharides	R.D. Preston	23, 153-159 (1964)

This is a monthly journal, in English, French or German. It publishes both review articles and short communications in all fields of the natural sciences. Edited by P. Huber, R. Matthey, A.v. Muralt, L. Ruzicka and H. Mislin and published by Birkhauser-Verlag, Basel.

Title	Authors	Reference
Peptolide (in German)	E. Schröder and K. Lübke	19, 57-67 (1963)
The venom of the Colombian arrow poison frog phyllobates bicolor (in English)	F. Märki and B. Witkop	19, 329-338 (1963)
The synthesis of novabiocin (in German)	B.P. Vaterlaus K. Koebel J. Kiss et al.	19, 383-391 (1963)
The filling mechanism of the swim bladder (Generation of high gas pressures through hairpin counter-current multiplication) (in English)	W. Kuhn A. Ramel H.J. Kuhn and E. Marti	19, 497-511 (1963)
X-ray crysttalographic measurements of high-polymers (in German)	P.H. Hermans	19, 553-564 (1963)
Stereospecific polymerization of cyclic monomers (in German)	G. Natta	19, 609-664 (1963)
The chemistry of certain imines related to the diterpene alkaloids (in English)	S.W. Pelletier	20, 1-56 (1964)
Chemical basis of heredity, the genetic code (in English)	S. Ochoa	20, 57-112 (1964)
Enzymes, drugs, and antibodies, some chemical factors (in English)	J.H. Turnbull	20, 113-168 (1964)
Human pituitary growth hormones (in English)	C. Haoli and W.-K. Liu	20, 169-240 (1964)
Origin and function of some methyl groups in branched fatty acids in plant sterols and in quinones of the vitamin K and ubiquinone group (in German)	E. Lederer	20, 473-528 (1964)
Bradykinin, kallidin and their synthetic analogues (in English)	E. Schröder and R. Hempel	20, 529-592 (1964)

FORTSCHRITTE DER ARZNEIMITTELFORSCHUNG
(Progress in Drug Research)

This is a continuing series of review articles, in German, French and English. It publishes selected subjects pertaining to drug research, medicinal, pharmaceutical and biochemical aspects thereof. Occasional articles on physical methods are included. Edited by E. Jucker and published by Birkhauser Verlag, Basel-Stuttgart.

Title	Authors	Reference
The effects of structural alteration on the anti-inflammatory properties of hydrocortisone (in English)	L.H. Sarett, A.A. Patchett and S. Steelman	5, 11-153 (1963)
Analgesia and addiction (in English)	L.B. Mellett and L.A. Woods	5, 155-267 (1963)
Phenothiazine and azaphenothiazine as drugs (in German)	E. Schenker and H. Herbst	5, 269-627 (1963)
Metabolism of drugs and other foreign compounds by enzymatic mechanisms (in English)	J.R. Gillette	6, 11-73 (1963)
Recent studies in the field of indole compounds (in English)	R.V. Heinzelman and J. Szmuszkovicz	6, 75-150 (1963)
Physico chemical methods in pharmaceutical chemistry. I. Spectrofluorometry (in English)	H.G. Leemann, K. Stich and M. Thomas	6, 151-278 (1963)
Biological activity of the terpenoids and their derivatives (in English)	M. Martin-Smith and T. Khatoon	6, 279-346 (1963)
Concerning new drugs (in German)	W. Kunz	6, 347-406 (1963)
Basic mechanisms of drug action (in English)	D.R.H. Gourley	7, 11-57 (1964)
The use of radio-isotopes in pharmaceutical research (in German)	K.E. Schulte and I. Mleinek	7, 59-131 (1964)
The development of antifertility substances (in English)	H. Jackson	7, 133-192 (1964)
Antibacterial chemotherapy of tuberculosis (in German)	F. Trendelenburg	7, 193-303 (1964)
The pharmacology of homologous series (in English)	H.R. Ing	7, 305-339 (1964)
Aminonucleoside nephroses (in German)	U.C. Dubach	7, 341-463 (1964)

FORTSCHRITTE DER CHEMIE ORGANISCHER NATURSTOFFE
(Progress in the Chemistry of Organic Natural Products)

This is an annual collection of review articles, in English, German or French, covering the recent advances in the chemistry of natural products. Edited by L. Zechmeister and published by Springer-Verlag, Wien.

Title	Authors	Reference
The biosynthesis of rubber	J. Bonner	21, 1-16 (1963)
The polyene antifungal antibiotics	W. Oroshnik and A.D. Metane	21, 17-79 (1963)
The chemistry of tetracycline (in German)	H. Muxfeldt and R. Bangert	21, 80-120 (1963)
Anthracyclinones and anthracyclines (Rhodomycinone, pyrromycinone and their glycosides) (in German)	H. Brockmann	21, 121-182 (1963)
Folic acid and folic acid enzymes (in German)	L. Jaenicke and C. Kutzbach	21, 183-274 (1963)
Chemistry of the natural rotenoids	L. Cromtie	21, 275-325 (1963)

FORTSCHRITTE DER CHEMISCHEN FORSCHUNG

This is a collection of review articles, in English, French or German, published at irregular intervals. Edited by E. Heilbronner, U. Hofmann, K. L. Schafer and G. Wittig and published by Springer-Verlag, Berlin-Heidelberg-New York.

Title	Authors	Reference
Fundamental problems of structural chemistry. Studies on the additivity of atomic distances (in German)	F. Rogowski	4, 1-50 (1963)
The chemistry of the transplutonium elements (in German)	F. Weigel	4, 51-137 (1963)
Naturally occurring acetylene derivatives (in German)	F. Bohlmann, H. Bornowski and Chr. Arndt	4, 138-272 (1963)
The chemistry of s-triazin (in German)	A. Kreutzberger	4, 273-300 (1963)
Chemistry of the Rashig's hydroxylamine synthesis, and related topics (in German)	F. Seel	4, 301-332 (1963)
New polarographic and voltametric processes for trace analysis (in German)	R. Neeb	4, 333-458 (1963)
Chemistry of silicone compounds (A development based on pyrochemical studies) (in German)	G. Fritz	4, 459-553 (1963)
The chemistry of the formation of heterocyclic nitrogen-containing thioxo compounds with carbon disulfide (in English)	J.F. Willems	4, 554-652 (1963)

FORTSCHRITTE DER HOCHPOLYMEREN-FORSCHUNG
(Advances in Polymer Science)

This is a collection of review articles, in English, German or French. It publishes review articles on selected subjects of polymer chemistry and has no fixed schedule of publication. Edited by J.D. Ferry, C.G. Overberger, G.V. Schulz, A.J. Staverman and H.A. Stuart and published by Springer-Verlag Berlin.

Title	Authors	Reference
The high resolution nuclear magnetic resonance spectroscopy of polymers (in English)	F.A. Bovey and G.V.D. Tiers	3, 139-195 (1963)
Intrinsic viscosities and unperturbed dimensions of long chain molecules (in English)	M. Kurata and W.H. Stockmayer	3, 196-312 (1963)
Fluorescence methods in polymer science (in English)	G. Oster and Y. Nishijima	3, 313-331 (1964)
The adsorption of macro-molecules from solution (in German)	F. Patat, E. Killmann and C. Schliebener	3, 332-393 (1964)
Glass transition in amorphous polymers. A phenomenological study (in French)	A.J. Kovacs	3, 394-507 (1964)
Recent advances in cationic polymerization (in English)	J.P. Kennedy and A.W. Langer, Jr.	3, 508-580 (1964)
The crystallization and melting of high polymers (in German)	H.G. Zachmann	3, 581-687 (1964)
Macroradical reactivity studied by electron spin resonance (in English)	S.E. Bresler and E.N. Kazbekov	3, 688-711 (1964)

JOURNAL OF CHEMICAL EDUCATION

This is a monthly journal, in English, of the division of Chemical Education of the American Chemical Society. Edited by W.F. Kieffer and published by the American Chemical Society.

Title	Authors	Reference
Conventions defining thermodynamic properties of aqueous ions and other chemical species	R.M. Noyes	40, 2-10 (1963)
A convenient synthetic technique to oxidize mercaptans to disulfides	T.J. Wallace, W. Bartok and A. Schrieshiem	40, 39 (1963)
Standard ionic crystal structures	W.G. Gehman	40, 54-60 (1963)
A modular approach to chemical instrumentation	E.N. Wise	40, 73-75 (1963)
German chemical and atomic energy documentation centers	H.A. Buch	40, 82-85 (1963)
Kinetics in the study of organic reaction mechanisms	R.H. DeWolfe	40, 95-98 (1963)
Some aspects of the history of chemistry in Russia	H.M. Leicester	40, 108-109 (1963)
Y. Hirshberg — In Memorium	E. Fischer	40, 112-113 (1963)
The physical and chemical character of graphite	P.A.H. Tee and B.L. Tonge	40, 117-123 (1963)
Ionic and molecular halides of the phosphorus family	R.R. Holmes	40, 125-130 (1963)
Intrinsic bond energies	S. Siegel and B. Siegel	40, 143-145 (1963)
Chemistry of diphosphorus compounds	J.E. Huheey	40, 153-158 (1963)
Organometallic derivatives of the transition elements	H.D. Kaesz	40, 159-168 (1963)
Chemistry of the diene-iron tricarbonyl complexes	R. Pettit, G. Emerson and J. Mahler	40, 175-181 (1963)
Multicenter and assisted mechanistic pathways in the reactions of organometallic compounds	R.E. Dessy and F. Paulik	40, 185-194 (1963)
Energy relations in teaching organic chemistry	N.L. Allinger	40, 201-202 (1963)

Ribonucleic acid, the simplest information-transmitting molecule	H. Fraenkel-Conrat	40, 216-222 (1963)
The early history of carbon-14	M.D. Kamen	40, 234-242 (1963)
Contour surfaces for atomic and molecular orbitals	E.A. Ogryzlo and G.B. Porter	40, 256-261 (1963)
The chemical behavior of ions in gases	J.L. Franklin	40, 284-293 (1963)
The valence-shell electron-pair repulsion (VSEPR) theory of directed valency	R.J. Gillespie	40, 295-301 (1963)
α-Pinene, a starting material for a sequence of organic experiments	X.A. Dominguez and G. Leal	40, 347-348 (1963)
The professional presentation of scientific papers	H.G. Cassidy	40, 373-376 (1963)
What is humic acid	C. Steelink	40, 379-384 (1963)
The bromonium ion	J.G. Traynham	40, 392-395 (1963)
Rules for molecular orbital structures	H. Meislich	40, 401-408 (1963)
Tangent-sphere models of molecules. I. Theory and construction	H.A. Bent	40, 446-452 (1963)
Plastic Dreiding models	L.F. Fieser	40, 457-459 (1963)
Conduction and semi-conduction	N.J. Juster	40, 489-496 (1963)
Cyclobutane chemistry. I. Structure and strain energy	A. Wilson and D. Goldhamer	40, 504-511 (1963)
Tangent-sphere models of molecules. II. Uses in teaching	H.A. Bent	40, 523-530 (1963)
Organic semiconductors	N.J. Juster	40, 547-555 (1963)
Significant structure theory of liquids	H. Eyring and R.P. Marchi	40, 562-572 (1963)
Terpenes to platinum: the chemical career of Ler Aleksandrovich Chugaev	G.B. Kauffman	40, 656-665 (1963)

Title	Author(s)	Vol.	Pages	Year
Molecular orbital theory for transition metal complexes	H.B. Gray	41	2-12	(1964)
Principles of chemical reaction	R.T. Sanderson	41	13-22	(1964)
Homogeneous catalysis	J.A. Leisten	41	23-27	(1964)
Ficin as a catalyst in organic syntheses	J.L. Abernathy and G.L. Leonardo	41	53-54	(1964)
Rocket propulsion: the chemical challenge	J.R. Dafler	41	58-64	(1964)
An introduction to the sequence rule. A system for the specification of absolute configuration	R.S. Cahn	41	116-125	(1964)
Leo Hendrik Baekeland (1863-1944)	J. Gillis (Translated by: R.E. Oesper)	41	224-226	(1964)
The ultraviolet spectra of aromatic hydrocarbons. Predicting substitution and isomerism changes	P.E. Stevenson	41	234-239	(1964)
Phthalocyanine compounds	F.H. Moser and A.L. Thomas	41	245-249	(1964)
Toxicity: Killer of great chemists	S. Soloveichik	41	282-284	(1964)
A brief history of polarimetry	R.E. Lyle and G.G. Lyle	41	308-313	(1964)
Boron-nitrogen heterocycles	D.A. Payne, Jr. and E.A. Eads	41	334-336	(1964)
The shape of the F-orbitals	H.G. Friedman, Jr. G.R. Choppin and D.G. Feuerbacher	41	354-358	(1964)
Principles of halogen chemistry	R.T. Sanderson	41	361-366	(1964)
Molecular charge transfer complexes (A group experiment in physical chemistry)	W.E. Wentworth G.W. Drake W. Hirsch and E. Chen	41	373-379	(1964)
Rotational and pseudo-rotational barriers in simple molecules	S.I. Miller	41	421-424	(1964)
Kinetics in gas-phase stirred flow reactors	W.C. Herndon	41	425-428	(1964)

Liquid field theory	F.A. Cotton	41,	466-476 (1964)
Activation of small molecules by coordination	M.M. Jones	41,	493-500 (1964)
The discovery of a new class of compounds: Carboranes	J. Bobinski	41,	500-501 (1964)
X-ray crystallography as a tool for structural chemists	W.M. Macintyre	41,	526-529 (1964)
Organic photochemistry and the excited state	P.A. Leermakers and G.F. Vesley	41,	535-541 (1964)
Recent advances in the base-catalyzed reactions of sulfur compounds	T.J. Wallace	41,	542-549 (1964)
The exposition of isotope effects on rates and equilibria	M.M. Kreevoy	41,	636-638 (1964)
Homolytic, cationotropic, and anionotropic reactions	B.E. Hoogenboom	41,	639-644 (1964)
An introduction to nonequilibrium thermodynamics	H.J.M. Hanley	41,	647-653 (1964)
Crystalline molecular sieves	D.W. Breck	41,	678-689 (1964)

JOURNAL DE CHIMIE PHYSIQUE ET DE PHYSICO-CHIMIE BIOLOGIQUE

This is a monthly journal, generally in French, publishing mainly original articles as well as symposia and selected review articles. Published by the Societe de Chimie Physique, Paris.

Title	Authors	Reference
Symposia: Physico Chemistry of the Separation of Isotopes, held June 4-8, 1962, Paris, France.		
The effect of isotopic substitution on the living cell (in English)	D. Rittenberg	60, 318-323 (1963)
The effect of the substitution of H_2O by D_2O on the activity of phosphopyruvic kinase in the presence of alkaline ions (in French)	R. Coelho and I. Penset-Härstrom	60, 324-327 (1963)
The effect of complete substitution of D for H in succinic acid dehydrogenase on its enzymatic activity (in English)	S.M. Rittenberg and E. Borak	60, 328-331 (1963)
Fluorescence and photosynthesis (in French)	J. Lavorel	60, 608-611 (1963)
Symposia: Molecular Interactions in the Liquid Phase, held June 9-7, 1963, Paris, France		
General survey of intermolecular forces (in English)	H.C. Longuet-Higgins	61, 13-19 (1964)
The interactions of electron donors and acceptors (in English)	R.S. Mulliken	61, 20-38 (1964)
Properties of pure liquids in relation to their cohesion (in French)	J. Barriol	61, 39-43 (1964)
Cohesion of liquid mixtures	J.A.A. Ketalaar	61, 44-49 (1964)
Intermolecular forces between molecules of different species (in English)	J.H. Hildebrand	61, 53-57 (1964)
Prefreezing effects in the viscosity and other properties of liquids (in English)	A.R. Ubbelohde	61, 58-66 (1964)
The solubility of hydrogen halides as a function of the basic strength of solvents (in English)	W. Gerrard	61, 73-80 (1964)
Molecular associations and acid base equilibria	P. Huyskens and T. Zeegers-Huyskens	61, 81-86 (1964)
Isotope effects in phase equilibria: A new tool for the study of intermolecular forces (in English)	J. Bigeleisen	61, 87-91 (1964)

Structure of liquids studied by means of x-rays	H. Curien	61,	92-96 (1964)
Cold neutron investigation of the structure of liquids (in English)	J.A. Janik	61,	97-104 (1964)
Hydrophobic bond. Activity and conformation of trypsin in dimethylsulfoxide-water systems (in English)	F.A. Bettelheim and P. Senatore	61,	105-111 (1964)
Variations of activity coefficients of phenyl and its alkylated derivatives in organic liquids as a function of their polarization	A.B. Lindenberg and M. Massin	61,	112-121 (1964)
Dipole moments and molecular complexes (in English)	J.W. Smith	61,	125-131 (1964)
Molecular conformation of some complexes formed by hydrogen bonding in pyrrol	H. Lumbroso	61,	132-138 (1964)
Static dielectric constants and structure of solutions of polar molecules in non-polar solvents	C. Brot	61,	139-145 (1964)
Study of the structure of liquids with hydrogen bonds by dispersion and dipolar absorption	M. Moriamez, L. Raczy, E. Constant and A. Lebrun	61,	146-162 (1964)
Study of some intermolecular associations of carboxylic acids by means of hertzian spectra of dipolar absorption	E. Constant and A. Lebrun	61,	163-173 (1964)
Relaxation times in nuclear magnetic resonance and their relation to adsorption, molecular association and polymerization	L. Giulotto	61,	177-181 (1964)
NMR and molecular interactions in the liquid phase	G. Mavel	61,	182-194 (1964)
Attempts to interpret quantitatively the chemical displacement of protons bound in a hydrogen bond, as a function of concentration	B. Lemanceau, C. Lussan, N. Souty and J. Biais	61,	195-198 (1964)
Solvent effects in proton nuclear magnetic resonance spectra of substituted benzenes (in English)	P. Diehl	61,	199-203 (1964)
Dynamic nuclear polarisation in liquids at high magnetic fields (in English)	K.H. Hausser	61,	204-209 (1964)
Dynamic polarization of liquids with the help of porous paramagnetic surfaces	A.P. Legrand, J. Auvray and J. Uebersfeld	61,	210-214 (1964)

The charge-transfer mechanism of hydrogen bonding as revealed from electronic absorption spectra (in English)	S. Nagakura	61, 217-221 (1964)
Influence of the hydrogen bond on $\pi \to \pi$ bands of organic molecules. Electronic theory	S. Besnainou M.R. Prat and S. Bratoz	61, 222-227 (1964)
The influence of the solvent on the electronic spectra of 2-hydroxy-1-naphthaldehydanil	E. Lippert J. Muszik and W. Voss	61, 228-229 (1964)
Ultraviolet spectra and empiric determination of the solvent polarity (Z constant)	E.M. Kosower	61, 230-235 (1964)
Ultrasonic absorption in solutions containing non-electrolytic dissociation equilibria (in English)	W. Maier	61, 239-244 (1964)
Infra-red spectra study of intermolecular actions in the liquid phase	M.-L. Josien	61, 245-256 (1964)
Infrared spectroscopy of weak charge-transfer complexes (in English)	E.E. Ferguson	61, 257-262 (1964)
The high resolution absorption spectroscopy of aromatic free radicals (in English)	G. Porter and B. Ward	61, 1517-1522 (1964)
ESR of organic free radicals	M. Soutif	61, 1549-1569 (1964)
Electronic structure of paramagnetic molecules	R. Lefebvre	61, 1592-1597 (1964)
Luminescence and the triplet state	A. Rousset	61, 1621-1630 (1964)
The first triplet state of benzene	S. Leach and R. Lopez-Delgado	61, 1636-1642 (1964)
Study of the second triplet state of aromatic molecules by ESR	J.H. van der Waals and M.S. de Groot	61, 1643-1654 (1964)
Environmental effects in the ESR spectra of organic triplet states (in English)	A.M. Trozzolo E. Wasserman and W.A. Yager	61, 1663-1665 (1964)
Theoretical studies of free radicals of biological interest	A. Pullman	61, 1666-1680 (1964)
Magnetic phenomena determined by optical excitation of aromatic amino acids	P. Douzou and M. Ptak	61, 1681-1685 (1964)

JOURNAL OF PHARMACEUTICAL SCIENCES

This is a monthly journal, in English. It is the scientific organ of the American Pharmaceutical Association, dedicated to research articles and some reviews. Edited by E.G. Feldman and published by the American Pharmaceutical Association, Washington, D.C.

Title	Authors	Reference
Radioisotopes in the pharmaceutical sciences and industry	J.E. Christian	50, 1-13 (1961)
Role of metal-binding in the biological activities of drugs	W.O. Foye	50, 93-108 (1961)
Kinetics of drug absorption, distribution, metabolism, and excretion	E. Nelson	50, 181-192 (1961)
Hypotensive veratrum ester alkaloids	S.M. Kupchan	50, 273-287 (1961)
Application of infrared spectrophotometry to pharmaceutical analysis	J. Carol	50, 451-463 (1961)
Biosynthesis of the ergot alkaloids	V.E. Tyler, Jr.	50, 629-640 (1961)
Stereochemistry	W.H. Hartung and J. Andrako	50, 805-818 (1961)
Sulfurous acid salts as pharmaceutical antioxidants	L.C. Schroeter	50, 891-900 (1961)
Influence of gibberellic acid on medicinal plants	L.A. Sciuchetti	50, 981-998 (1961)
Antibiotics 1956-1961	R. Pratt	51, 1-27 (1962)
Colorants for pharmaceuticals	C.J. Swartz and J. Cooper	51, 89-99 (1962)
Structure-activity relationships of drugs affecting the lungs	D.M. Aviado	51, 191-201 (1962)
Crystallography. Part I.	J.A. Biles	51, 499-509 (1962)
Crystallography. Part II.	J.A. Biles	51, 601-617 (1962)
Gas chromatography and its application to pharmaceutical analysis	E. Brochmann-Hanssen	51, 1017-1031 (1962)
Pharmaceutics of penicillin	M.A. Schwartz and F.H. Buckwalter	51, 1119-1128 (1962)

Plastics in pharmaceutical practice and related fields. Part I.	J. Autian	<u>52</u>,	1-23 (1963)
Plastics in pharmaceutical practice and related fields. Part II.	J. Autian	<u>52</u>,	105-122 (1963)
X-ray emission spectroscopy in pharmaceutical analysis	G.J. Papariello and W.J. Mader	<u>52</u>,	209-217 (1963)
Synthesis of tetracycline analogs	G.C. Barrett	<u>52</u>,	309-330 (1963)
Metabolism of phenothiazine drugs	J.L. Emmerson and T.S. Miya	<u>52</u>,	411-419 (1963)
Column partition chromatography in pharmaceutical analysis	J. Levine	<u>52</u>,	1015-1031 (1963)
Thiophene compounds of biological interest	W.L. Nobles and C.D. Blanton, Jr.	<u>53</u>,	115-129 (1964)
Naturally occurring coumarins and related physiological activities	T.O. Soine	<u>53</u>,	231-264 (1964)
Photometric titrations	S.P. Eriksen and K.A. Connors	<u>53</u>,	465-479 (1964)
1,4-Benzodiazepines	S.J. Childress and M.I. Gluckman	<u>53</u>,	577-590 (1964)
Electron spin resonance spectroscopy	L.D. Tuck	<u>53</u>,	1437-1445 (1964)

This is a monthly journal, in English, containing both original articles and reviews. Edited by G. Brownlee and published by the Council of the Pharmaceutical Society of Great Britain.

Title	Authors	Reference
The estimation of penicillins and penicillin destruction	J.M.T. Hamilton-Miller, J.T. Smith and R. Knox	15, 81-91 (1963)
The relation between chemical structure and biological activity	J.M. van Rossum	15, 285-316 (1963)
Phosphatidylethanolamine and lysophosphatidylethanolamine	D.C. Robins	15, 701-722 (1963)
A molecular basis for drug action	E.J. Ariëns and A.M. Simonis	16, 137-157 (1964)
A molecular basis for drug action (The interaction of one or more drugs with different receptors)	E.J. Ariëns and M.F. Sugrue	16, 569-595 (1964)

JOURNAL OF POLYMER SCIENCE

This is a monthly journal, mainly in English, containing original articles and symposia in the field of polymer chemistry. Edited by R.M. Fuoss, J.J. Hermans, H. Marks, H.W. Melville, C.G. Overberger and G. Smets and published by Interscience Publishers, Inc. (a division of John Wiley and Sons), New York.

Title	Authors	Reference
1st Biannual American Chemical Society Polymer Symposium, held at Michigan State University, East Lansing, Michigan, June 20-22, 1962	H.W. Starkweather, Jr. (Editor)	
Crystallization in polymers	P.H. Lindenmeyer	1, 5-39 (1963)
Motion in the solid state of high polymers	B. Wunderlich	1, 41-64 (1963)
Some contributions to the study of vinyl polymerization in the crystalline state	H. Morawetz	1, 65-82 (1963)
Double chain polymers and nonrandom crosslinking	J.F. Brown, Jr.	1, 83-97 (1963)
A dynamic mechanical study of rubber-modified polystyrenes	S.G. Turley	1, 101-116 (1963)
Properties of polyethylene-acrylic acid graft copolymers	J.K. Rieke and G.M. Hart	1, 117-133 (1963)
The stereoregular polymerization of vinyl ethers with transition metal catalysts	E.J. Vandenberg	1, 207-236 (1963)
Ziegler polymerization of olefins. I. Dependence on structure of the metal alkyl and the transition metal compound	J. Boor, Jr.	1, 237-255 (1963)
Ziegler polymerization of olefins. II. Nature of catalytic sites	J. Boor, Jr.	1, 257-279 (1963)
Kinetics and mechanism of ethylene polymerization by the Ziegler-Martin catalyst	T.P. Wilson and G.F. Hurley	1, 281-304 (1963)
Mechanisms of homogeneous anionic polymerization by alkyllithium initiators	M. Morton L.J. Fetters and E.E. Bostick	1, 311-323 (1963)
Alkali-metal polymerization. Homopolymerization and the attempted copolymerization of t-butyl vinyl ketone	C.G. Overberger and A.M. Schiller	1, 325-337 (1963)
Quantitative studies of elementary reactions in electron-transfer initiations	M. Szwarc	1, 339-354 (1963)

Proceedings of the 4th Cellulose Conference, held at Syracuse, New York, October 18-19, 1962.	R.H. Marchessault (Editor)			
Solution properties of birch xylan. I. Measurement of molecular weight	R.G. LeBel D.A.I. Goring and T.E. Timell	2,	9-28	(1963)
Solution properties of birch xylan. II. Fractionation and configuration	R.G. LeBel and D.A.I. Goring	2,	29-48	(1963)
Four polymer-homologous series of oligosaccharides from a 4-O-methylglucuronoxylan	R.H. Marchessault and T.E. Timell	2,	49-61	(1963)
Structural studies on the water-soluble arabino galactans of mountain and european larch	J.K.N. Jones and P.E. Reid	2,	63-71	(1963)
Investigation into the efficacy of water-soluble arabino galactan as a plasma substitute	R.E. Semple	2,	73-78	(1963)
Ozonization of cellulose and wood	C. Schuerch	2,	79-95	(1963)
Some considerations on the kinetics of the acid hydrolysis of poly- and oligosaccharides	A. Meller	2,	97-107	(1963)
Effect of deacetylation and nitration on normal glycosidic linkages in cellulose	T.E. Timell	2,	109-116	(1963)
Viscosity and sedimentation in dilute solutions of cellulosic macromolecules	J.J. Hermans	2,	117-128	(1963)
Flow of gels of cellulose microcrystals. I. Random and liquid crystalline gels	J. Hermans, Jr.	2,	129-144	(1963)
Flow of gels of cellulose microcrystals. II. Effect of added electrolyte	M.R. Edelson and J. Hermans, Jr.	2,	145-152	(1963)
Radiation induced graft polymerization of styrene in wood	K.V. Ramalingam G.N. Werezak and J.W. Hodgins	2,	153-167	(1963)
Grafting polymers onto cellulose by high-energy radiation. II. Effect of swelling agents on the gamma-ray induced direct radiation grafting of styrene onto cellulose	R.Y.-M. Huang and W.H. Rapson	2,	169-188	(1963)
Method for the investigation of carbohydrate graft copolymers	C.P.J. Glaudemans and E. Passaglia	2,	189-197	(1963)

Symposium on Molecular Architecture of Wood	W.A. Cote (Editor)	2,	200-286 (1963)
Morphology of Polymers, American Chemical Society Symposium, held at Los Angeles, California, April 4-5, 1963.	T.G. Rochow (Editor)		
Studies of matrix rigidity to determine intimate morphology	W.O. Statton	3,	3-8 (1963)
Morphology of large molecules in polyethylene	V.G. Peck and L.D. Moore, Jr.	3,	9-19 (1963)
The molecular weight of amorphous polymers by electron microscopy	M.J. Richardson	3,	21-29 (1963)
The use of viscosity data to assess molecular entanglement in dilute polymer solutions	T. Gillespie	3,	31-37 (1963)
The morphology of synthetic latexes	E.B. Bradford and J.W. Vanderhoff	3,	41-64 (1963)
Colloidal particles in the thermosetting resins	E.H. Erath and M. Robinson	3,	65-76 (1963)
Morphology of molded melamine-formaldehyde	H.P. Wohnsiedler	3,	77-89 (1963)
The morphology of polymer fracture surfaces	J.P. Berry	3,	91-101 (1963)
Morphological aspects of polymerization in the solid state	D.G. Grabar and C.S. Hsia Chen	3,	105-107 (1963)
The morphology of crystalline polymers, with especial reference to single crystals grown from the molten state	N.K.J. Symons	3,	109-116 (1963)
Kinetics of spherulite formation	F.P. Price	3,	117-119 (1963)
Mechanisms and kinetics of spherulitic crystallization in high polymers	H.D. Keith	3,	121-122 (1963)
Fracture studies of isothermally bulk-crystallized linear polyethylene	F.R. Anderson	3,	123-134 (1963)
The morphology of pressure-sensitive adhesive films	C.W. Hock	3,	139-149 (1963)
Morphology of polyphase solids	M.C. Botty	3,	151-162 (1963)

Title	Authors	Vol.	Pages	Year
International Symposium on Macromolecular Chemistry, held under the auspices of the International Union of Pure and Applied Chemistry, in Paris, July 1-6, 1963.	M. Magat (Editor)			
Principles of conversional polymerization of unsaturated hydrocarbons	B.A. Krentsel, L.G. Sidorova and A.V. Topchiev	4,	3-9	(1963)
Elementary reactions of the polymerization of propylene at the surface of $TiCl_3$ (in French)	K. Vesely, J. Ambroz and O. Hamrik	4,	11-19	(1963)
Optically active vinyl polymers. VIII. Some aspects of the stereospecific polymerization of racemic α-olefins	P. Pino, F. Ciardelli and G. Paolo	4,	21-36	(1963)
Kinetic study of the heterogeneous polymerization of styrene and/or α-d-styrene by Ziegler-Natta catalysis	C.G. Overberger and P.A. Jarovitzky	4,	37-48	(1963)
The role of aluminum alkyl chlorides in polymerization of propylene with titanium chloride catalysts	A.D. Caunt	4,	49-69	(1963)
Polymerization of ethylene and propylene with Ziegler-Natta catalysts	H. Schnecko, M. Reinmoller, K. Weirauch and W. Kern	4,	71-80	(1963)
Kinetic studies in ethylene polymerization with Ziegler-type catalysts	A. Schindler	4,	81-96	(1963)
New catalysts of the polymerization of ethylene under usual and unusual conditions (in French)	K.A. Kochechkov, O.A. Paleev, T.I. Sogolova et al.	4,	97-102	(1963)
Study of the role of hydrogen in the polymerization of propylene by the use of hydrogen labelled with tritium (in French)	G. Bourat, J. Ferrier and A. Perez	4,	103-107	(1963)
The role of hydrogen in Ziegler-Natta polymerizations	A.S. Hoffman, B.A. Fries and P.C. Condit	4,	109-126	(1963)
Physical restrictions on the rate of polymerizations by some stereospecific catalysts	P.E. Allen, D. Gill and C.R. Patrick	4,	127-140	(1963)
Optically active polymers: autocatalytic phenomena in asymmetrical synthesis (in French)	M. Farina, G. Natta and G. Bressan	4,	141-146	(1963)

Stereospecific polymerization of aldehydes by metalloorganic compounds	H. Sobue and H. Kubota	4,	147-155 (1963)
A kinetic investigation of the formation of amorphous and crystalline polyethylidene from the gold-catalyzed decomposition of diazoethane in ethyl ether solutions	L. Trossarelli, M. Guaita, G. Pegone and A. Priola	4,	157-165 (1963)
Recent progress in the mechanism of the formation of polyethylidene from diazoethane	A.G. Nasini and L. Trossarelli	4,	167-171 (1963)
Directed growth of chains in anionic-coordination polymerization	S.S. Medvedev and A.R. Gantmakher	4,	173-195 (1963)
Mechanism of stereospecific polymerization of styrene with alkali metal alkyls	D. Braun and W. Kern	4,	197-209 (1963)
Determination of active centers in stereospecific diene polymerization	W. Gooper, D.E. Eaves, G.D.T. Owen and G. Vaughan	4,	211-232 (1963)
Polymerization of butadiene by triisobutyl etherate and titanium tetraiodide	J.F. Henderson	4,	233-247 (1963)
Measure of the polymerization rates of isoprene in the presence of metallic lithium (in French)	F. Schue, A. Hinschberger and J. Marchal	4,	249-257 (1963)
Measure of the polymerization rates of 2,3-di-methyl-1,3-butadiene in the presence of metallic lithium (in French)	F. Schue, C. Ortlieb, M. Brini et al.	4,	259-265 (1963)
Synthesis of optically active polymers by asymmetric catalysts. I. Mechanism of propylene oxide polymerization induced by asymmetric catalyst	T. Tsuruta, S. Inoue, M. Ishimori and N. Yoshida	4,	267-279 (1963)
Mechanism of stereoregular polymerization of acetaldehyde	J. Furukawa, T. Saegusa and H. Fujii	4,	281-287 (1963)
Crystalline hydrocarbon polymers by cationic hydride shift mechanism	J.P. Kennedy, L.S. Minckler, Jr., G.G. Wanless and R.M. Thomas	4,	289-296 (1963)
On the mechanism of an anionic polymerization of isoprene by organolithium compounds (in French)	A. Guyot and P.Q. Tho	4,	299-309 (1963)

Polymerization of propylene oxide catalyzed by trimethyl aluminum	R.O. Colclough and K. Wilkinson	4, 311-332 (1963)
Stereospecific polymerization of methyl methacrylate by means of tertiary alcoholates	J. Trekoval and D. Lim	4, 333-343 (1963)
On the two-state mechanism for homogeneous ionic polymerization	B.D. Coleman and T.G. Fox	4, 345-360 (1963)
Stereospecific polymerization of vinyl monomers by ionic mechanism in homogeneous systems	T. Higashimura, T. Watanabe, K. Suzuoki and S. Okamura	4, 361-374 (1963)
Study of the stereospecific polymerization of isoprene by organolithium compounds (in French)	B. Francois, V. Sinn and J. Parrod	4, 375-385 (1963)
Some particular kinds of cobalt catalysts in the polymerization of butadiene (in French)	C. Longiave and R. Castelli	4, 387-398 (1963)
Cobalt catalysts for preparing syndiotactic 1,2-polybutadiene	E. Susa	4, 399-410 (1963)
Polymerization of propylene to syndiotactic polymers (in French)	A. Zambelli, G. Natta and I. Pasquon	4, 411-426 (1963)
Nonbonding interactions in the free propagating methyl methacrylate radical	C.E.H. Bawn, W.H. Janes and A.M. North	4, 427-441 (1963)
Theory of the thermodynamic properties of solutions of graft and block copolymers	M.L. Huggins	4, 445-452 (1963)
Contribution to the thermodynamics of very dilute solutions of copolymers (in French)	M. Lautout-Magat	4, 453-471 (1963)
Study of grafted copolymers and sequences in solution (in French)	Y. Gallot, E. Franta, P. Rempp and H. Benoit	4, 473-489 (1963)
Fractionation of a grafted copolymer, polyisobutylene-polystyrene, prepared by radio chemical technique (in French)	A. Chapiro, P. Cordier, J. Jozefowicz and J. Sebban-Danon	4, 491-506 (1963)
Configuration of macromolecular chains of copolymeric sequences in concentrated solution. Example of the copolymer sequences polystyrolene-polyethylene and polyoxypropylene-polyoxyethylene in solution in solvents favorable to one or the other of the sequences (in French)	A.E. Skoulios, G.T. Souladze and E. Franta	4, 507-518 (1963)

Analysis of copolymers by means of density gradient centrifugation	J.J. Hermans and H.A. Ende	4, 519-527 (1963)
Graft and block copolymers of some vinyl aromatic hydrocarbons	A. Rembaum, J. Moacanin and E. Cuddihy	4, 529-549 (1963)
Preparation and characterization of some cellulose graft copolymers. Part II.	V. Stannett, J.D. Wellons and H. Yasuda	4, 451-562 (1963)
Breakage of a covalence by physical means in a polymeric sequence (in French)	M. Brendlé, G. Riess and A. Banderet	4, 563-575 (1963)
Properties of copolymers obtained by the grafting of vinylic monomers on cellulose (in French)	Kh. U. Usmanov and U. Azizov	4, 579-587 (1963)
Graft copolymers of polyethylene and acrylic acid. II.	J.K. Rieke, G.M. Hart and F.L. Saunders	4, 489-604 (1963)
Dynamic mechanical properties of some graft copolymers	M. Baccaredda, E. Butta and V. Frosini	4, 605-613 (1963)
Grafts of nylon and unsaturated acids	E.E. Magat, I.K. Miller, D. Tanner and J. Zimmerman	4, 615-629 (1963)
Behaviour of the chemical functions introduced by grafting at the surface of solid polymers. I. Study of the structural modifications caused by polyacrylic acid grafted to the surface of a polyethylene polymer (in French)	C. Job and P. LeBel	4, 631-647 (1963)
Rheological study of copolymers grafted to vinyl polychloride and of copolymer styrolene anhydride (in French)	P. LeBel and C. Job	4, 649-672 (1963)
Reactive fiber. II. Chemical reactivities of cellulose fiber grafted with glycidyl methacrylate	Y. Iwakura, T. Kurosaki, K. Uno and Y. Imai	4, 673-698 (1963)
Internal pressure of block copolymers	C. Rossi, U. Bianchi and E. Bianchi	4, 699-705 (1963)
Elastomeric polycarbonate block copolymers	E.P. Goldberg	4, 707-730 (1963)
Polyallomers: Synthesis and properties	H.J. Hagemeyer, Jr. and M.B. Edwards	4, 731-742 (1963)

Polyester-urethane block terpolymers	C.M. Cusano E.P. Dunigan and P. Weiss	4, 743-752 (1963)
Grafting as a method of surface modification of hetero-chain polymers	V.V. Korshak K.K. Mozgova and M.A. Shkolina	4, 753-764 (1963)
International Symposium on Macromolecular Chemistry, held under the auspices of the International Union of Pure and Applied Chemistry, Paris, July 1-6, 1963.	M. Magat (Editor)	
Effect of phase trnasitions on polymerization of monomers below their melting point	V.A. Kargin V.A. Kabanov and I.M. Papissov	4, 767-787 (1963)
Polymerization in the crystalline state. IV. Calcium acrylate and barium methacrylate	J.B. Lando and H. Morawetz	4, 789-803 (1963)
Polymerization in the crystalline state. V. Oriented chain growth in the thermally initiated polymerization of p-acetamidostyrene and p-benzamidostyrene	S.Z. Jakabhazy H. Morawetz and N. Morosoff	4, 805-826 (1963)
Cationic polymerization of trioxane in solid phase	S. Okamura E. Kobayashi M. Takeda et al.	4, 827-838 (1963)
Structure of polymers formed by radiation-induced solid-phase polymerization of cyclic monomers	K. Hayashi H.M. Nishii and S. Okamura	4, 839-848 (1963)
Effect of polymorphism in solid-state polymerization	C.S. Hsia Chen and D.G. Grabar	4, 849-868 (1963)
Copolymerization in solid solutions	C.S. Hsia Chen and D.G. Grabar	4, 869-880 (1963)
Radiopolymerization of solid acrylonitrile: effect of of crystallization conditions and phase transitions (in French)	R. Bensasson A. Dworkin and R. Marx	4, 881-895 (1963)
Radiation-induced solid-phase polymerization. I. Polymerization of acrylonitrile	I.M. Barkalov V.I. Goldanskii N.S. Enikolopyan et al.	4, 897-908 (1963)
Radiation-induced solid-phase polymerization. II. Polymerization of vinyl acetate	I.M. Barkalov V.I. Goldanskii N.S. Enikolopyan et al.	4, 909-921 (1963)

Investigations in the field of radiation-induced solid-state polymerization	Gy. Hardy K. Nyitrai J. Varga et al.	4, 923-934 (1963)
Polymer fractionation studies in the solid-state polymerization of acrylamide initiated by gamma radiation	B. Baysal	4, 935-941 (1963)
Radiation-induced stereospecific polymerization of isocyanate in the solid state	H. Sobue Y. Tabata M. Hiraoka and K. Oshima	4, 943-951 (1963)
Production of polyoxymethylene of high molecular weight and high crystallinity from trioxane	M. Baccaredda E. Butta and P. Giusti	4, 953-965 (1963)
Kinetics of vinylic polymerization with per salts in non-aqueous media (in French)	O.F. Solomon and M. Dimonie	4, 969-976 (1963)
Cyclopolymerization of isoprene in the presence of $AlC_2H_5Cl_2$	I. Kössler M. Štolka and K. Mach	4, 977-985 (1963)
Nitrogen-containing radical-ions as initiators for vinyl polymerization	H. Ringsdorf	4, 987-997 (1963)
Low temperature polymerization of chlorosubstituted aldehyde	D.E. Ilyina B.A. Krentsel and G.E. Semenido	4, 999-1007 (1963)
Polymerization of nitriles and pyridine	V.A. Kabanov V.P. Zubov V.P. Kovaleva and V.A. Kargin	4, 1009-1026 (1963)
Mechanochemical reactions of polymerization and degradation at low temperatures	N.A. Plate and V.A. Kargin	4, 1027-1041 (1963)
New process for the preparation of silanic polymers containing carbon and silicium atoms in the principal chains (in French)	N.S. Nametkine and V.M. Vdovine	4, 1043-1051 (1963)
An effective catalyst for the polymerization of vinyl-silicic compounds (in French)	N.S. Nametkine A.V. Topchiev and S.G. Dourgarian	4, 1053-1059 (1963)
Suppression of the initiation in a polycondensation of the second type: Opening of the β-lactones by the betaines (in French)	Y. Etienne and R. Soulas	4, 1061-1074 (1963)

Low-temperature solution polycondensation	P.W. Morgan	4, 1075-1096 (1963)
A new class of cocatalysts in the anionic polymerization of caprolactane (in French)	Ch. Mermoud	4, 1097-1103 (1963)
Synthesis of new nitrogen- and oxygen-containing polymers with conjugated bonds from inorganic salts and carbamide	I.M. Paushkin and A.F. Lunin	4, 1105-1107 (1963)
Initiation of butadiene popcorn polymerization	G.H. Miller R.R. Eliason and G.O. Pritchard	4, 1109-1115 (1963)
Studies on the formation of non-homogeneous polymers capable of growth (popcorn polymers) (in German)	J.W. Breitenbach	4, 1117-1126 (1963)
Radiation-induced ionic polymerization of styrene; Effects of additives	A. Charlesby and J. Morris	4, 1127-1133 (1963)
Radiochemical polymerization of isobutene in the presence of solid additives (in French)	C. David F. Provoost and G. Verduyn	4, 1135-1149 (1963)
Radiation-induced polymerization at high pressures	L.A. Wall and D.W. Brown	4, 1151-1160 (1963)
Polymerization of chloroprene. I.	A.K. Banbrook R.S. Lehrle and J.C. Robb	4, 1161-1171 (1963)
Grafting of styrene onto teflon and polyethylene by preirradiation	J. Dobó A. Somogyi and T. Czvikovszky	4, 1173-1193 (1963)
Comparison of radiation- and peroxide-initiated grafting of styrene to polyethylene film	W.K.W. Chien and H.Z. Friedlander	4, 1195-1209 (1963)
Grafting of styrene-acrylonitrile mixtures on the polyvinylchloride by direct radio-chemical methods (in French)	A. Chapiro A.-M. Jendrychowska-Bonamour	4, 1211-1222 (1963)
Grafting on polypropylenes of different microstructures	F. Geleji and L. Odor	4, 1223-1232 (1963)
Radiation-induced graft copolymerization onto cellulose and polyvinyl alcohol fibers with binary mixtures of comonomers	I. Sakurada T. Okada S. Hatakeyama and F. Kimura	4, 1233-1249 (1963)
Graft polymerization in cellulose materials. Part I. Cation-exchange membranes from paper and acrylic acid	G.N. Richards and E.F.T. White	4, 1251-1260 (1963)

Polymerization under conditions of rapid decomposition of the indicator	J. Coupek M. Kolinsky and D. Lim	4, 1261-1273 (1963)
Some examples of polymerization in mesomorphic phase (in French)	J. Herz F. Reiss-Husson P. Rempp and V. Luzzati	4, 1275-1290 (1963)
Organized polymerization. I. Olefins on a clay surface	H.Z. Friedlander	4, 1291-1301 (1963)
Synthesis and study of the photoelectric properties of polyazines and Schiff polybases	A.V. Topchiev V.V. Korshak U.A. Popov and I.D. Rosenstein	4, 1305-1313 (1963)
Synthesis and some electrophysical properties of polymers with system of conjugated bonds	V.V. Korshak S.L. Sosin and A.M. Sladkov	4, 1315-1326 (1963)
Conductance mechanism in organic semiconductor polymers	A.V. Airapetyants R.M. Voitenko B.E. Davydov and B.A. Krentsel	4, 1327-1333 (1963)
Conducting polymers from cyclopentadiene	P.E. Blatz	4, 1335-1346 (1963)
Polymers with long chains of doubly conjugated bonds: Preparation and properties (in French)	J.-P. Roth P. Rempp and J. Parrod	4, 1347-1366 (1963)
On the nature of paramagnetic centers detected by ESR in conjugated polymers (in French)	M. Nechtschein	4, 1367-1376 (1963)
Polydicyanoacetylene: Preparation and properties	M. Benes J. Peska and O. Wichterle	4, 1377-1383 (1963)
Catalysis on polymers presenting properties of electronic paramagnetic resonance (in French)	F. Dawans J. Gallard Ph. Teyssie and Ph. Traynard	4, 1385-1400 (1963)
Modification of electronic properties of some synthetic polymers	M. Kryszewski and M. Skorko	4, 1401-1416 (1963)
Relationship between luminescence and semiconducting properties of some synthetic polymers	M. Kryszewski H. Kurczewska and A. Szymanski	4, 1417-1427 (1963)

Study of the electronic conductability of desoxyribonucleic acid films (in French)	M. Hanss P. Douzou and C. Sadron	$\underline{4}$, 1429-1435 (1963)
On the eventual role of electronic doublets in the quantic behaviour of bipolymers (in French)	P. Douzou J.C. Francq M. Ptak and C. Sadron	$\underline{4}$, 1437-1445 (1963)
On electrical asymmetry at the junction of cationic and anionic permselective membranes	H.Z. Friedlander	$\underline{4}$, 1447-1456 (1963)
Polyvinylanthraquinone redox resins (Electron exchange polymers)	G. Manecke and W. Storck	$\underline{4}$, 1457-1466 (1963)
Formation of molecular complexes on high polymers (in French)	G. Smets V. Balogh and Y. Castille	$\underline{4}$, 1467-1480 (1963)
New ferrous-containing polymers based on ferrocene and their electrophysical properties	I.M. Paushkin L.S. Polak T.P. Vishnyakova et al.	$\underline{4}$, 1481-1494 (1963)
Recent results of stereospecific polymerization by heterogeneous catalysis	F. Danusso	$\underline{4}$, 1497-1509 (1964)
Kinetics and mechanism of heterogeneous stereospecific α-olefin polymerization	H.W. Coover, Jr.	$\underline{4}$, 1511-1528 (1964)
Stereospecific polymerization in homogeneous media (in French)	W. Kern and D. Braun	$\underline{4}$, 1529-1549 (1964)
Solid phase polymerizations (in French)	A. Chapiro	$\underline{4}$, 1551-1570 (1964)
Some novel initiators of vinyl polymerization	C.H. Bamford	$\underline{4}$, 1571-1587 (1964)
Grafted and sequenced copolymers in solution (in French)	H. Benoit	$\underline{4}$, 1589-1600 (1964)
Solid-state properties of graft copolymers	V.A. Kargin	$\underline{4}$, 1601-1632 (1964)
Structure and electrical properties of some synthetic solid polymers	W.O. Baker	$\underline{4}$, 1633-1650 (1964)
A Conference on Rheo-optics of Polymers, held under the Auspices of the Polymer Research Institute, Amherst, Mass., August 24, 1963.	R.S. Stein (Editor)	
Streaming birefringence of polymer solutions	W. Philippoff	$\underline{5}$, 1-9 (1964)

Streaming birefringence as a rheological research tool	H. Wayland	5,	11-36 (1964)
Birefringent and rheologic properties of milling yellow suspensions	F.N. Peebles, J.W. Prandos and E.H. Honeycutt, Jr.	5,	37-53 (1964)
Particle size distribution in dilute liquid / liquid dispersions by light scattering	E.E. Lindsey, D.C. Chappelear, D.M. Sullivan and V.A. Augstkalns	5,	55-66 (1964)
On phenomenological rheo-optic constitutive relation	E.H. Dill	5,	67-74 (1964)
Stress-strain-time-birefringence relations in photoelastic plastics with creep	C.L. Amba-Rao	5,	75-86 (1964)
Dynamic birefringence of amorphous polymers	B.E. Read	5,	87-100 (1964)
Temperature dependence of orientation birefringence of polymers in the glassy and rubbery states	R.D. Andrews and T.J. Hammack	5,	101-112 (1964)
Application of equivalent model method to dynamic rheo-optical properties of crystalline polymers	M. Takayanagi, S. Uemura and S. Minami	5,	113-122 (1964)
Dynamic birefringence of several high polymers	R. Yamada and C. Hayashi	5,	123-137 (1964)
Dynamic birefringence of polyolefins	K. Sasaguri and R.S. Stein	5,	139-152 (1964)
Rheo-optical properties of polymers. X. Relaxation	D.G. Legrand and W.R. Haaf	5,	153-161 (1964)
Morphological and rheological studies of polyethylene by light scattering	R.S. Moore and S. Matsuoka	5,	163-177 (1964)
Studies of rates of spherulite deformation by low-angle light scattering	R. Erhardt, K. Sasaguri and R.S. Stein	5,	179-190 (1964)

Thermal Analysis of High Polymers. American Chemical Society Symposium, held at New York, September 11-12, 1963. Sponsored by the Polymer Chemistry Division. — B. Ke (Editor)

Differential thermal analysis of synthetic fibers	R.F. Schwenker, Jr. and R.K. Zuccarello	6,	1-16 (1964)

Differential thermal analysis of linear polyesters	R.M. Schulken, Jr. R.E. Boy, Jr. and R.H. Cox	6,	17-25 (1964)
Thermal analysis of polyurethane elastomers	P.E. Slade, Jr. and L.T. Jenkins	6,	27-32 (1964)
The specific heat of poly-1-butene	H. Wilski and T. Grewer	6,	33-41 (1964)
Differential thermal analysis of polybutene-1	B.H. Clampitt and R.H. Hughes	6,	43-51 (1964)
A differential thermal analysis study of the effects of thermal history on polyethylene	H.W. Holden	6,	53-64 (1964)
Effect of flame retardants on pyrolysis and combustion of α-cellulose	W.K. Tang and W.K. Neill	6,	65-81 (1964)
Calorimetric determinations of the crystallization and melting processes in polymers (in German)	F.H. Müller and H. Martin	6,	83-91 (1964)
Crystallization of polypropylene measured by differential thermal analysis	H.J. Donald E.S. Humes and L.W. White	6,	93-99 (1964)
Differential calorimetric measurements on various chlorinated polyethylenes in connection with x-ray measurements (in German)	E. Hellmuth H.G. Kilian and F.H. Muller	6,	101-108 (1964)
Thermodynamic properties of lexan polycarbonate from 110-560°K.	J.M. O'Reilly F.E. Karasz and H.E. Bair	6,	109-115 (1964)
Calorimetric measurements of thermal effects on stretching (in German)	F.H. Müller and N. Weimann	6,	117-124 (1964)
Studies of the thermal behavior of polymers by torsional braid analysis	J.K. Gillham and A.F. Lewis	6,	125-136 (1964)
Dynamic differential thermal analysis of the glass transition interval	B. Wunderlich and D.M. Bodily	6,	137-148 (1964)
Specific heat of synthetic high polymers. XI. The glass transition in isotactic polypropylene	I. Abu-isa V.A. Crawford A.R. Haly and M. Dole	6,	149-155 (1964)
Thermal conductivity of high polymers	K. Eiermann	6,	157-165 (1964)

Anisotropy of thermal conductivity in stretched amorphous linear polymers and in strained elastomers	J. Hennig and W. Knappe	6,	167-174 (1964)
Thermogravimetry of polymer pyrolysis kinetics	H.C. Anderson	6,	175-182 (1964)
Kinetics of thermal degradation of char-forming plastics from thermogravimetry. Application to a phenolic plastic	H.L. Friedman	6,	183-195 (1964)
Form III to form II phase transition of polybutene-1	C. Geacintov, R.S. Schotland and R.B. Miles	6,	197-207 (1964)

MEDICINAL CHEMISTRY

This is a series of reviews prepared under the auspices of the Division of Medicinal Chemistry of the American Chemical Society. Edited by E.E. Campaigne and W.H. Hartung and published by John Wiley and Sons, Inc., New York-London.

Title	Authors	Reference	
Non-barbiturate hypnotics	K.W. Wheeler	6,	1-245 (1963)
Spinal cord depressant drugs derived from polyhydroxy alcohols	E.J. Pribyl	6,	246-289 (1963)
X-ray contrast media	J.O. Hoppe	6,	290-349 (1963)

DIE NATURWISSENSCHAFTEN

This is a bi-weekly journal, mostly in German, containing original articles and reviews in all fields of the natural sciences. Edited by E. Lamla and published by Springer-Verlag, Berlin.

Title	Authors	Reference
Transfer of genetic information (in German)	H.G. Wittmann	<u>50</u>, 76-88 (1963)
The cycle of ^{14}C in nature (in German)	K.O. Munnich	<u>50</u>, 211-218 (1963)
Hydrogen bonds as means of chemical transport (in German)	M. Eigen	<u>50</u>, 426-437 (1963)
Concerning serologically effective protein resins and enzyme resins (in German)	G. Manecke	<u>51</u>, 25-34 (1964)
Honey-comb formation by the honey-bee <u>apis mellifica</u> L. (in German)	H. Rembold	<u>51</u>, 49-54 (1964)
The mechanism of biological X-rays action (in English)	O. Warburg with K. Gawehn A.W. Geissler <u>et al.</u>	<u>51</u>, 373-376 (1964)
Compounds of the noble gases (in German)	H. Schwarz	<u>51</u>, 397-403 (1964)
Protein synthesis and ribonucleic acids in nucleus free reticulocites (in German)	H.-G. Schweiger	<u>51</u>, 521-533 (1964)

NUCLEONICS

This is a monthly journal, in English, containing original articles, technical information and reviews in all areas of the nuclear sciences. The pagination is independent for each number so that both volume, month and page are pertinent. Edited by J.D. Lutz and published by McGraw-Hill, Inc., New York.

Title	Authors	Reference
Radiation Effects on Man		
Somatic effects	L.D. Hamilton	21, (2) 48-53 (1963)
Hereditary effects	W.J. Schull	21, (2) 54-57 (1963)
Radiation-protection standards	L.S. Taylor	21, (2) 58-60 (1963)
Radiotracers in Medicine		
Tailoring the isotope to the need	L.G. Stang, Jr. and P. Richards	22, (1) 46-49 (1964)
Optimization of a scanning method using Tc^{99m}	P.V. Harper, R. Beck, D.C. Charleston and K.A. Lathrop	22, (1) 50-54 (1964)
Biological aspects in the choice of scanning agents	M. Blau and M.A. Bender	22, (1) 55-57 (1964)
Studying organ circulation, function and structure with radioisotopes	G.V. Taplin	22, (1) 58-60 (1964)

This is a general scientific review journal, in French, originally published since 1840 as " La Revue Scientifique" and renamed in 1960 " Nucleus ". Edited by L. Longchambon and published by Dunod, Paris.

Title	Authors	Reference
Catalysis and stereospecific polymerization. I.	G. Natta	4, 97-108 (1963)
Catalysis and stereospecific polymerization. II.	G. Natta	4, 211-224 (1963)
Recent progress in photosynthesis	C. Mentzner	5, 204-206 (1964)
Molecular biophysics	C. Sadron	5, 246-261 (1964)
The synthesis of bombybol	A. Butenandt and E. Hecker	5, 325-332 (1964)

ÖSTERREICHISCHE CHEMIKER-ZEITUNG

This is a monthly journal, in German, containing communications, technical information and review articles. Edited by A. Siegel and published by Springer-Verlag, Wien.

Title	Authors	Reference
Emulsion polymerization of vinyl acetate	G. Müller	64, 33-35 (1963)
Cellulose and artificial fibres	R. Reichherzer	64, 35-38 (1963)
Chemisorption, desorption and mechanism of catalytical reactions on metallic catalysts	A. Krause	64, 142-144 (1963)
Reactions of organic compounds on electrodes	N. Konopik	64, 168-177 (1963)
The problem of the sulfur-sulfur bond	M. Schmidt	64, 236-246 (1963)
Chemistry of hot atoms (chemical transformations by nuclear processes)	N. Getoff	64, 329-334 (1963)
The role of free radicals in oxidative phenol-coupling	J. Spona	65, 47-55 (1964)
New findings in the chemistry of allylic compounds	H. Schmid	65, 109-116 (1964)
Reactions similar to those occuring in muscle, of transformation of chemical to mechanical energy	W. Kuhn	65, 137-143 (1964)
The formation of organic compounds from carbonic acid in aqueous solutions under the influence of cobalt-60 gamma radiation and ultra-violet light	N. Getoff	65, 143-156 (1964)
New findings and developments in the field of high-fast viscose fibres	K. Götze	65, 209-218 (1964)
Coordination chemistry of crystal violet	V. Gutmann	65, 273-275 (1964)
Determination of aridity in concentrated salt solutions and organic solvents	K. Schwabe	65, 339-357 (1964)

ORGANIC REACTIONS

This is a collection of review articles, in English, on a number of the major reactions of organic chemistry. This series was initiated by Professor Roger Adams. Edited by A.C. Cope and published by John Wiley and Sons, Inc., New York-London.

Title	Authors	Reference		
Hydration of olefins, dienes and acetylenes via hydroboration	G. Zweifel and H.C. Brown	13,	1-54	(1963)
Halocyclopropanes from halocarbenes	W. Parham and E.E. Schweizer	13,	55-90	(1963)
Free radical additions to olefins to form carbon-carbon bonds	C. Walling and E.S. Huyser	13,	91-149	(1963)
Formation of carbon-hetero atom bonds by free radical chain additions to carbon-carbon multiple bonds	F.W. Stacey and J.F. Harris, Jr.	13,	150-376	(1963)
The Chapman rearrangement	J.W. Schulenberg and S. Archer	14,	1-51	(1964)
α-Amidoalkylations at carbon	H.E. Zaugg and W.B. Martin	14,	52-269	(1964)
The Wittig reaction	A. Maercker	14,	270-490	(1964)

PROCEEDINGS OF THE CHEMICAL SOCIETY

This is a monthly journal, in English, containing review articles, short communications and technical information. Edited by R.S. Cahn and published by the Chemical Society (London) Burlington House. After 1964, material published in <u>Proceedings</u> will be published in <u>Chemistry in Britain</u>.

Title	Authors	Reference
Amino-acid sequences in the active centres of certain enzymes (Pedler Lecture)	F. Sanger	76-83 (1963)
The strange history of intensive drying	W.V. Farrar	125-130 (1963)
Stereoselectivity in the reactions of cyclic compounds (Tilden Lecture)	H.B. Henbest	159-165 (1963)
The biosynthesis of alkaloids (Tilden Lecture)	A.R. Battersby	189-200 (1963)
Contributions of x-ray analysis to natural-product chemistry (Presidential Address)	J.M. Robertson	229-238 (1963)
The mechanism of the enzymic decarboxylation of acetoacetic acid (Centenary Lecture)	F.H. Westheimer	253-261 (1963)
The biogenesis of phenolic alkaloids (Hugo Muller Lecture)	D.H.R. Barton	293-298 (1963)
Chemistry at Cambridge from 1901-1910	A.J. Berry and E.A. Moelwyn-Hughes	357-363 (1963)
The Steacie memorial lecture	H.E. Gunning	73-79 (1964)
The group displacement law	J.A. Cranston	104-107 (1964)
A glow in the dark - the rationale of phosphorylation (Tilden Lecture)	V.M. Clark	129-135 (1964)
Some basic problems of solid-state chemistry (Liversidge Lecture)	J.S. Anderson	166-173 (1964)
Activated molecules (Tilden Lecture)	A.F. Trotman-Dickenson	249-256 (1964)
Molecular rearrangements of terpenes (Simonsen Lecture)	G. Ourisson	274-282 (1964)
Applications of optical rotatory dispersion and circular dichroism in stereochemistry (Centenary Lecture)	C. Djerassi	314-330 (1964)
Niels Bohr memorial lecture	G. Thompson	351-354 (1964)

This is a collection of review articles, in English. Edited by F.A. Cotton and published by Interscience Publishers (a division of John Wiley and Sons, Inc.) New York-London.

Title	Authors	Reference
Preparation and properties of primary, secondary, and tertiary phosphines	L. Maier	5, 27-210 (1963)
Polarographic behavior of coordination compounds	A.A. Vicek	5, 211-384 (1963)
The coupling of vibrational and electronic motions in degenerate and non-degenerate electronic states of inorganic and organic molecules. Part III. Non-degenerate electronic states	A.D. Liehr	5, 385-430 (1963)
Reaction of metal halides with ammonia and aliphatic amines	G.W.A. Fowles	6, 1-36 (1964)
The magnetic properties of transition metal complexes	B.N. Figgis and J. Lewis	6, 37-239 (1964)
The coordination model for non-aqueous solvent behavior	R.S. Drago and K.F. Purcell	6, 271-322 (1964)

PROGRESS IN MEDICINAL CHEMISTRY

This is an annual collection of review articles, in English. Edited by G.P. Ellis and G.B. West and published by Butterworths, London.

Title	Authors	Reference
Some chemical aspects of neuromuscular block	J.B. Stenlake	3, 1-51 (1963)
The chemotherapy of trypanosomiasis	L.P. Walls	3, 52-88 (1963)
Antitussive drugs	C.I. Chappel and C. von Seemann	3, 89-145 (1963)
The chemistry and pharmacology of the rauwolfia alkaloids	R.A. Lucas	3, 146-186 (1963)
Anticonvulsant drugs	A. Spinks and W.S. Waring	3, 261-331 (1963)
Local anaesthetics	S. Wiedling and C. Tegnér	3, 332-398 (1963)

PROGRESS IN NUCLEIC ACID RESEARCH

This is an annual collection of review articles, in English. Edited by J.N. Davidson and W.E. Cohn and published by Academic Press, New York-London.

Title	Authors	Reference
Messenger ribonucleic acid	F. Lipmann	<u>1</u>, 135-161 (1963)
Some problems concerning the macromolecular structure of ribonucleic acids	A.S. Spirin	<u>1</u>, 301-345 (1963)
The structure of DNA as determined by x-ray scattering techniques	V. Luzzati	<u>1</u> 347-368 (1963)
Molecular mechanisms of radiation effects	A. Wacker	<u>1</u>, 369-399 (1963)
Nucleic acids and information transfer	L.F. Cavalieri and B.H. Rosenberg	<u>2</u>, 1-18 (1963)
Nuclear ribonucleic acid	H. Harris	<u>2</u>, 19-59 (1963)
Plant virus nucleic acids	R. Markham	<u>2</u> 61-81 (1963)
The nucleases of escherichia coli	I.R. Lehman	<u>2</u>, 83-123 (1963)
Specificity of chemical mutagenesis	D.R. Krieg	<u>2</u>, 125-168 (1963)
Column chromatography of oligonucleotides and polynucleotides	M. Staehelin	<u>2</u>, 169-195 (1963)
Mechanism of action and application of azapyrimidines	J. Skoda	<u>2</u>, 197-219 (1963)
The function of the pyrimidine base in the ribonuclease reaction	H. Witzel	<u>2</u>, 221-258 (1963)
Preparation and fractionation of nucleic acids	K.S. Kirby	<u>3</u> 1-31 (1964)
Cellular sites of RNA synthesis	D.M. Prescott	<u>3</u>, 33-57 (1964)
Ribonucleases in takadiastase: properties, chemical nature and applications	F. Egami, K. Takahashi and T. Uchida	<u>3</u>, 59-101 (1964)
Chemical effects of ionizing radiations on nucleic acids and related compounds	J.J. Weiss	<u>3</u>, 103-144 (1964)

PROGRESS IN NUCLEIC ACID RESEARCH (Contd.)

Title	Author	Reference
The regulation of RNA synthesis in bacteria	F.C. Neidhardt	3, 145-181 (1964)
Actinomycin and nucleic acid function	E. Reich and I.H. Goldberg	3, 183-234 (1964)
De Novo protein synthesis in Vitro	B. Nisman	3, 235-297 (1964)
Free nucleotides in animal tissues	P. Mandel	3, 299-334 (1964)

PROGRESS IN ORGANIC CHEMISTRY

This is a collection of review articles, in English. Edited by J. Cook and W. Carruthers and published by Butterworth, Inc., Washington, D.C.

Title	Authors	Reference
Vegetable tannins	E. Haslam and R.D. Haworth	6, 1-37 (1964)
Bisbenzylisoquinoline alkaloids	M.F. Grundon	6, 38-85 (1964)
Polyacetylenes and related compounds in nature	J.D. Bu'Lock	6, 86-134 (1964)
The chemistry of the phenoxazones	W. Schafer	6, 135-163 (1964)
Carbenes	W. Kirmse	6, 164-213 (1964)
Molecular orbital studies of organic equilibria and reaction rates	S.F. Mason	6, 214-247 (1964)

PROGRESS IN PHYSICAL ORGANIC CHEMISTRY

The purpose of this series is to provide critical, authoritative reviews on all phases of physical organic chemistry. The first volume appeared in 1963. Edited by S.G. Cohen, A. Streitwieser, Jr. and R.W. Taft and published by John Wiley and Sons, Inc., New York-London.

Title	Authors	Reference
Ionization potentials in organic chemistry	A. Streitweiser, Jr.	1, 1-30 (1963)
Nucleophilic aromatic substitution reactions	S.D. Ross	1, 31-74 (1963)
Ionization and dissociation equilibria in solution in liquid sulfur dioxide	N.N. Lichtin	1, 75-108 (1963)
Secondary isotope effects	E.A. Halevi	1, 109-221 (1963)
Quantitative comparisons of weak organic bases	E.M. Arnett	1, 223-403 (1963)
The properties and reactivity of methylene: derived principally from its gas-phase reactions	J.A. Bell	2, 1-61 (1964)
Mechanism and catalysis of simple carbonyl group reactions	W.P. Jencks	2, 63-128 (1964)
Carbonium ions	N.C. Deno	2, 129-193 (1964)
Theoretical interpretations of the Hammett and derivative structure-reactivity relationships	S. Ehrenson	2, 195-251 (1964)
Electrophilic aromatic substitution reactions	E. Berliner	2, 253-321 (1964)
An examination of structure-reactivity relationships	C.D. Ritchie and W.F. Sager	2, 323-400 (1964)

PROGRESS IN REACTION KINETICS

This is a collection of review articles. in English. Edited by G. Porter and published by Pergamon Press, Oxford.

Title	Authors	Reference
The rate constants of halogen atom reactions	G.C. Fettis and J.H. Knox	$\underline{2}$, 1-38 (1964)
Mercury photosensitized reactions	R.J. Cvetanovic	$\underline{2}$, 39-130 (1964)
The reactions of methylene and some simple carbenes	H.M. Frey	$\underline{2}$, 131-164 (1964)
The kinetics of cis-trans isomerization	R.B. Cundall	$\underline{2}$, 165-215 (1964)
The kinetics of propagation of anionic polymerization and copolymerization	M. Szwarc and J. Smid	$\underline{2}$, 217-284 (1964)
Rate constants of protolytic reactions in aqueous solution	M. Eigen, W. Kruse, G. Maass and L. De Maeyer	$\underline{2}$, 285-318 (1964)
The rates of reactions of some haem compounds	Q.H. Gibson	$\underline{2}$, 319-335 (1964)
Kinetic treatment of consecutive processes	R.M. Noyes	$\underline{2}$, 337-362 (1964)

PROGRESS IN THE CHEMISTRY OF FATS AND OTHER LIPIDS

This is a collection of review articles, in English, published by Pergamon Press, Oxford-New York and edited by R.T. Holman, W.O. Lundberg and T. Malkin.

Title	Authors	Reference		
Plasmalogens	E. Klenk and H. Bebuch	6,	3-29	(1963)
The synthesis of phospholipids	E. Baer	6,	33-86	(1963)
Fatty acid oxidation	D.E. Green	6,	88-115	(1963)
Fatty acid biosynthesis	D.M. Gibson	6,	118-136	(1963)
The chemistry of serum lipoproteins	N.K. Freeman, F.T. Lindgren and A.V. Nichols	6,	216-250	(1963)
Lipoproteins of the fowl-serum, egg and intracellular	O.A. Schjeide	6,	253-289	(1963)
Ozonolysis of fatty acids and their derivatives	R.G. Kadesch	6,	292-312	(1963)
Recent developments in the glyceride structure of vegetable oils	H.J. Dutton and C.R. Scholfield	6,	313-339	(1963)
The higher saturated branched chain fatty acids	S. Abrahamsson, S. Ställberg-Stenhagen and E. Stenhagen	7,	1-157	(1963)
Gas chromatography of lipids	E.C. Horning, A. Karmen and G.C. Sweeley	7,	167-246	(1963)
Antioxidant effects in biochemistry and physiology	J.G. Bieri	7,	247-266	(1963)
The coenzyme Q group (Ubiquinones)	F.L. Crane	7,	267-289	(1963)

This is the official journal of the International Union of Pure and Applied Chemistry (I.U.P.A.C.), in English, French or German. It publishes both the main invited lectures of symposia sponsored by I.U.P.A.C. as well as the recommendations of the Unions various commissions. Two volumes are published annually, since 1960-61. Edited by B.C.L. Weedon and published for I.U.P.A.C. by Butterworths, London. Unless otherwise indicated, articles are in English.

Title	Authors	Reference
Plenary Lectures Presented at the 7th International Conference on Coordination Chemistry, held in Stockholm and Uppsala, Sweden, 25-29 June, 1962.		
Metal complexes of peptides and proteins	F.R.N. Gurd	6, 49-59 (1963)
Some important aspects of the chemistry of iso-heteropolyanions	P. Souchay	6, 61-84 (1963)
Physical properties and structure of donor-acceptor complexes	N.N. Greenwood	6, 85-96 (1963)
Fast elementary steps in chemical reaction mechanisms	M. Eigen	6, 97-115 (1963)
Plenary Lectures Presented at the International Symposium on Pharmaceutical Chemistry, held in Florence, Italy, 17-19 September, 1962.		
The co-ordination of chemical processes in the living cell	T. Bücher	6, 209-226 (1963)
Drug-receptor interaction	H.R. Ing	6, 227-232 (1963)
Some recently developed fractionation procedures and their application to peptide and protein hormones	J. Porath	6, 233-244 (1963)
Structure-function relations in the corticotropin series	K. Hofmann	6, 245-264 (1963)
Chemical structure and biological activity in the field of polypeptide hormones	R. Schwyzer	6, 265-295 (1963)
Peptide-type antibiotics	J.C. Sheehan	6, 297-304 (1963)
Synthetic approach to the structure-activity relation of some antibiotics	M.M. Shemyakin and M.N. Kolosov	6, 305-325 (1963)
Iron-containing antibiotics and microbic growth factors	V. Prelog	6, 327-338 (1963)
Chemical and toxicological studies with cyclopeptides of _Amanita phalloides_	T. Wieland	6, 339-349 (1963)

PURE AND APPLIED CHEMISTRY (Contd.)

Title	Author	Reference
Attempts at chemotherapy of neoplastic and related diseases	F. Bergel	6, 351-368 (1963)
Photoreactions between flavin coenzymes and skin-photosensitizing agents	L. Musajo	6, 369-384 (1963)
Recent developments in psycho-pharmacology (in French)	D. Bovet	6, 385-407 (1963)
Recent pharmaceutical research on hydrazine derivatives	E. Jucker	6, 409-433 (1963)
Chemical structure and biological activity of catecholamines	P. Pratesi	6, 435-449 (1963)
Recent findings on apocynea alkaloids (Funtumia, malouetia, paravallaris, holarrhena, picralima and vinca) (in French)	M.-M. Janot	6, 451-470 (1963)
Recent researches on indole alkaloids	A.R. Battersby	6, 471-481 (1963)

Special and Introductory Lectures Presented at the 2nd International Symposium on the Chemistry of Natural Products, held in Prague, Czechoslovakia, August 27-September 2, 1962.

Title	Author	Reference
Perspectives in the biogenesis and chemistry of terpenes (in German)	L. Ruzicka	6, 493-523 (1963)
Progress in the total synthesis of steroids	I.V. Torgov	6, 525-544 (1963)
Conformation and reactivity of medium-sized ring compounds	V. Prelog	6, 545-560 (1963)
The total synthesis of tetracycline	R.B. Woodward	6, 561-573 (1963)
Mass spectrometric investigations in the steroid, terpenoid and alkaloid fields	C. Djerassi	6, 575-599 (1963)
The biogenesis of certain alkaloids	R. Robinson	6, 601-620 (1963)
The highly-oxygenated diterpenoid alkaloids	L. Marion	6, 621-633 (1963)
Recent alkaloid structures of the aquamicine type (in French)	M.-M. Janot	6, 635-662 (1963)
Photochemical transformations of natural products	D.H.R. Barton	6, 663-677 (1963)
Some aspects of the chemotaxonomy	H. Erdtman	6, 679-708 (1963)

The chemistry of the aphid colouring matters	A. Todd	6,	709-717 (1963)

Special Lectures Presented at the International Symposium on Molecular Structure and Spectroscopy, held in Tokyo, Japan, 10-15 September, 1962.

Molecular vibrations and physicochemical problems	H.W. Thompson	7,	13-21 (1963)
Recent results of chemical interest from microwave spectroscopy	E.B. Wilson, Jr.	7,	23-31 (1963)
Interference spectroscopy and some of its applications in the far infrared and in the near infrared	G. Sutherland	7,	33-43 (1963)
Formation and dissociation of excited dimers	T. Forster	7,	73-78 (1963)
Electronic spectra and electron-transfer interaction between electron donor and acceptor	S. Nagakura	7,	79-92 (1963)
N.M.R. Studies of conformational equilibria in substituted ethanes	H.S. Gutowsky	7,	93-102 (1963)
Molecular spectra of some organic sulphur compounds	A. Mangini	7,	103-109 (1963)
Infrared spectral perturbations in matrix experiments	G.C. Pimentel and S.W. Charles	7,	111-123 (1963)
Force constants of small molecules	T. Shimamouchi	7,	131-145 (1963)

Congress Lectures Presented at the 19th International Congress of Pure and Applied Chemistry, held in London, U.K., 10-17 July, 1963.

Structure of carbanions	D.J. Cram	7,	155-172 (1963)
On small rings containing triple bonds. Another chemistry of the "as if" (in German)	G. Wittig	7,	173-191 (1963)
Equilibration and vapour-phase halogenation of benzene derivatives	E.C. Kooyman	7,	193-202 (1963)
Rearrangements of free alkyl radicals and alkyl cations in solution	O.A. Reutov	7,	203-227 (1963)
Isothiocyanates of natural derivation	A. Kjaer	7,	229-245 (1963)
On the structures of some biologically active microbial metabolites (in French)	E. Lederer	7,	247-268 (1963)

The use of the Schmidt reaction in the elaboration of selected alkaloids containing a seven-membered ring	S. Uyeo	7, 269-283 (1963)
The structure, stereochemistry and absolute configuration of anhydroryanodine	K. Wiesner	7, 285-296 (1963)
Studies on the synthesis of corrins	A. Eschenmoser	7, 297-316 (1963)
Recent studies on the synthesis of homocyclic systems	W.S. Johnson	7, 317-334 (1963)
Organic chemistry in peptide synthesis	J. Rudinger	7, 335-362 (1963)
Recent advances in the chemistry of large-ring conjugated systems	F. Sondheimer	7, 363-388 (1963)
Variation of the half-wave potential of organic compounds with pH	P.J. Elving	7, 423-454 (1963)
Modern progress and problems in the determination of trace elements in pure substances	I.P. Alimarin	7, 455-472 (1963)
Masking and promotion of reactions in quantitative analysis	A. Ringbom	7, 473-488 (1963)
The biosynthesis of antibiotics	A.J. Birch	7, 527-537 (1963)
The vitamin B_{12} coenzyme	A. Dolphin A.W. Johnson R. Rodrigo and N. Shaw	7, 539-549 (1963)
Constitution of rifamycins	V. Prelog	7, 551-564 (1963)
The sclerotiorin group of fungal metabolites: their structure and biosynthesis	W.B. Whalley	7, 565-587 (1963)
Fungal metabolism of certain aromatic compounds related to lignin	M.E.K. Henderson	7, 589-602 (1963)
Protein synthesis in yeast	A.H. Cook	7, 621-637 (1963)
The structure and organization of the polysaccharides of yeast	D.H. Northcote	7, 669-675 (1963)

Plenary Lectures Presented at the Symposium on Thermodynamics and Thermochemistry, held in Lund, Sweden, 18-23 July, 1963.

Excursion in chemical thermodynamics, from the past into the future	F.D. Rossini	<u>8</u>,	95-112 (1964)
Key heat of formation data	H.A. Skinner	<u>8</u>,	113-130 (1964)
The calorimetry of combustions and related reactions: inorganic reactions	C.E. Holley, Jr.	<u>8</u>,	131-142 (1964)
The calorimetry of combustions and related reactions: organic compounds	J.D. Cox	<u>8</u>,	143-156 (1964)
Calorimetry of non-reacting systems with particular emphasis on solution and mixing processes	M.L. McGlashan	<u>8</u>,	157-169 (1964)
On the thermochemistry of stepwise complex formation	I. Leden	<u>8</u>,	171-178 (1964)
Biochemical calorimetry	I. Wadsö	<u>8</u>,	179-185 (1964)
Progress in the calorimetry and thermodynamics of phase and ordering transitions	E.F. Westrum, Jr.	<u>8</u>,	187-214 (1964)

Proceedings of the Symposium on Isotope Mass Effects in Chemistry and Biology, held in Vienna, Austria, 9-13 December, 1963.

Correlation of kinetic isotope effects with chemical bonding in three-centre reactions	J. Bigeleisen	<u>8</u>,	217-223 (1964)
Validity of some approximation procedures used in the theoretical calculation of isotope effects	M. Wolfsberg and M.J. Stern	<u>8</u>,	225-242 (1964)
Solvent isotope effect in the H_2O - D_2O mixtures	A.J. Kresge	<u>8</u>,	243-258 (1964)
Deuterium isotope effects in two-phase liquid systems containing water	C.U. Linderstrøm-Lang	<u>8</u>,	259-272 (1964)
Deuterium solvent isotope effects on acid-base equilibria in dioxan-water mixtures	V. Gold and B.M. Lowe	<u>8</u>,	273-279 (1964)
Conversion of oxygen into CO_2, as part of the isotopic determination of oxygen (in French)	H. Hering H. LeBail J. Sutton et al.	<u>8</u>,	281-286 (1964)
Carbon isotope effects in the pyrolytic decomposition of manganous oxalate	P.E. Yankwich and P.D. Zavitsanos	<u>8</u>,	287-304 (1964)

Large-scale bacteriogenic fractionation of sulphur isotopes	M. L. Jensen and N. Nakai	8, 305-315 (1964)
Variations in the ratio $^{48}Ca/(total\ Ca)$ in the natural environment	J. T. Corless and J. W. Winchester	8, 317-323 (1964)
Secondary isotope effects as probes for force constant changes	M. Wolfsberg and M. J. Stern	8, 325-338 (1964)
Secondary isotope effects on pi-complex formation	E. A. Halevi and B. Ravid	8, 339-345 (1964)
Steric origin of some secondary isotope effects	V. F. Raaen and C. J. Collins	8, 347-355 (1964)
Detection and computation of isotope fractionation in the adsorption chromatography of dual-labelled compounds	P. D. Klein D. W. Simborg and P. A. Szczepanik	8, 357-370 (1964)
Tritium and deuterium Arrhenius parameter effects in a base-promoted elimination reaction: evidence for tunnelling	V. J. Shiner, Jr. and B. Martin	8, 371-378 (1964)
Carbon-14 and tritium isotope effects in Hofmann degradation of quarternary amines	H. Simon and G. Mullhofer	8, 379-384 (1964)
Tritium and nitrogen-15 isotope effects in the decomposition of p-nitrophenethyltrimethyl-ammonium iodide	E. M. Hodnett and J. J. Sparapany	8, 385-392 (1964)
Intramolecular proton exchange in aromatic carbonium ions: isotope effect in hexamethylbenzene	E. L. Mackor and C. MacLean	8, 393-404 (1964)
Rate of tritium exchange in diethyl methyl-d_3-malonate-t and diethyl malonate-d t in buffered aqueous solutions	O. Gjurović-Deletis S. Borčić and D. E. Sunko	8, 405-408 (1964)
Application of the successive labelling technique to some carbon, nitrogen and chlorine isotope effect studies of organic reaction mechanisms	A. Fry	8, 409-419 (1964)
Kinetic hydrogen isotope effects in the hydration of isobutene	V. Gold and M. A. Kessick	8, 421-425 (1964)
Abnormal tritium isotope effect in the addition of hydrogen atoms to liquid unsaturated hydrocarbons	J. Y. Yang and J. G. Burr	8, 427-433 (1964)
Competitive deuteration of toluene and toluene-α, α, α-d_3	W. M. Lauer and K. C. Senan	8 435-439 (1964)

Effect of deuterium substitution on solvolysis rates of (methylcyclopropyl)-carbinyl derivatives	M. Nikoletić S. Borćić and D.E. Sunko	8,	441-448 (1964)
Deuterium isotope effects as criteria of mechanism in the reactions of organophosphorus compounds	D. Samuel	8,	449-457 (1964)
Kinetic deuterium isotope effects in enzymic formate oxidation	H. Aebi	8,	459-469 (1964)
Isotope effects in fully deuterated hexoses, proteins and nucleic acids	J.J. Katz H.L. Crespi and A.J. Finkel	8,	471-481 (1964)
Mitosis and macromolecule synthesis in cells exposed to D_2O	P.R. Gross W. Spindel and G.H. Cousineau	8,	483-497 (1964)
Dose-determined effects of HT_3DR as DNA label upon the liver cell replication time and pattern in the growing rat	J. Post and J. Hoffman	8,	499-505 (1964)
Comparison of effects of D_2O and x-ray irradiation on lipid composition of tissue culture cells	G.H. Rothblat R.W. Hartzell, Jr. and D. Kritchevsky	8,	507-513 (1964)

Special Lectures Presented at the 3rd International Symposium on the Chemistry of Natural Products, held in Kyoto, Japan, 12-18 April, 1964.

Chemistry of some dimeric indole alkaloids	G. Büchi	9,	21-34 (1964)
Some studies in the biogenesis of plant products	D.H.R. Barton	9,	35-47 (1964)
The structure of tetrodotoxin	R.B. Woodward	9,	49-74 (1964)
Intramolecular rearrangements in peptide systems: hydroxy- and amino-acyl incorporation into peptides	M.M. Shemyakin and V.K. Antonov	9,	75-94 (1964)
High resolution mass spectrometry of natural products	K. Biemann	9,	95-118 (1964)
Specification of the stereospecificity of some oxido-reductases by diamond lattice sections	V. Prelog	9,	119-130 (1964)
Progress in the synthesis of polycyclic natural products	G. Stork	9,	131-144 (1964)
Peptides of amanita phalloides	Th. Wieland	9,	145-157 (1964)
Isotope labelling and mass spectrometry of natural products	C. Djerassi	9,	159-178 (1964)

Recent advances in x-ray analysis of natural product structures	J.M. Robertson	9, 179-191 (1964)

Special Lectures Presented at the Symposium on Organo-Phosphorus Compounds, held in Heidelberg, Germany, 20-22 May, 1964.

The preparation of organo-phosphorus compounds from the element	M. Grayson	9, 193-204 (1964)
Synthesis of organophosphorus compounds by means of p-substituted metallophosphines (in German)	K. Issleib	9, 205-223 (1964)
Synthesis and properties of optically active tertiary phosphenes (in German)	L. Horner	9, 225-244 (1964)
Variation on a subject from Staudinger: a contribution to the history of formation of olefins by organic phosphor carbonyls (in German)	G. Wittig	9, 245-254 (1964)
The Wittig reaction	S. Trippett	9, 255-269 (1964)
Stereospecific carbonyl olefination with phosphorylids	L.D. Bergelson and M.M. Shemyakin	9, 271-283 (1964)
New reactions of phosphene alkenes and their synthetic utility (in German)	H.J. Bestmann	9, 285-306 (1964)
The Michaelis-Arbusow and Perkow reactions (in German)	B.A. Arbusow	9, 307-335 (1964)
Condensations of carbonyl compounds with phosphite esters	F. Ramirez	9, 337-369 (1964)
The nature of the chemical bonding in organo-phosphorus compounds	R.F. Hudson	9, 371-386 (1964)

Plenary Lectures Presented at the International Symposium on Organic Photochemistry, held in Strasbourg, France, 20-24 July, 1964.

Some aspects of transitions between electronic levels	W.A. Noyes, Jr. and I. Unger	9, 461-472 (1964)
Chemiluminescence in solution	E.J. Brown	9, 473-479 (1964)
Photochemical isomerization of polyenic systems (in French)	M. Mousseron	9, 481-492 (1964)

Report on recent photochemical investigations	H.E. Zimmerman	9,	493-498 (1964)
Reactivity of excited states of aromatic ketones	G. Porter and P. Suppan	9,	499-505 (1964)
Mechanism and stereoselectivity of photosensitized oxygen transfer reactions	K. Gollnick and G.O. Schenck	9,	507-525 (1964)
Photolysis of the diazirines	H.M. Frey	9,	527-537 (1964)
Photochemistry of dienes	W.G. Dauben and W.T. Wipke	9,	539-553 (1964)
Photochemical transformation of α,β-epoxyketones and related carbonyl systems	O. Jeger, K. Schaffner and H. Wehrli	9,	555-565 (1964)
Photochemistry of the organic solid state	M.D. Cohen	9,	567-574 (1964)
Stereospecificity in photochemical reactions of ketones	R.C. Cookson	9,	575-584 (1964)
Photochemistry of unsaturated nitro compounds	O.L. Chapman, A.A. Griswold, E. Hoganson et al.	9,	585-590 (1964)
Photochemical reactions of carbonyl compounds in solution: Paterno-Buchi reaction	N.C. Yang	9,	591-596 (1964)
Photochemical reactions of dicarbonyl compounds	P. de Mayo	9,	597-606 (1964)
Photochemistry of non-conjugated ketones in solution	G. Quinkert	9,	607-621 (1964)

QUARTERLY REVIEWS

This is a review journal, in English, containing articles in all fields of chemistry. Edited by R.S. Cahn and published by the Chemical Society (London), Burlington House.

Title	Authors	Reference
Griseofulvin	J.F. Grove	17, 1-19 (1963)
Optical rotatory power	S.F. Mason	17, 20-66 (1963)
Electron-spin resonance spectra of aromatic radicals and radical-ions	A. Carrington	17, 67-99 (1963)
Radiation chemistry of hydrocarbons	F. Williams	17, 101-132 (1963)
The metal carbonyls	E.W. Abel	17, 133-159 (1963)
An outline of acylation	D.P.N. Satchell	17, 160-203 (1963)
Thermochemical properties of phosphorus compounds	S.B. Hartley, W.S. Holmes, M.F. Mole et al.	17, 204-223 (1963)
Methods of studying chemical kinetics in flames	G. Dixon-Lewis and A. Williams	17, 243-263 (1963)
Bond-energy term values in hydrocarbons and related compounds	H.A. Skinner and G. Pilcher	17, 264-288 (1963)
The macrolide antibiotics	M. Berry	17, 343-361 (1963)
Far-infrared spectroscopy	J.L. Wood	17, 362-381 (1963)
The acceptor properties of quadripositive silicon, germanium, tin, and lead	I.R. Beattie	17, 382-405 (1963)
The Wittig reaction	S. Trippett	17, 406-440 (1963)
The chemistry of the inorganic azides	P. Gray	17, 441-473 (1963)
The absolute intensities of infrared absorption bands	D. Steele	18, 21-44 (1964)
Neighbouring group participation	B. Capon	18, 45-111 (1964)
Semiconductivity of the phthalocyanenes	C.J. Hoffman	18, 113-121 (1964)

Chemical activation	B.S. Rabinovitch and M.C. Flowers	18,	122-167 (1964)
Phosphonitrilic derivatives and related compounds	N.L. Paddock	18,	168-210 (1964)
The cyclisation of olefinic acids to ketones and lactones	M.F. Ansell and M.H. Palmer	18,	211-225 (1964)
Deviations from the Arrhenius equation	J.R. Hulett	18,	227-242 (1964)
Radical rearrangement in the gas-phase oxidation and related processes	A. Fish	18,	243-269 (1964)
Synthesis of sesquiterpenes	J.M. Mellor and S. Munavalli	18,	270-294 (1964)
Prediction of the strengths of organic bases	J. Clark and D.D. Perrin	18,	295-320 (1964)
Molecular interactions in clathrates: a comparison with other condensed phases	W.C. Child, Jr.	18,	321-346 (1964)
Quinone methides	A.B. Turner	18,	347-360 (1964)
Inorganic nitrates and nitrato-compounds	B.D. Field and C.J. Hardy	18,	361-388 (1964)
Substituent interactions in ortho-substituted nitrobenzenes	J.D. Loudon and G. Tennant	18,	389-413 (1964)

RADIATION RESEARCH

This is a monthly journal, in English. It is the official organ of the Radiation Research Society. Edited by T.C. Evans and published by Academic Press, Inc., New York-London.

Title	Authors	Reference
Transfer of radiation-induced unpaired spins from proteins to sulfur compounds	T. Henriksen, T. Sanner and A. Pihl	18, 163-176 (1963)
On the mechanism of radiation protection by cysteamine: An investigation by means of electron spin resonance	M.G. Ormerod and P. Alexander	18, 495-509 (1963)
Electron spin resonance studies of radiation effects on polyamino acids	R.C. Drew and W. Gordy	18, 552-579 (1963)
Kinetics of enzyme inactivation by ionizing radiation	T. Sanner and A. Pihl	19, 12-26 (1963)
The method of a competitive scavenger in radiation chemistry	V.N. Shubin and P.I. Dolin	19, 345-358 (1963)

These are the proceedings of the Exciton Model Symposium sponsored by the Radiation Research Society Meeting, held at Colorado Springs, May 22, 1962.

Title	Authors	Reference
Introductory Remarks	M. Kasha	20, 53-54 (1963)
Energy transfer mechanisms and the molecular exciton model for molecular aggregates	M. Kasha	20, 55-71 (1963)
Energy transfer phenomena and dissociation processes in electronically excited molecules	J.L. Magee	20, 71-76 (1963)
Wannier excitons in simple van der Waals crystals	R.S. Knox	20, 77-82 (1963)
Energy exchange processes	D.S. McClure	20, 83-86 (1963)
Theory of the interaction of localized electronic excitations	W.T. Simpson	20, 87-100 (1963)
Absorption of light by a rigid crystal	C.A. Mead	20, 101-106 (1963)
The problem of defects in relation to the optical properties of organic molecular crystals	R.M. Hochstrasser	20, 107-117 (1963)
Optical absorption by a pair of ions	D.L. Dexter	20, 118-119 (1963)

The symmetry and spectral properties of helical polynucleotides	W. Rhodes	20,	120-132 (1963)
The exciton contribution to the optical rotation of polymers	I. Tinoco, Jr.	20,	133-139 (1963)
Exciton theory by second quantization	R. Hoffmann	20,	140-148 (1963)
On the theories of hypochromism in polynucleotides	O. Sinanoglu	20,	149-153 (1963)
Vibronic states of dimers	R. E. Merrifield	20,	154-158 (1963)
		21,	1-643 (1964)
		22,	1-734 (1964)
		23,	1-652 (1964)

RECENT PROGRESS IN HORMONE RESEARCH

These are the proceedings of the Laurentian Hormone Conference, an annual conference on hormone research. Edited by G. Pincus and published by Academic Press, Inc., New York.

Title	Authors	Reference
The characterization of pituitary hormones by starch gel electrophoresis	K.A. Ferguson and A.L.C. Wallace	19, 1-55 (1963)
Studies of human steroidal hormones by gas chromatographic techniques	E.C. Horning, T. Luukkainen, E.O.A. Haahti et al.	19, 57-106 (1963)
Metabolic effects of adrenocortical steroids in vivo and in vitro: relationship to antiinflammatory action	E.M. Glenn, W.I. Miller and C.A. Schlagel	19, 107-199 (1963)
Carbonic anhydrase in steroid-responsive tissues	G. Pincus and G. Bialy	19, 201-250 (1963)
Androgen biosynthesis and related studies	R.I. Dorfman, E. Forchielli and M. Gut	19, 251-273 (1963)
Studies on the secretion and interconversion of the androgens	R.L. Vande Wiele, P.C. MacDonald, E. Gurpide and S. Lieberman	19, 275-310 (1963)
The control of aldosterone secretion	J.R. Blair-West, J.P. Coghlan, D.A. Denton et al.	19, 311-383 (1963)
The human sebaceous gland: its regulation by steroidal hormones and its use as an end organ for assaying androgenicity in vivo	J.S. Strauss and P.E. Pochi	19, 385-444 (1963)
On the hormonal regulation of carbohydrate metabolism; studies with C^{14} glucose	R.C. De Bodo, R. Steele, N. Altszuler et al.	19, 445-488 (1963)
Functional characterization and metabolic pathways of the pancreatic islet tissue	A. Lazarow	19, 489-546 (1963)
The metabolic errors in certain types of familial goiter	J.B. Stanbury	19, 547-577 (1963)
Production and properties of transplantable tumors of the thyroid gland in the Fischer rat	S.H. Wollman	19, 579-618 (1963)

Andrenocortical factors associated with adaptation of vertebrates to marine environments	W.N. Holmes J.G. Phillips and I.C. Jones	19,	619-672 (1963)
Chemical communication among animals	E.O. Wilson and W.H. Bossert	19,	673-716 (1963)
Some studies with antisera to growth hormone, ACTH and TSH	E.E. McGarry J.C. Beck L. Ambe and R. Nayak	20,	1-31 (1964)
Calciphyloxis and the parathyroid glands	H. Selye G. Gabbiani and B. Tuchweber	20,	33-58 (1964)
Parathyroids, calcitonin, and control of plasma calcium	D.H. Copp	20,	59-88 (1964)
Hypothalmic factors releasing pituitary hormone	R. Guillemin	20,	89-130 (1964)
The neuroendocrine regulation of hypophyseal luteinizing hormone secretion	S.M. McCann and V.D. Ramirez	20,	131-181 (1964)
Some effects of hormones on the metabolism of adipose tissue	E.G. Ball and R.L. Jungas	20,	183-214 (1964)
The effects of corticosteroids upon connective tissue and lysosomes	G. Weissmann and L. Thomas	20,	215-245 (1964)
Testicular hormones and the synthesis of ribonucleic acids and proteins in the prostate gland	H.G. Williams-Ashman S. Liao R.L. Hancock et al.	20,	247-301 (1964)
Ovarian steroid synthesis and secretion in vivo	R.V. Short	20,	303-340 (1964)
Steroid secretions of the normal and polycystic ovary	V.B. Mahesh and R.B. Greenblatt	20,	341-394 (1964)
Effects of steroids on pituitary gonadotropin and fertility	F.J. Saunders	20,	395-434 (1964)
Hormone antagonists: inhibitors of specific activities of estrogen and androgen	L.J. Lerner	20,	435-490 (1964)
A quantitative study of inhibition and recovery of spermatogenesis	D.J. Patanelli and W.O. Nelson	20,	491-543 (1964)
Kinetics of the germinal epithelium in man	C.G. Heller and Y. Clermont	20,	545-575 (1964)

This is a quarterly review journal, in English. Edited by W.H. Powers and published by Wayne State University Press for the Kresge-Hooker Science Library Associates.

Title	Authors	Reference
The alkylation of saturated and unsaturated ketones	J.-M. Conia	24, 43-62 (1963)
Steric effects in some free radical reactions	E.C. Kooyman	24, 93-102 (1963)
Paramagnetic resonance of phosphorescent organic molecules	C.A. Hutchison, Jr.	24, 105-127 (1963)
Chemical compounds of the noble gases	C.L. Chernick	24, 139-155 (1963)
Organic functional group analysis by photometric titration	C.A. Reynolds	24, 157-165 (1963)
The role of chelating agents in spectrophotometric analysis	D.F. Boltz	24, 167-178 (1963)
The mechanisms of oxidation reduction reactions in solutions	C.H. Brubaker, Jr.	24, 181-189 (1963)
Optical rotatory dispersion and circular dichroism of transition metal complexes	F. Woldbye	24, 197-223 (1963)
A review of recent advances in the chemistry of the diterpenoid alkaloid atisine	R.E. Ireland	24, 225-235 (1963)
Exchange reactions of multidentate ligand complexes	D.W. Margerum	24, 237-251 (1963)
Cyclizations by radical reactions	M. Julia	25, 3-29 (1964)
The chemistry of the vic-dioximes	C.V. Banks	25, 85-103 (1964)
The significance of complexes of macrocyclic ligands and their synthesis by ligand reactions	D.H. Busch	25, 107-126 (1964)
Electrophilic aliphatic substitution	C.K. Ingold	25, 145-158 (1964)
Biochemical dimensions of photosynthesis (An introduction)	D.W. Krogmann	25, 161-164 (1964)
Two light reactions in photosynthesis	G.E. Hoch	25, 165-180 (1964)

Studies on photosynthesis using cell free preparations of blue green algae	W.A. Susor W.C. Duane and D.W. Krogmann	25,	197-208 (1964)
The energy conversion reactions of photosynthesis	B. Vennesland	25,	211-224 (1964)
Selective inhibitors of photosynthesis	N.E. Good and S. Izawa	25,	225-236 (1964)
On the mechanism of photophosphorylation in chloroplasts	M. Avron	25,	237-246 (1964)
Functional lipids in the photosynthesis of higher plants	W.L. Ogren J.J. Lightbody and D.W. Krogmann	25,	247-257 (1964)
Respiration during photosynthesis	H.V. Marsh, Jr. J.M. Gamiche and M. Gibbs	25,	259-271 (1964)

REVIEWS OF PURE AND APPLIED CHEMISTRY

This is a quarterly review journal, in English. Edited by J.W. Loder and published by The Royal Australian Chemical Institute, Melbourne, Australia.

Title	Authors	Reference		
Thin layer chromatography	J.H. Russel	13,	15-29	(1963)
Synthesis of purines from imidazoles	J.H. Lister	13,	30-47	(1963)
Water molecules in hydrated organic crystals	J.R. Clark	13,	50-90	(1963)
Chemistry of arylaluminium compounds	J.R. Surtees	13,	91-109	(1963)
Adsorption-flocculation reactions of macromolecules	V.K. La Mer and T.W. Healy	13,	112-133	(1963)
Chemistry and biochemistry of β-1,3-glucans	A.E. Clarke and B.A. Stone	13,	134-156	(1963)
Chemistry of organic surface coatings	D.H. Solomon	13,	171-188	(1963)
Genetic coding in plant and bacterial viruses	R.H. Symons	13,	211-246	(1963)
Long-range H^1-H^1 spin-spin coupling in nuclear magnetic resonance spectroscopy	S. Sternhell	14,	15-46	(1964)
Chemistry of _Flindersia_ species	E. Ritchie	14,	47-56	(1964)
Recent advances in the chemistry of dyes and pigments	K.G. Neill	14,	67-80	(1964)
Macromolecular systems incorporating reversible electrochemical groups	A.S. Lindsey	14,	109-126	(1964)
Recent advances in the chemistry of simple oxygen heterocycles	R.D. Topsom	14,	127-146	(1964)
5-Hydroxymethylfurfural	C.J. Moye	14,	161-170	(1964)

THE ROYAL INSTITUTE OF CHEMISTRY JOURNAL

This is a monthly journal, in English. Since 1955, review articles and conferences have been published in full. Edited by F.W. Gibbs and published by The Royal Institute of Chemistry, London.

Title	Authors	Reference
Colorimetric reagents	W.C. Johnson	<u>87</u>, 39-44 (1963)
Flash photolysis	R.N. Dixon	<u>87</u>, 75-80 (1963)
The powder method in x-ray diffraction analysis	C.K. Prout	<u>87</u>, 155-159 (1963)
Units and standards of measurement	P.H. Bigg	<u>87</u>, 407-413 (1963)
Radiocarbon dating	H. Barker	<u>88</u>, 34-38 (1964)
The bond theory of organometallic compounds	D.A. Brown	<u>88</u>, 65-70 (1964)
Recent advances in polarography	H.M. Davis	<u>88</u>, 104-109 (1964)
Archaeometry — the physical sciences applied to archaeology	E.T. Hall	<u>88</u>, 146-148 (1964)
Masers and lasers	T.P. Melia	<u>88</u>, 149-153 (1964)
Selection of personnel for research	E.S. Hiscocks	<u>88</u>, 187-193 (1964)
Aromatic five-membered ring compounds	D. Lloyd	<u>88</u>, 304-309 (1964)
Biosynthesis of the macrolide antibiotics	H. Grisebach and W. Hofheinz	<u>88</u>, 332-340 (1964)
Gas-liquid chromatography	D.R. Browning	<u>88</u>, 376-381 (1964)
Recent developments in collagen research	A.G. Ward	<u>88</u>, 406-413 (1964)

RUSSIAN CHEMICAL REVIEWS

This is a monthly publication, in Russian (Uspekhi Khimii) of the Academy of sciences of the USSR. The english translation is published by The Chemical Society, London. The pagination below refers to the English version.

Title	Authors	Reference		
Mechanisms of the hydrolysis of aromatic sulphonyl chlorides and of alkyl and aryl arenesulphonates	R.V. Vizgert	32,	1-20	(1963)
The chemistry of some 1-substituted alka-1,3-dienes	I.I. Guseinov and G.S. Vasil'ev	32,	20-30	(1963)
Free radical reactions in the electrolysis of organic compounds	A.P. Tomilov and M. Ya. Fioshin	32,	30-44	(1963)
Metal complexes of hydroxy-acids	I.V. Pyatnitskii	32,	44-58	(1963)
Advances in the chemistry of aromatic diazonium compounds	B.I. Belov and V.V. Kozlov	32,	59-75	(1963)
Phenol-dienone rearrangement in the reactions of phenols	V.V. Ershov A.A. Volod'kin and G.N. Bogdanov	32,	75-93	(1963)
Hydration and solvation of acids and salts undergoing extraction	Yu. A. Zolotov	32,	107-116	(1963)
Advances in the synthesis of perfumes possessing an isoprenoid structure	V.N. Belov and N.I. Skvortsova	32,	121-140	(1963)
Secondary reactions in radical polymerisation	L.M. Pyrkov and S. Ya. Frenkel	32,	140-157	(1963)
Advances in the preparation, investigation and use of quinolinium compounds	G.T. Pilyugin and B.M. Gutsulyak	32,	167-188	(1963)
Mechanism of the catalytic redistribution of hydrogen in unsaturated cyclic hydrocarbons	V.M. Gryaznov	32,	188-200	(1963)
Infrared spectra of salts and complexes of carboxylic acids and some of their derivatives	L.L. Shevchenko	32,	201-207	(1963)
The electrochemistry of semi-conductors	V.A. Myamlin and Yu. V. Pleskov	32,	207-223	(1963)

Title	Authors	Vol.	Pages	Year
Polyynes	A.M. Sladkov and Yu.P. Kudryavtsev	32,	229-243	(1963)
Inorganic cyclic silicon-containing compounds and their organic derivatives	K.A. Andrianov, I. Haiduc and L.M. Khananashvili	32,	243-268	(1963)
Perchloric acid	A.A. Zinov'ev	32,	268-282	(1963)
The natural pigments	N.S. Vul'fson	32,	283-290	(1963)
Chemistry of alloxazines and isoalloxazines	V.M. Berezovskii and T.V. Eremenko	32,	290-307	(1963)
Catalytic hydrogenation and hydrogenolysis of furan compounds	I.F. Bel'skii and N.I. Shuikin	32,	307-321	(1963)
Analysis of trace impurities by gas chromatography	M.S. Vigdergauz, M.I. Afanas'ev and K.A. Gol'bert	32,	330-339	(1963)
Hot synthesis of compounds labelled with radioactive tritium	An.N. Nesmeyanov and V.V. Pozdeev	32,	341-353	(1963)
Thin-layer chromatography	A.A. Akhrem and A.I. Kuznetsova	32,	366-386	(1963)
The thiylation of olefins	E.N. Prilezhaeva and M.F. Shostakovskii	32,	399-426	(1963)
Status of the theory of electronic orbitals in molecules and some of its problems	N.D. Sokolov	32,	436-444	(1963)
Reactions of olefins in the presence of metal oxides	Ya.T. Eidus and B.K. Nefedov	32,	445-460	(1963)
Polyfluorinated linear bifunctional compounds (containing like functions) as potential monomers	I.L. Knunyants, Li Chih-yuan and V.V. Shokina	32,	461-476	(1963)
Chromatography of proteins on cellulose ion exchangers	A.Ya. Nikolaev	32,	476-490	(1963)
Organic derivatives of laminar silicates	E.V. Kukharskaya and A.D. Fedoseev	32,	490-495	(1963)
Chemical characteristics, structure, and properties of liquid crystal	V.A. Usol'tseva and I.G. Chistyakov	32,	495-509	(1963)

Vibrational spectra of compounds containing elements of the carbon subgroup	N.A. Chumaevskii	32,	509-522 (1963)
Stereochemistry of the perhydroindene systems	E.P. Serebryakov and V.F. Kucherov	32,	523-535 (1963)
Pyrazolidine-3,5-diones, their synthesis and pharmacological importance	A.M. Khaletskii and B.L. Moldaver	32,	535-550 (1963)
Interaction of dyes with substances of high molecular weight	M.V. Savost'yanova	32,	550-569 (1963)
Recent progress in the synthesis of monocyclic and polycyclic aromatic hydrocarbons	V.R. Skvarchenko	32,	571-589 (1963)
Effect of irradiation on the adsorption and catalytic properties of semiconductors and dielectrics	V.G. Baru	32,	590-604 (1963)
Chemical changes during β-decay processes	V.D. Nefedov V.M. Zaitsev and M.A. Toropova	32,	604-619 (1963)
Advances in the use of infrared spectroscopy for the characterization of OH bonds	G.V. Yukhnevich	32,	619-633 (1963)
Syntheses and reactions in the cyclobutane series	M. Yu. Lukina	32,	635-651 (1963)
Modern views on the mechanism of the cationic polymerisation of vinyl monomers	B.L. Erusalimskii	32,	651-666 (1963)
Chelate polymers	V.V. Rode E.G. Rukhadze and A.P. Terent'ev	32,	666-682 (1963)
Allene hydrocarbons	A.A. Petrov and A.V. Fedorova	33,	1-13 (1964)
Mechanism of the oxidative thermal degradation and of the stabilisation of polymers	M.B. Neiman	33,	13-27 (1964)
1,3-Dichlorobut-2-ene and new preparations based on it	V.I. Isagulyants and G.T. Esayan	33,	27-38 (1964)
Theory of catalytic hydrogen waves in organic polarography	S.G. Mairanovskii	33,	38-55 (1964)
New aspects of the chemistry of complexes of selenium	D.I. Ryabchikov and I.I. Nazarenko	33,	55-64 (1964)

Addition of alcohols and mercaptans to compounds with triple bonds	M.F. Shostakovskii A.V. Bogdanova and G.I. Plotnikova	33,	66-77 (1964)
Chemistry of 19-norsteroids	A.A. Akhrem and and Yu. A. Titov	33,	77-92 (1964)
2,4,6-Trichloro-1,3,5-triazine (cyanurylchloride) and its future applications	V.I. Mur	33,	92-103 (1964)
Self-diffusion and interdiffusion in polymer systems	S.S. Voyutskii and V.L. Vakula	33,	103-117 (1964)
Electrochemical and physicochemical properties of aluminium compounds in non-aqueous solutions	Yu. M. Kessler N.M. Alpatova and O.R. Osipov	33,	119-138 (1964)
Electrically conducting non-aqueous systems formed from non-conducting components	S.T. Miskidzh'yan	33,	138-151 (1964)
Modern views on the plasticising of polymers	V.A. Voskresenskii and E.M. Orlova	33,	151-158 (1964)
Monoalkylhydrazines	A.N. Kost and R.S. Sagitullin	33,	159-176 (1964)
Hexachlorobutadiene	L.M. Kogan	33,	176-188 (1964)
Advances in the chemistry of thiamine	A.M. Yurkevich	33,	188-199 (1964)
The kinetics of ion exchange on sulphonated cation exchangers	A.B. Sigodina N.I. Nikolaev and N.N. Tunitskii	33,	199-212 (1964)
Alpha radiolysis of aqueous solutions	M.V. Vladimirova	33,	212-220 (1964)
Advances in the chemistry of divinylacetylene	S.A. Vartanyan	33,	243-258 (1964)
The principles of biocatalyst modelling	L.A. Nikolaev	33,	275-286 (1964)
Chemical effects of nuclear transformations in solids (Chemical consequences of nuclear recoil)	A.N. Murin R.V. Bogdanov and S.M. Tomilov	33,	295-310 (1964)
Commercial methods of synthesis of α,β-unsaturated aldehydes and ketones	G.S. Mironov and M.L. Farberov	33,	311-319 (1964)
Catalytic reduction of dinitriles	L. Kh. Freidlin and T.A. Sladkova	33,	319-330 (1964)

Some aspects of high-temperature chemistry from the viewpoint of chemical thermodynamics	V.A. Kireev	33,	330-342 (1964)
Process in the synthesis of macrocyclic and other perfumery materials	V.N. Belov and N.I. Skvortsova	33,	251-266 (1964)
The chemistry of the odour of foodstuffs	R.V. Golovnya G.A. Mironov and S.D. Sokolov	33,	366-285 (1964)
Chemical properties and molecular structure of derivatives of sym-heptazine [1,3,4,6,7,9b-heptazaphenalene, tri-1,3,5-triazine]	A.I. Finkel'shtein and N.V. Spiridonova	33,	400-405 (1964)
Diene synthesis with dienophiles containing hetero-atoms	Yu.A. Arbuzov	33,	407-424 (1964)
Progress in the chemistry of the aminodeoxy-sugars	V.I. Veksler	33,	424-438 (1964)
The catalytic synthesis of carboxylic acids and their esters from carbon monoxide, alkenes, and alcohols	Ya.T. Eidus and K.V. Puzitskii	33,	438-451 (1964)
Unsaturated organoboron compounds	A.V. Topchiev A.A. Prokhorova and M.V. Kurashev	33,	453-462 (1964)
The prolongation of the action of pharmaceutical preparations by mixing or combining them with polymers	K.P. Khomyakov A.D. Virnik and Z.A. Rogovin	33,	462-467 (1964)
Thiolactones	M.G. Lin'kova N.D. Kuleshova and I.L. Knunyants	33,	493-508 (1964)
The chemistry of 1,3,4-oxadiazole derivatives	E.P. Nesynov and A.P. Grekov	33,	508-514 (1964)
The formation of biochemically important compounds in the prebiological stage of the earth's development	A.G. Pasynskii and T.E. Pavlovskaya	33,	514-524 (1964)
Polymeric boron compounds	V.A. Zamyatina and N.I. Bekasova	33,	524-532 (1964)
Compounds with conjugated double bonds	V.V. Pen'kovskii	33,	532-549 (1964)
Present-day methods of synthesising macrocyclic compounds	L.I. Belen'kii	33,	551-571 (1964)
Gas-liquid chromatography of lipids	A.G. Vereshchagin	33,	578-590 (1964)
Oscillographic polarography in organic chemistry	G.K. Budnikov	33,	590-608 (1964)

SCIENCE

This is a weekly publication of the American Association for the Advancement of Science, in English. It publishes letters, (review) articles and reports. Edited by P.H. Abelson.

Title	Authors	Reference
Low-energy electron diffraction	A.V. MacRae	139, 379-388 (1963)
On the genetic code	F.H.C. Crick	139, 461-464 (1963)
Permissible exposure to ionizing radiation	K.Z. Morgan	139, 565-571 (1963)
Nuclear activation analysis	R.E. Wainerdi and N.P. DuBeau	139, 1027-1033 (1963)
Hazards of new drugs	W. Modell	139, 1180-1185 (1963)
Myoglobin and the structure of proteins	K.C. Kendrew	139, 1259-1266 (1963)
Involvement of RNA in the synthesis of proteins	J.D. Watson	140, 17-26 (1963)
Accuracy of radiocarbon data	W.F. Libby	140, 278-280 (1963)
Early history of carbon-14	M.D. Kamen	140, 584-590 (1963)
Information retrieval	R.R. Shaw	140, 606-609 (1963)
Foreign literature of chemistry	J.L. Wood, K.L. Coe and G.O. Platau	140, 610-613 (1963)
Bent chemical bonds	W.H. Flygare	140, 1179-1185 (1963)
Chemical insect attractants	M. Jacobson and M. Beroza	140, 1367-1373 (1963)
Carbamyl phosphate	M.E. Jones	140, 1373-1379 (1963)
Separation and fractionation of macromolecules and particles	A. Tiselius, J. Porath and P.-A. Albertsson	141, 13-20 (1963)
Metabolic control of timing	S.B. Hendricks	141, 21-27 (1963)
Genetic control of hemoglobin synthesis	W.E. Nance	141, 123-130 (1963)

SCIENCE (Contd.)

Title	Author(s)	Citation
Informational retrieval systems	J.A. Swets	141, 245-250 (1963)
Cross-linking of polymers by radiation	G. Adler	141, 321-329 (1963)
Chemical kinetics in shock tubes	S.H. Bauer	141, 867-879 (1963)
Biological implications of gas chromatography	A. Karmen	142, 163-172 (1963)
Ultrahigh vacuum instrumentation	T.A. Vanderslice	142, 178-184 (1963)
Origin of the electron microscope	M.M. Freundlich	142, 185-188 (1963)
Scientific instruments in space exploration	R.L. Heacock	142, 188-195 (1963)
Measurement of optical activity: new approaches	B. Carroll and I. Blei	142, 200-208 (1963)
Intermetallic compounds	W.W. Scanlon	142, 1265-1269 (1963)
Correlation of structure and function in enzyme action	D.E. Koshland, Jr.	142, 1533-1541 (1963)
Organic photochemistry	G.S. Hammond and N.J. Turro	142, 1541-1553 (1963)
Free radicals and unstable molecules	S.N. Foner	143, 441-450 (1964)
Crystalline deamino-oxytocin	D. Jarvis and V. du Vigneaud	143, 545-548 (1964)
Biosynthesis of unsaturated fatty acids in microorganisms	J. Erwin and K. Bloch	143, 1008-1012 (1964)
Alpha-chymotrypsin and the nature of enzyme catalysis	C. Niemann	143, 1287-1296 (1964)
Molecular theories of memory	W. Dingman and M.B. Sporn	144, 26-29 (1964)
Mossbauer effect in chemistry and solid state physics	G.K. Wertheim	144, 253-259 (1964)
Differential dialysis	L.C. Craig	144, 1093-1079 (1964)
Tarichatoxin - tetrodotoxin: a potent neurotoxin	H.S. Mosher, F.A. Fuhrman, H.D. Buchwald and H.G. Fischer	144, 1100-1110 (1964)

SCIENCE (Contd.)

Biological complexity and radio-sensitivity	H.S. Kaplan and L.E. Moses	145,	21-25 (1964)
Enzymatic alteration of nucleic acid structure	P.R. Srinivasan and E. Borek	145,	548-553 (1964)
The chemistry of noble gas compounds	H.H. Hyman	145,	773-783 (1964)
Lipopolysaccharide of the gram-negative cell wall	M.J. Osborn, S.M. Rosen, L. Rothfield, L.D. Zeleznick and B.L. Horecker	145,	783-789 (1964)
RNA Codewords and protein synthesis	M. Nirenberg and P. Leder	145,	1399-1407 (1964)
The hydrated electron	E.J. Hart	146,	19-25 (1964)
Nuclear magnetic resonance spectroscopy in superconducting magnetic fields	F.A. Nelson and H.E. Weaver	146,	223-232 (1964)
Infrared spectroscopy and catalysis research	R.P. Eischens	146,	486-493 (1964)
Macromolecules	H. Mark	146,	1023-1024 (1964)

SURVEY OF PROGRESS IN CHEMISTRY

This is a collection of review articles, in English. Edited by A.F. Scott and published by Academic Press, New York and London.

Title	Authors	Reference		
New research tools of chemists	R. Schaeffer	1,	1-33	(1963)
High temperature reactions	A.W. Searcy	1,	35-79	(1963)
The implications of some recent structures for chemical valence theory	R.E. Rundle	1,	81-131	(1963)
Metallocenes	W.F. Little	1,	133-210	(1963)
Oxidation reduction mechanisms in organic chemistry	K.B. Wiberg	1,	211-248	(1963)
The chemistry of biological transfer reactions	W.P. Jencks	1,	249-300	(1963)
The structure of the Grignard reagent and the mechanism of its reaction	R.M. Salinger	1,	301-324	(1963)
Mechanisms of substitution reactions of metal complexes	F. Basolo	2,	1-55	(1964)
Fast reactions in solutions	E.M. Eyring	2,	56-89	(1964)
Equilibria in concentrated mineral acid solutions	N.C. Deno	2,	155-187	(1964)
Nucleophilic displacement at the sulfur-sulfur bond	R.C. David	2,	189-238	(1964)
The mechanisms of some photochemical reactions of organic molecules	J. Saltiel	2,	239-328	(1964)

This is a basic review journal for chemistry. Articles are published in Swedish or English. Those in the Swedish language generally have an English summary. A review of the Nobel prize in chemistry for each year is included. Edited by V. Runnstrom-Reio and published by Svenska Kemistamfundet.

Title	Authors	Reference
The Nobel prize in chemistry for 1962. (in Swedish)	B. Strandberg	75, 1-10 (1963)
The Nobel prize in physiology and medicine for 1962 (in Swedish)	G. Frick	75, 11-16 (1963)
Electron transport-coupled phosphorylation in biological, especially photosynthetic systems (in Swedish)	H. Baltscheffsky	75, 17-24 (1963)
The Kolbe electrolytic synthesis (in Swedish)	L. Eberson	75, 115-124 (1963)
Elasticity and swelling of macromolecular materials (in English)	W. Prins	75, 125-133 (1963)
Isomerism, mesomerism and resonance (in Swedish)	L. Melander	75, 153-160 (1963)
Studies on complex dibenzofurans (in English)	N.E. Stjernström	75, 184-198 (1963)
S-Phosphorylated thiols (Studies on their synthesis, chemistry and biochemistry) (in English)	S. Akerfeldt	75, 231-246 (1963)
Preparative methods utilizing peroxide reactions catalyzed by metal ions (in English)	S.-O. Lawesson and G. Sosnovsky	75, 343-366 (1963)
Ethylene oxide — A small molecule of great importance (in Swedish)	B. Weibull	75, 385-397 (1963)
Electronic spectra of organic compounds (in Swedish)	J. Sandstrom	75, 446-459 (1963)
Isocyanides in organic syntheses (in Swedish)	K. Sjöberg	75, 493-506 (1963)
Ethylamine derivatives containing halogy mercapto, acetothiolo and acetoxy groups. A study of their synthesis and reactivities (in English)	B. Hansen	75, 511-537 (1963)
Syntheses with peresters catalysed by metal ions (in English)	S.-O. Lawesson and G. Sosnovsky	75, 568-578 (1963)
The coordination chemistry of the metal dioximes (in English)	D. Dyrssen	75, 618-637 (1963)

Organotin compounds (in Swedish)	H. Meyer	76, 8-20	(1964)
The future of plastics (in Swedish)	J. Cedwall, P. Flodin, S. Kjellström and N. Sundén	76, 21-30	(1964)
The Nobel prize in chemistry for 1963 (in Swedish)	B. Rånby	76, 57-66	(1964)
How long is a carbon-carbon bond — and why? (in Swedish)	I. Fischer-Hjalmars	76, 77-86	(1964)
Biological nitrogen fixation (The effect of irradiated nitrogen gas on nitrogen fixation by azotobacter (in English)	B. Zacharias	76, 87-96	(1964)
Studies on the chemistry of terpenes (in English)	T. Norin	76, 97-120	(1964)
Quantum chemistry and organic chemistry (in Swedish)	G. Bergson	76, 145-155	(1964)
Tables for Craig distribution and for simple liquid/liquid extraction (in Swedish)	T. Alm and N. Löfgren	76, 156-158	(1964)
Thiol and disulphide groups in enzymes (in Swedish)	S. Åkerfeldt	76, 186-194	(1964)
Enzymes in analytical chemistry (in Swedish)	H.G. Boman	76, 234-242	(1964)
Living polymers and their applications in polymer chemistry (in English)	M. Szwarc	76, 243-264	(1964)
The paper chromatographic identification of aromatic compounds related to metabolites of fungal and mammalian origin (in English)	L. Reio	76, 265-289	(1964)
Extraction of water and acids by long-chain tertiary amines and some related problems (in English)	E. Högfeldt	76, 290-308	(1964)
Microbial production of amino acids (in Swedish)	R. Kihlberg	76, 321-337	(1964)
The importance of fundamental research in thermodynamics and thermochemistry for the chemical industries (in English)	F.D. Rossini	76, 338-353	(1964)
Study of the hydrolysis equilibria of cations by EMF methods (in English)	G. Biedermann	76, 362-384	(1964)
Chemical data for science and industry (in English)	G. Waddington	76, 413-418	(1964)

Thermodynamic data compilation and review at the National Bureau of Standards (in English)	G.T. Armstrong	76,	419-427 (1964)
Polyene macrolide antibiotics: The structures of fungichromin, pimaricin and filipin (in English)	O. Ceder	76,	428-441 (1964)
Organic inhibitors for crystal growth, corrosion and pickling inhibitors (in Swedish)	L.O. Hardelius	76,	498-509 (1964)
On the use of computors for chemical problems (in Swedish)	C-E. Froberg	76,	588-593 (1964)
N,N'-bis-(β,β,β-Trinitroethyl)-urea (in Swedish)	A. Wetterholm	76,	628-634 (1964)
Separation of hydroxy-acids with anion exchangers (in Swedish)	O. Samuelson	76,	635-645 (1964)
Isotopic exchange reactions studied by electro-deposition (Non-rapid and moderately rapid iodide exchanges) (in English)	P. Beronius	76,	646-656 (1964)
The 1964 Nobel prize in chemistry (in Swedish)	S. Abrahamsson	76,	677-683 (1964)
Chemistry of the noble gases (in Swedish)	N-G. Vannerberg	76,	693-699 (1964)
Quantum chemistry of sulphur compounds (in English)	R. Zahradnik	76,	700-710 (1964)

TALANTA

This is an international journal of analytical chemistry, in English, German and French. Edited by C.L. Wilson and published by Pergamon Press, Oxford-New York.

Title	Authors	Reference
A nomogram for acid-base titrations (in English)	E. Wänninen	$\underline{10}$, 221-229 (1963)
Application of Feigl's reactions in biochemistry and agricultural chemistry (in English)	A. Bondi	$\underline{10}$, 679-684 (1963)
The accuracy of gas chromatography (in English)	A.F. Williams and W.J. Murray	$\underline{10}$, 937-959 (1963)
The determination of tertiary amines in the presence of primary and secondary amines (in English)	M. Miller and D.A. Keyworth	$\underline{10}$, 1131-1138 (1963)
Precipitation of metal chelates from homogeneous solution (in English)	F.H. Firsching	$\underline{10}$, 1169-1175 (1963)
The status of and trends in analytical chemistry (in English)	I.M. Kolthoff	$\underline{11}$, 75-84 (1964)
The boron-carbon-hydrogen system (in English)	I. Shapiro	$\underline{11}$, 211-220 (1964)
Analytical applications of radioactive vitamin B_{12} (in English)	C. Rosenblum	$\underline{11}$, 255-270 (1964)
Some sulphur-containing organic compounds as reagents for the photometric determination of selenium (in English)	A.I. Busev	$\underline{11}$, 485-493 (1964)
Wet oxidation of organic compositions (mixed nitric and hydrochloric acids with ammonium perchlorate as oxygen donors) (in English)	G.F. Smith	$\underline{11}$, 633-640 (1964)
Recent developments in the ring oven technique (in English)	H. Weisz	$\underline{11}$, 1041-1060 (1964)
Radiometric titrations (in English)	T. Braun and J. Tolgyessy	$\underline{11}$, 1277-1312 (1964)
Recent uses of liquid ion exchangers in inorganic analysis (in English)	H. Green	$\underline{11}$, 1561-1580 (1964)

This is an international journal, in English, French or German. It is devoted to the publication of original articles, reviews and symposia in all fields of organic chemistry. Published by Pergamon Press, Oxford.

Title	Authors	Reference
Control of the steric course of the Wittig reaction (Stereochemical studies and synthetic applications)	L.D. Bergelson and M.M. Shemyakin	19, 149-159 (1963)
The total synthesis of strychnine	R.B. Woodward M.P. Cava W.D. Ollis et al.	19, 247-288 (1963)
The effect of hydrogen bonding on the electronic spectra of organic molecules	W.A. Lees and the late A. Burawoy	19, 419-438 (1963)
The dipole moments of some aryl methyl sulphides (Evidence for the expansion of the valence shell of sulphur)	V. Baliah and M. Uma	19, 455-464 (1963)
The free radical mechanism of nitration	A.I. Titov	19, 557-580 (1963)
A method for assessing electrophilic substituent effects in the gas-phase, and the correlation with reactivity in the condensed-phase	R. Taylor and G.G. Smith	19, 937-947 (1963)
The thermochemistry of sulphur-containing molecules and radicals. I. Heats of combustion and formation	H. Mackle and P.A.G. O'Hare	19, 961-971 (1963)
The mechanism of the Etard reaction	I. Necsoiu A.T. Balaban I. Pascaru et al.	19, 1333-1142 (1963)
The thermochemistry of sulphur-containing molecules and radicals. II. (The dissociation energies of bonds involving sulphur: the heats of formation of sulphur-containing radicals	H. Mackle	19, 1159-1170 (1963)
A bond energy scheme. II. Strain and conjugation energies in cyclic compounds	J.D. Cox	19, 1175-1184 (1963)
An examination of the principle of the additivity of substituent group influence in benzoic acids	J.F.J. Dippy and S.R.C. Hughes	19, 1527-1530 (1963)
The molecular geometry of some strained five- and six-membered rings	L.C.G. Goaman and D.F. Grant	19, 1531-1537 (1963)

The M.O. theory of bent bonds in strained cyclic hydrocarbons	D. Peters	19, 1539-1546 (1963)
The effect of ring size on acidity and rate of C-alkylation of cyclic β-ketoesters	S.J. Rhoads and A.W. Decora	19, 1645-1659 (1963)
Nuclear magnetic resonance of natural products (VI) (Triterpenes derived from betuline) (in French)	J.M. Lehn and A. Vystrcil	19, 1733-1745 (1963)
Conjugation and non-bonded interaction	I. Fischer-Hjalmars	19, 1805-1815 (1963)
Synthesis of higher alicyclic compounds from thiophene derivatives	Ya. L. Gol'dfarb S.Z. Taits and L.I. Belen'kii	19, 1851-1866 (1963)
Dipole moments of some substituted acetophenones (Influence of ortho substituents on the position of the acetyl group)	V. Baliah and K. Aparajithan	19, 2177-2183 (1963)
Nitro groups as proton acceptors in hydrogen bonding	W.F. Baitinger P. von R. Schleyer T.S.S.R. Murty and L. Robinson	20, 1635-1647 (1964)
Identification of the structure of cyclopropanes. Infrared and NMR spectroscopic studies of cyclopropanes (in German)	H. Weitkamp and F. Korte	20, 2125-2135 (1964)
New microwave procedure for determining dipole moments and relaxation times	W.F. Hassell M.D. Magee S.W. Tucker and S. Walker	20, 2137-2155 (1964)
The ultraviolet spectrum of alkyl benzenes (Evidence for carbon-hydrogen and carbon-carbon bond hyperconjugation) (Baker-Nathan effects)	M. Ballester and J. Riera	20, 2217-2227 (1964)
Mass spectroscopic fragmentation of steroid alkaloids (in German)	H. Budzikiewicz	20, 2267-2278 (1964)
Reactions with lead tetraacetate I. (Oxidation of saturated aliphatic alcohols Part I)	V.M. Mičović R.I. Mamuzic D. Jeremič and M. Lj. Mihailović	20, 2279-2287 (1964)
Charge distributions in carbonyl and thiocarbonyl compounds (LCAO calculations with the ω-technique)	M.J. Janssen and J. Sandström	20, 2339-2349 (1964)
Infrared absorption of the azo (-N=N-) in azoic dyes (in French)	P. Bassignana and C. Cogrossi	20, 2361-2371 (1964)

VITAMINS AND HORMONES

This is an annual collection of review articles, in English. Edited by R.S. Harris, I.G. Wool and J.A. Loraine and published by Academic Press, New York-London.

Title	Authors	Reference
Intrinsic factor	L. Ellenbogen and D.R. Highley	21, 1-49 (1963)
Vitamin A, vitamin D, cartilage, bones and teeth	F.C. McLean and A.M. Budy	21, 51-68 (1963)
Modified thiamine compounds	C. Kawasaki	21, 69-111 (1963)
Interrelations between vitamin B_6 and hormones	J.M. Hsu	21, 113-134 (1963)
Corticosteroids and enzyme activity	F. Rosen and C.A. Nichol	21, 135-214 (1963)
The endocrine system and the stomach	G.P. Crean	21, 215-280 (1963)
Biochemistry of biotin	S.P. Mistry and K. Dakshinamurti	22, 1-55 (1964)
The biochemistry of the inositol lipids	J.N. Hawthorne	22, 57-79 (1964)
The biochemistry of progesterone	K. Fotherby	22, 153-204 (1964)
Synthesis and labeling of the vitamin B_6 group	J.M. Osbond	22, 367-397 (1964)
γ-Aminobutyric acid (γABA), vitamin B_6, and neuronal function A speculative synthesis	E. Roberts J. Wein and D.G. Simonsen	22, 503-559 (1964)
Recent developments in the analysis for vitamin B_6 in foods	E.W. Toepfer and M.M. Polansky	22, 825-832 (1964)
Methods for the determination of vitamin B_6 in biological materials	C.A. Storvick and J.M. Peters	22, 833-854 (1964)

ZEITSCHRIFT FUR ANALYTISCHE CHEMIE

This journal, mainly in German, contains original articles, abstracts of current papers and review papers (including symposia) on the progress of analytical chemistry in all its phases. Several issues during 1963-1964 contain a large proportion of current papers. Edited by W. Fresenius and published by Springer-Verlag, Berlin-Gottingen-Heidelberg.

Title	Authors	Reference
Modern analysis as an aid in food sciences (in German)	L. Acker	<u>192</u>, 27-43 (1963)
Microanalysis by X-rays (in French)	R. Castaing	<u>192</u>, 51-64 (1963)
Development of luminescence analysis in the last 30 years (in German)	J. Eisenbrand	<u>192</u>, 83-91 (1963)
Analysis of pesticide residues (in German)	H. Frehse and H. Niessen	<u>192</u>, 94-136 (1963)
Development and state of vacuum spectroscopy (in German)	G. Graue, R. Marotz and S. Eckhard	<u>192</u>, 137-156 (1963)
Development of gas analysis methods (in German)	H. Kienitz	<u>192</u>, 160-189 (1963)
Analysis of synthetic fibers (in German)	W. Kupper	<u>192</u>, 219-248 (1963)
Nomenclature of dithiocarbamic acids and their salts (in German)	W. Haas and K. Irgolič	<u>193</u>, 248-259 (1963)
Analytical chemistry of thiocarbamides. Qualitative determination of thiourea and derivatives (in German)	P.C. Gupta	<u>196</u>, 412-433 (1963)

These are the proceedings of a symposium on Modern Methods in the Analysis of Organic Compounds, held in Eindhoven, Holland, 20-23 May, 1964.

Title	Authors	Reference
I. Importance of analysis in organic chemistry (in German)	J.F. Arens	<u>205</u>, 1-12 (1964)
Analysis of elements and functional groups. Present state of organic elemental analysis (in German)	W. Schöniger	<u>205</u>, 13-28 (1964)
Ultramicro and trace-analysis of organic materials. Part 4. Determination of the chlorine content of non-volatile organic substances in 1-10 μg. samples. (in German)	G. Schwab and G. Tölg	<u>205</u>, 29-40 (1964)
Part 5. Determination of nitrogen content of non-volatile organic compounds in 5-20 μg. samples (in German)	G. Tölg	<u>205</u>, 40-50 (1964)

Title	Authors	Vol.	Pages	Year
Six-minute determination of carbon and hydrogen with titrimetric finish (in German)	L. Blom and M.H. Kraus	205,	50-80	(1964)
Three decimal place hydrogen and carbon determination in mineral oil samples (in German)	J. Boes P. Gouverneur	205,	58-66	(1964)
Rapid conbustion apparatus with an electric end-point reading for the determination of carbon and hydrogen in organic substances (in German)	F. Salzer	205,	66-80	(1964)
Routine analysis with a C - N - H automatic-analysis machine (in German)	H. Weitkamp R. Mayntz and F. Korte	205,	81-89	(1964)
Calorimetric sub-micro-determination of bromine in organic compounds	T.R.F.W. Fennell and J.R. Webb	205,	90-94	(1964)
Group reactions in organic analysis (in German)	S. Veibel	205,	94-109	(1964)
II and III. Mass spectrometry and X-ray structural analysis. Present state of mass spectrometry in organic analysis (in German)	E. Stenhagen	205,	109-124	(1964)
Application of a mass spectrometer with high resolving power in organic analysis (in German)	D. Henneberg	205,	124-133	(1964)
Structural determination of organic molecules by x-ray analysis (in German)	D. Henneberg	205,	134-148	(1964)
Methods in automatic structural analysis (in German)	W. Hoppe	205,	148-153	(1964)
New automatic single-crystal x-ray diffractometer according to W. Hoffe (in German)	E.A. Mayer	205,	153-165	(1964)
IV. Constitution of organic compounds and molecular spectroscopy. Spectroscopic methods in determinations of constitution (in German)	G. Kresze	205,	165-179	(1964)
Precision analysis of multicomponent mixtures in quartz spectrophotometric region (in German)	I.S. Herschberg	205,	180-194	(1964)
Applications of proton magnetic resonance to functional-group analysis (in German)	N. Van Meurs	205,	194-202	(1964)
Modern analytical methods, especially gas chromatography, with stereoisomeric alicyclic compounds (in German)	W. Hückel and C.-M. Jennewein	205,	202-216	(1964)
Correlation between the constitution of aliphatic acids and the kinetics of their dissociation and recombination (in German)	H.W. Nürnberg and H.W. Dürbeck	205,	217-237	(1964)

V. Separation processes. New methods in ion-exchange chromatography (in German)	D. Jentzsch G. Oesterhelt E. Rodel and H.G. Zimmermann	205, 237-243 (1964)
Some rapid identification methods in gas chromatography	F.H. Huyten and G.W.A. Rijders	205, 244-262 (1964)
Possibility and limitation of the identification of materials from their behaviour in isothermal-extraction gas-chromatography (in German)	J.F.K. Huber and A.I.M. Keulemans	205, 263-274 (1964)
Permanent columns for liquid-chromatography (in German)	R. Dijkstra J. van Duijn and E.A.M.F. Dahmen	205, 274-284 (1964)
Direct and automatic coupling of thin-layer chromatography to gas chromatography (in German)	R. Kaiser	205, 284-298 (1964)
Selection of the stationary phase for thin-layer chromatography (in German)	P.J. Schorn	205, 298-308 (1964)
Chromatography on silver nitrate-silica gel thin layers (in German)	F.C. Den Boer	205, 308-313 (1964)
Analysis of mixtures of polyphenols by thin-layer chromatography. Part 2. Quantitative results (in German)	F.J. Ritter P. Canonne and F. Geiss	205, 313-324 (1964)
Separation of sterol acetates by thin-layer chromatography in reversed-phase systems and on silica gel G-silver nitrate layers	J.W. Copius-Peereboom	205, 325-332 (1964)
Mechanism of peak-broadening in paper chromatography	C.L. De Ligny and D. Bax	205, 333-341 (1964)
VI. Analysis of synthetics. Analysis in polymer research	P.W.O. Wijga	205, 342-357 (1964)
Pyrolysis-gas chromatography with linearly-programmed temperature in pocked and tubular columns. Thermal degradation of polyolefins. Part 1.	E.W. Cieplinski L.S. Ettre B. Kolb and G. Kemmner	205, 357-371 (1964)
Infrared determination of unsaturated bonds in polyethylene	R.J. DeKock P.A.H.M. Hol and H. Bos	205, 371-381 (1964)
VII. Clinical analysis: analysis of natural substances. New developments in clinical analysis	E.J. Van Kampen	205, 381-393 (1964)

ZEITSCHRIFT FUR ANALYTISCHE CHEMIE (Contd.)

Title	Authors	Reference
Improved methods in the quantitative high-voltage electrophoresis in agar-agar layers (in German)	K. Dose and S. Risi	205, 394-403 (1964)
Determination of amino-acids by measurements of electrical conductivity (in German)	A. Niemann	205, 403-406 (1964)
Amino-acids and peptides (in German)	F. Weygand	205, 406-416 (1964)
Determination of creatinine in serum and urine (in German)	K. Beyermann	205, 416-438 (1964)
Determination of acetoacetic acid in 0.1 ml. of blood. Separation of acetone and its determination with salicylaldehyde (in German)	F. Bahner	205, 438-442 (1964)
Preparative thin-layer chromatographic separation of phosphatides, and thin-layer chromatographic detection of their hydrolysis products (in German)	O.W. Thiele and W. Wober	205, 442-445 (1964)
Reaction of nucleotides and nucleic acids with diazonium salts (in German)	H. Kossel	205, 445-455 (1964)
Analysis of pharmaceuticals in biological liquids by the tropaeolin method (in German)	A. Haussler and P. Hajdu	205, 455-460 (1964)
Gas-chromatographic separation of tropane alkaloids	H. Frauendorf and H. Vogel	205, 460-470 (1964)

ZEITSCHRIFT FUR VITAMIN-, HORMON-, UND FERMENTFORSCHUNG

This is a journal publishing original articles and reviews mainly in German. Edited by R. Abderhalden and published by Verlag-Urban and Schwarzenberg G.M.B.H., Wien-Innsbruck.

Title	Authors	Reference
Paper chromatography of phenolic steroids (in German)	R. Knuppen	12, 355-401 (1963)
Chemistry and pharmacology of corticoids (in German)	K. Junkmann and H. Witzel	13, 49-494 (1964)

PART II
REVIEWS IN SYMPOSIA, COLLECTIVE VOLUMES AND NON-PERIODICAL PUBLICATIONS
Pages 149-218

ADVANCES IN CHEMISTRY

This is a collection of the proceedings of national and international conferences sponsored by the American Cjemical Society. Edited by R.F. Gould and published by the American Chemical Society, Applied Publications Division, Washington, D.C.

Title	Authors	Reference
REACTIONS OF COORDINATED LIGANDS AND HOMOGENEOUS CATALYSIS. This is the collection of papers comprising the symposium sponsored by the Division of Inorganic Chemistry at the 141st Meeting of the American Chemical Society, Washington, D.C, March 22-24, 1962. D.H. Busch, Symp. Chairman.		
Reactions of ligands in metal complexes	D.H. Busch	37, 1-18 (1963)
Metal ion catalysis of nucleophilic organic reactions in solution	M.L. Bender	37, 19-36 (1963)
Metal ion catalysis in biological systems	G.L. Eichhorn	37, 37-55 (1963)
Ring substitution reactions of metal-cyclopentadienyls and metal-arenes	M.D. Rausch	37, 56-77 (1963)
The chemistry of quasiaromatic metal chelates	J.P. Collman	37, 78-98 (1963)
Synthesis and reactions of iron isonitrile complexes	W.Z. Heldt	37, 99-115 (1963)
Effect of coordination on the reactivity of aromatic ligands (General patterns of reactivity)	M.M. Jones	37, 116-124 (1963)
The reactions of coordinated ligands. V. Metal complexes of β-mercaptoamines and their reactions with alkyl halides	D.H. Busch, J.A. Burke, Jr., D.C. Jicha, M.C. Thompson and M.L. Morris	37, 125-142 (1963)
Reactivity of hydroxyl groups in the bis(2-hydroxyethyliminodiacetato)chromium(III) ion	R.A. Krause and S.D. Goldby	37, 143-149 (1963)
Phosphorus-fluorine chemistry. II. Coordination compounds of zero-valent nickel and molybdenum with fluorine-containing phosphine ligands	R. Schmutzler	37, 150-160 (1963)
Metal chelate compounds as acid catalysts in solvolysis reactions	A.E. Martell	37, 161-173 (1963)
Effects of metal ions on imidazole catalysis of the mutarotation of glucose	N.C. Li and L. Jean	37, 174-180 (1963)
Cobalt-catalyzed cleavage of N-hydroxyethylenediamine to ethylene diamine in the presence of oxygen	D. Huggins and W.C. Drinkard	37, 181-191 (1963)

Interaction of amines and β-ketoimine copper (II) compounds	D.F. Martin	37,	192-200 (1963)
Catalytic hydrogenation by pentacyanocobaltate (II)	J. Kwiatek I.L. Mador and J.K. Seyler	37,	201-215 (1963)
Iron-catalyzed autoxidation of mercaptoacetate (Facilitation of a two-electron transfer through coordination)	D.L. Leussing and T.N. Tischer	37,	216-225 (1963)
Redox reactions of ligands. I. Oxidation of oxalato complexes of chromium (III) by cerium (IV) in aqueous sulfuric acid	J.E. Teggins M.T. Wang and R.M. Milburn	37,	226-242 (1963)
Reactions of some niobium (V) and tantalum (V) halides with pyridine	R.E. McCarley B.G. Hughes J.C. Boatman and B.A. Torp	37,	243-255 (1963)

SALINE WATER CONVERSION. A symposium presented at the 137th National Meeting of the American Chemical Society before the Division of Water and Waste Chemistry, 1960-1963. Volume 38.

NONSTOICHIOMETRIC COMPOUNDS. A symposium sponsored by the Division of Inorganic Chemistry at the 141st Meeting of the American Chemical Society, Washington, D.C., March 21-23, 1962. R. Ward, Symp. Chairman. (Only selected articles from the symposium are given below).

Current problems in nonstoichiometry	J.S. Anderson	39,	1-22 (1963)
Nonstoichiometry in metal hydrides	G.G. Libowitz	39,	74-86 (1963)
Neutron and x-ray diffraction studies of nonstoichiometric metal hydrides	S.S. Sidhu N.S. Satyamurthy F.P. Campos and D.D. Zauberis	39,	87-98 (1963)
Nonstoichiometric hydrides (Interstitial-atom, proton, and hydride-anion models)	T.R.P. Gibb, Jr.	39,	99-110 (1963)
Solutions of hydrogen in palladium	J.G. Aston and P. Mitacek, Jr.	39,	111-121 (1963)
Nonstoichiometry in intermetallic compounds	G.R.B. Elliott and J.F. Lemons	39,	144-152 (1963)
Nonstoichiometric phases in the $MS-M_2S_3$ sulfide systems	J. Flahaut L. Domange M. Patrie et al.	39,	179-190 (1963)

ADVANCES IN CHEMISTRY (Contd.)

Investigations of nonstoichiometric sulfides. I. Titanium sulfides, TiS_2 and Ti_2S_3	J. Benard and Y. Jeannin	39,	191-203 (1963)
Investigations of nonstoichiometric sulfides. II. The system In_2S_3MgS	J. Benard G. Duchefdelaville and M. Huber	39,	204-209 (1963)
Nonstoichiometry in clathrate compounds	L.A.K. Staveley	39,	218-223 (1963)

MASS SPECTRAL CORRELATIONS.
Edited by F.W. McLafferty; 40, 117 pp. (1963)

NEW APPROACHES TO PEST CONTROL AND ERADICATION. A symposium sponsored by the Pesticides Subdivision of the Division of Agricultural and Food Chemistry at the 142nd Meeting of the American Chemical Society, Atlantic City, N.J., September 11, 1962. S.A. Hall, Symp. Chairman.

Recent progress in the chemistry of insect sex attractants	M. Jacobson	41,	1-10 (1963)
Synthetic chemicals as insect attractants	M. Beroza and N. Green	41,	11-30 (1963)
The male annihalation technique in the control of fruit flies	L.D. Christenson	41,	31-35 (1963)
Chemosterilants as a potential weapon for insect control	C.N. Smith	41,	36-41 (1963)
Chemosterilants for the control of houseflies	G.C. LaBrecque	41,	42-46 (1963)
Aziridine chemosterilants (Sulfur-containing aziridines)	A.B. Borkovec and C.W. Woods	41,	47-55 (1963)
Antifeeding compounds for insect control	D.P. Wright, Jr.	41,	56-63 (1963)
The status of bacillus thurungiensis	A.M. Heimpel	41,	64-74 (1963)

SYMPOSIUM ON BORON-NITROGEN CHEMISTRY.
Edited by K. Niedenzu; 42, 330 pp. (1963).

CONTACT ANGLE WETTABILITY, AND ADHESION. Papers presented at the Kendall Award Symposium honoring William A. Zisman. Sponsored by the Division of Colloid and Surface Chemistry at the 144th Meeting of the American Chemical Society, Los Angeles, California, April 2-3, 1963. Edited by F.M. Fowkes; 43, 389 pp. (1964).

ADVANCES IN CHEMISTRY (Contd.)

RICHARD J. BLOCK MEMORIAL SYMPOSIUM.
Amino Acids and Serum Proteins. Edited by
J.A. Stekol; 44, 154 pp. (1964).

MOLECULAR MODIFICATION IN DRUG DESIGN.
This is a collection of papers based on the symposium sponsored by the Division of Medicinal Chemistry at the 145th Meeting of the American Chemical Society, New York, Sept. 9-10, 1963. F.W. Schueler, Symp. Chairman.

Molecular modification in modern drug research	M. Tishler	45,	1-14	(1964)
The synthetic penicillins	J.C. Sheehan	45,	15-24	(1964)
Molecular modification in the development of newer anti-infective agents (The sulfa drugs)	G. Zbinden	45,	25-38	(1964)
Impact of the synthetic anti-infectives on the therapy of bacterial infection	L.P. Garrod	45,	39-49	(1964)
Hypertension, an important disease of regulation	I.H. Page	45,	50-66	(1964)
Antihypertensive therapy	E.D. Freis	45,	67-76	(1964)
Molecular modifications among antihypertensive agents	C.J. Cavallito	45,	77-86	(1964)
Some results of molecular modifications of diuretics	J.M. Sprague	45,	87-101	(1964)
Sulfonylureas, science and serendipity	F.G. McMahon	45,	102-113	(1964)
Some rationals for the development of sntidepressant drugs	J.H. Biel	45,	114-139	(1964)
Molecular modifications in the development of phenothiazine drugs	M. Gordon, P.N. Craig and C.L. Zirkle	45,	140-147	(1964)
Role of synthetic drugs in therapy of mental illness	D.G. Friend	45,	148-161	(1964)
Narcotic antagonists as analgesics (Laboratory aspects)	A. Archer, L.S. Harris, N.F. Albertson, B.F. Tullar and A.K. Pierson	45,	162-169	(1964)
Narcotic antagonists as analgesics (Clinical aspects)	A.S. Keats and J. Telford	45,	170-176	(1964)

ADVANCES IN CHEMISTRY (Contd.)

Clinical control of fertility	G. Pincus	<u>45</u>,	177-189 (1964)
Synthetic progestational agents	J.C. Babcock	<u>45</u>,	190-203 (1964)
Androgenic-anabolic steroids	A. Segaloff	<u>45</u>,	204-220 (1964)
Perspectives in drug therapy	F.W. Schueler	<u>45</u>,	221-222 (1964)

PATENTS FOR CHEMICAL INVENTIONS.
A symposium sponsored by the Division of Chemical Literature and the Division of Industrial and Engineering Chemistry at the 145th Meeting of the American Chemical Society, New York, Sept. 9 and 13, 1963. Edited by E.J. Lawson and E.A. Godula; <u>46</u>, 117 pp. (1964)

FUEL SYSTEM CELLS.
A symposium sponsored by the Division of Fuel Chemistry. Edited by G.J. Young and H.R. Linden; <u>47</u>, 360 pp. (1964).

ADVANCES IN ENZYMIC HYDROLYSIS OF CELLULOSE AND RELATED MATERIALS

These are the proceedings of a symposium sponsored by The American Chemical Society and co-sponsored by The Army Research Office, 1962. Edited by E.T. Reese and published by Pergamon Press (The Macmillan Company) New York. A bibliography for the years 1950-1961 is included.

Title	Authors	Reference
Structural features of cellulose that influence its susceptibility to enzymatic hydrolysis	E.B. Cowling	1-32 (1963)
The effect of cellulolytic enzymes on some properties of cotton fibers	K. Selby	33-49 (1963)
Inhibition of cellulases and β-glucosidases	M. Mandels and E.T. Reese	115-157 (1963)
Endwise degradation of cellulose	K.W. King	159-170 (1963)

AUSGEWÄHLTE PHYSIKALISCHE METHODEN DER ORGANISCHEN CHEMIE

This is a collective volume, in German, on selected physical methods in organic chemistry. Edited by G. Geiseler and published by Akademie-Verlag-Berlin.

Title	Authors	Reference
Thermodynamic equilibrium in chemical reactions	G. Geiseler	1, 1-52 (1963)
Calorimetry	M. Rätzsch	1, 53-150 (1963)
Kinetics of chemical reactions	R. Kuschmiers	1, 151-200 (1963)
Distillation	G. Zimmermann	1, 201-285 (1963)
Gas chromatography	C.E. Döring	1, 287-352 (1963)
Capillary chromatography	H.R. Schütte	1, 353-402 (1963)
Paper chromatography	H.R. Schütte	1, 403-430 (1963)
Paper electrophoresis	H.R. Schütte	1, 431-447 (1963)
Liquid-phase partition methods	E. Stöckel	1, 449-501 (1963)
Thermal diffusion	W. Thierfelder	1, 503-528 (1963)
Determination of melting points and purity of substances	P. Laue	1, 529-549 (1963)
Purification of organic substances by zone melting and related methods	K.O. Bindernagel	1, 551-586 (1963)

BIOGENESIS OF NATURAL COMPOUNDS

This is a collective volume, in English, covering the biogenesis of several classes of natural compounds. Edited by P. Bernfeld and published by Pergamon Press (The Macmillan Co.) New York - London.

Title	Authors	Reference
The biosynthesis of amino acids	J.R. Mattoon	1-30 (1963)
The biogenesis of purine and pyrimidine nucleotides	A.J. Guarino	31-82 (1963)
The biogenesis of the lipids	K.P. Strickland	83-154 (1963)
The biosynthesis of steroids	E. Staple	155-181 (1963)
The biogenesis of heme, chlorophylls, and bile pigments	L. Bogorad	183-231 (1963)
The biogenesis of carbohydrates	P. Bernfeld	233-336 (1963)
The biogenesis of proteins	K. Moldave	337-379 (1963)
The biosynthesis of nucleic acids	R. Mantsavinos and S. Zamenhof	381-426 (1963)
The biogenesis of conjugation and detoxication products	R.T. Williams	427-474 (1963)
The biosynthesis of carotenoids and vitamin A	C.O. Chichester and T.O.M. Nakayama	475-508 (1963)
The biosynthesis of the water-soluble vitamins	V.H. Cheldelin and A. Baich	509-561 (1963)
The biosynthesis of phenolic plant products	T.A. Geissman	563-616 (1963)
The biosynthesis of tannins	S.G. Humphries	617-639 (1963)
The biogenesis of terpenes in plants	H.J. Nicholas	641-691 (1963)
The biogenesis of lignins	F.F. Nord and W.J. Schubert	693-726 (1963)
Rubber biogenesis	J. Bonner	727-735 (1963)
Alkaloid biogenesis	E. Leete	739-796 (1963)
The biosynthesis of fungal metabolites	W.B. Whalley	797-829 (1963)

BIOMEDICAL APPLICATIONS OF GAS CHROMATOGRAPHY

This is a collection of papers presented during the Fifth Annual Gas Chromatography Institute held at Canisius College, Buffalo, New York, 1963. Edited by H.A. Szymanski and published by Plenum Press, New York.

Title	Authors	Reference	
A brief introduction to the theory and practice of gas chromatography	H.A. Szymanski	1-37	(1964)
The gas chromatography of amines, alkaloids, and amino acids	H.-M. Fales and J.J. Pisano	39-87	(1964)
Separation, identification, and estimation of steroids	W.J.A. Vanden-Heuvel and E.C. Horning	89-150	(1964)
Gas chromatography of bile acids	J. Sjövall	151-167	(1964)
Gas chromatography of carbohydrates	W.W. Wells, C.C. Sweeley and R. Bentley	169-223	(1964)
Gas chromatography of urinary aromatic acids	C.M. Williams and C.C. Sweeley	225-269	(1964)
Analysis of fatty acids and derivatives by gas chromatography	W.R. Supina	271-305	(1964)
Determination of volatile organic anesthetics in blood	H.J. Lowe and L.M. Beckham	307-324	(1964)

BIOMEDICAL FRONTIERS IN MEDICINE

This is a collective volume, edited by H. Busch and published by Little, Brown and Co.

Title	Authors	Reference
Biochemical basis of genetic aberrations	H. Busch	3-39 (1963)
Genetically determined disorders of carbohydrated metabolism	W.L. Nyhan	40-91 (1963)
Genetically determined disorders of metabolism of amino acids and proteins	W.L. Nyhan	92-150 (1963)
Biochemical basis of newer diagnostic methods in clinical medicine	O. Bodansky	151-209 (1963)
Biochemistry of cancer	H. Busch	210-244 (1963)
Cancer chemotherapy with purine antimetabolites	R.E. Parks, Jr.	245-273 (1963)
Selected aspects of biochemical pathology	E. Farber	274-314 (1963)
Biochemistry of penicillin	R.E. Parks, Jr.	315-340 (1963)

BORON, METALLO-BORON COMPOUNDS AND BORANES

This is a collective volume, edited by R.M. Adams and published by Interscience Publishers Inc., New York.

Title	Authors	Reference
The hydroboron ions (Ionic boron hydrides)	R.M. Adams and A.R. Siedle	373-506 (1964)
The boranes or boron hydrides	R.M. Adams	507-692 (1964)
Toxicology of boron compounds	G.J. Levinskas	693-737 (1964)

CATALYSIS AND CHEMICAL KINETICS

This is a collection of review articles in the fields of catalysis and kinetics covering the post-war period in Poland. Published by Academic Press Inc., New York.

Title	Authors	Reference	
The potential theory of the adsorption of gases and vapours on adsorbents with energy-heterogeneous surfaces	M.M. Dubinin	17-31	(1964)
New data on the multiplet theory of catalysis	A.A. Balandin	33-58	(1964)
Hydrolytic hydrogenation of polysaccharides and the multiplet theory of catalysis	N.A. Vasyunina	75-83	(1964)
The theory of enzymatic and asymmetric catalysis	E.I. Klabunovskii	85-91	(1964)
Some applications of electrical conductivity measurements to the investigation of catalytic processes on semiconducting oxide catalysts	A. Bielanski	93-127	(1964)
Problems of chemical catalysis in connection with the Fischer-Tropsch synthesis of hydrocarbons	Z. Sokalski	129-170	(1964)
Studies on kinetics and selectivity in dehydrogenation processes of alcohols	E. Treszczanowicz	171-185	(1964)
The magnetic properties and structure of some metallic contacts	W. Trzebiatowski	187-200	(1964)
The rise of catalytic reactions involving hydrogen peroxide in the study of the formation of complexes and the development of very sensitive analytical methods	K.B. Yatsimirskii	201-206	(1964)
Mechanism and kinetics of certain redox systems in solutions	B. Jeżowska-Trzebiatowska	217-240	(1964)
Kinetics of reaction at the interface liquid-liquid or liquid-gas	E. Józefowicz	241-250	(1964)

CHELATING AGENTS AND METAL CHELATES

This is a collective volume, in English. Edited by F.P. Dwyer and D.P. Mellor and published by Academic Press, New York-London.

Title	Authors	Reference
The nature of the metal-ligand bond	D.P. Craig and R.S. Nyholm	51-93 (1963)
Bidentate chelates	C.M. Harris and S.E. Livingstone	95-141 (1963)
Design and stereochemistry of multidentate chelating agents	H.A. Goodwin	143-181 (1963)
Optical phenomena in metal chelates	A.M. Sargeson	183-235 (1963)
Oxidation-reduction potentials as functions of donor atom and ligand	D.A. Buckingham and A.M. Sargeson	237-282 (1963)
Metal chelates of ethylenediaminetetraacetic acid and related substances	F.L. Garvan	283-333 (1963)
Enzyme-metal ion activation and catalytic phenomena with metal complexes	F.P. Dwyer	335-382 (1963)
Metal chelates in biological systems	A. Shulman and F.P. Dwyer	383-439 (1963)
Physical and coordination chemistry of the tetrapyrrole pigments	J.E. Falk and J.N. Phillips	441-490 (1963)

THE CHEMISTRY OF CATIONIC POLYMERIZATION

This is a collective volume, edited by P.H. Plesch and published by Pergamon Press, Oxford-London-New York-Paris.

Title	Authors	Reference	
Carbonium ions	M.L. Burstall and F.E. Treloar	3-42	(1963)
Organic reactions related to cationic polymerization	M.L. Burstall	45-97	(1963)
A comparison of the radical, cationic and anionic mechanisms of addition polymerization	P.E.M. Allen and P.H. Plesch	101-139	(1963)
Isobutene	P.H. Plesch	143-208	(1963)
Aliphatic mono-olefins other than isobutene	C.M. Fontana	211-233	(1963)
Styrene	A.R. Mathieson	237-303	(1963)
Aromatic compounds other than styrene	S. Bywater	307-347	(1963)
Polyenes	W. Cooper	351-374	(1963)
Vinyl ethers	D.D. Eley	377-399	(1963)
Epoxides	A.M. Eastham	403-430	(1963)
Cyclic oxygen compounds other than epoxides	J.B. Rose	433-450	(1963)
Miscellaneous oxygen compounds	A. Schrage	453-476	(1963)
Sulphur compounds	J. Lal	479-512	(1963)
Nitrogen compounds	G.D. Jones	515-547	(1963)
Co-polymerization	R.B. Cundall	551-595	(1963)
Cationic reactions of polymers and cationic graft polymerization	G. Smets and M. van Beylen	599-609	(1963)
Cationic polymerizations induced by high energy radiation	S.H. Pinner	613-671	(1963)
Experimental techniques	P.H. Plesch	675-688	(1963)

THE CHEMISTRY OF HETEROCYCLIC COMPOUNDS

This is a collection of review monographs covering the entire field of heterocyclic chemistry. A. Weissberger is consulting editor for the whole series, which is published by Interscience Publishers, Inc., New York.

Title	Authors	Reference
The cyanine dyes and related compounds	F.M. Hamer	18, 1-790 (1964)
Heterocyclic compounds with three- and four-membered rings		
Ethylene oxides	A. Rosowsky	19, 1-523 (1964)
Aziridines	P.E. Fanta	19, 524-575 (1964)
Ethylene sulfides	D.D. Reynolds and D.L. Fields	19, 576-623 (1964)
Oxaziranes	W.D. Emmons	19, 624-646 (1964)
Thietane and its derivatives	Y. Etienne, R. Soulas and H. Lumbroso	19, 647-728 (1964)
β-Lactones	Y. Etienne and N. Fischer	19, 729-884 (1964)
Trimethyleneimines	J.A. Moore	19, 885-977 (1964)
Four-membered rings containing two heteroatoms	W.D. Emmons	19, 978-982 (1964)
Oxetanes	S. Searles, Jr.	19, 983-1068 (1964)
Pyrazolines, pyrazolidones and derivatives	R.H. Wiley and P.F. Wiley	20, 1-539 (1964)

CHEMISTRY OF ORGANIC SULFUR COMPOUNDS IN PETROLEUM AND PETROLEUM PRODUCTS

These are the proceedings of the Second Scientific Session, held in Ufa, and sponsored by the Bashkir Branch, Academy of Sciences, USSR, July 9-12, 1956. This is the first volume of a large publication, the aim of which is to make known results of research carried out in the Soviet Union in the field of chemistry and technology of organic compounds of sulfur and nitrogen. Edited by R.D. Obolentsev. Translated by and published for the National Science Foundation, Washington, D.C. by the Israel Program for Scientific Translations.

Title	Authors	Reference
Organic sulfur compounds of petroleum origin	R.D. Obolentsev	6-14 (1963)
Installation for the chromatographic separation of organic sulfur compounds from petroleum distillates	N.M. Ivanova, Ch.Kh. Mirkhaidarova and Ya.I. Nel'kenbaum	24-29 (1963)
Action of organic sulfur compounds on metals in the fuel system of gas-turbine engines	Ya.B. Chertkov and V.N. Zrelov	57-65 (1963)
Synthesis of compounds of the aliphatic series from thiophene and its homologs	S.Z. Taits	66-71 (1963)
Determination of the degree of purity of synthetic organic sulfur compounds	R.D. Obolentsev, S.V. Netupskaya, N.M. Pozdeev and E.V. Vafina	80-88 (1963)
Differential polarography for determination of elemental sulfur and disulfides in certain hydrocarbon solutions	R.D. Obolentsev and A.A. Ratovskaya	104-109 (1963)

CHEMISTRY OF WOOD

This is a collective volume, edited by B.L. Browning and published by Interscience Publishers Inc. (a division of John Wiley and Sons) New York-London.

Title	Authors	Reference	
The supply and uses of wood	B.L. Browning	1-6	(1963)
The structure of wood	I.H. Isenberg	7-55	(1963)
The composition and chemical reactions of wood	B.L. Browning	57-101	(1963)
Cellulose	E.H. Immergut	103-190	(1963)
The hemicelluloses (Addendum)	C. Schuerch and N.S. Thompson	191-247	(1963)
Wood lignins	K.V. Sarkanen	249-311	(1963)
Extraneous components of wood	M.A. Buchanan	313-367	(1963)
The chemistry of developing wood	R.E. Kremers	369-404	(1963)
The wood-water relationship	B.L. Browning	405-439	(1963)
Manufacture of wood pulp	N. Sanyer and G.H. Chidester	441-534	(1963)
Wood as a chemical raw material	J.F. Harris, J.F. Saeman and E.G. Locke	535-585	(1963)
The chemistry of bark	W. Jensen, K.E. Fremer, P. Sierilä and V. Wartiovaara	587-666	(1963)

These are the proceedings of the First Australian Conference on Electrochemistry, held in Sydney, 13-15 February and in Hobart, 18-20 February, 1963. Sponsored by The Royal Australian Chemical Institute, The University of New South Wales and The University of Tasmania. Edited by J.A. Friend and F. Gutmann and published by Pergamon Press, Oxford - New York.

Title	Authors	Reference
Ions in non-aqueous solvents (Chairmans Address)	H.N. Parton	455-469 (1963)
The voltage-current characteristics of benzene	E.O. Forster	470-480 (1963)
Triple ion equilibria in acetic acid	R.J.L. Martin	491-499 (1963)
The investigation of the kinetics of the electrode reactions of organic compounds by potentiostatic methods	M. Fleischmann, I.N. Petrov and W.F.K. Wynne-Jones	500-517 (1963)
The electrochemical aspects of some biochemical systems. II. Further investigations on the glucose-glucose oxidase systems	M.J. Allen and M. Nicholson	820-831 (1963)

ELECTRONIC ASPECTS OF BIOCHEMISTRY

These are the proceedings of the International Symposium held at Ravello, Italy, September 16-18, 1963. Sponsored by NATO. Edited by B. Pullman and published by Academic Press, New York-London.

Title	Authors	Reference	
Triplet states in DNA	I. Isenberg and R. Bersohn	3-5	(1964)
The use of fluorescence in studies of the structure and interactions of proteins	H. Edelhoch and R.F. Steiner	7-22	(1964)
The contribution of charge and energy migration phenomena to radiation inactivation mechanisms in enzymes	L.G. Augenstein and R. Mason	23-28	(1964)
Excitation migration and biological systems	J. Avery	29-41	(1964)
Luminescence of crystalline cell-nuclear chemicals under ionizing radiation	R.L. Lehman and R. Wallace	43-75	(1964)
Theory of the optical and other properties of biopolymers: Applicability and elimination of the first-neighbor and dipole-dipole approximations	D.F. Bradley, S. Lifson and B. Honig	77-91	(1964)
Mossbauer resonance studies in iron porphyrin and heme protein	J.E. Maling and M. Weissbluth	93-104	(1964)
Electrical conduction in solid proteins	D.D. Eley and R.B. Leslie	105-117	(1964)
Electron transport through conjugated systems	L.E. Orgel	119-120	(1964)
Impurity effects in organic semiconductors	H.A. Pohl	121-126	(1964)
Radio resistance of a series of aromatic molecules in the solid state	C. Williams-Dorlet, J. Duchesne and M. Lacroix	127-130	(1964)
Radioresistance of solid conjugated molecules	B. Pullman	131-134	(1964)
Molecular aspects of mutations	A. Pullman	135-152	(1964)
Electrons, arrows, and orbitals in biochemistry	H.C. Longuet-Higgins	153-154	(1964)
Some aspects of the mechanism of carcinogenesis	E. Boyland	155-165	(1964)

Some aspects of DNA replication; incorporation errors and proton transfer	P.-O. Löwdin	167-201 (1964)
Some new results in the quantum-mechanical calculation of DNA	J. Ladik	203-219 (1964)
Theoretical procedures for the study of biochemical σ-systems	G. Del Re	221-235 (1964)
Electronic aspects of biochemical hydroxylation	S. Diner	237-281 (1964)
Quantum theory of the chemical reactivity of conjugated molecules	O. Chalvet R. Daudel and F. Peradejordi	283-292 (1964)
Intermolecular forces in biological systems	L. Salem	293-299 (1964)
Solvent effects on Van der Waals dispersion attractions particularly in DNA	O. Sinanoğlu S. Abdulnur and N.R. Kestner	301-311 (1964)
Intermolecular charge fluctuation interaction in molecular biology	H. Jehle W.C. Parke and A. Salyers	313-333 (1964)
Possible electronic aspects of the biochemistry of muscle contraction	J. Gergely	335-345 (1964)
Reversible molecular reactions and biochemical mechanisms	P. Douzou	347-364 (1964)
Aromatic side chain effects on polypeptide structure	M. Goodman A.M. Felix and G. Schwartz	365-378 (1964)
Flavin free radicals	A. Ehrenberg	379-396 (1964)
Effects of hydroxyl free radicals on pyrimidine bases	C. Nofre and A. Cier	397-414 (1964)
Elucidation of induction and repression mechanisms in enzyme synthesis by analysis of model systems with the analog computer	F. Heinmets	415-479 (1964)
Biochemical macromolecules considered as specific information generators at very low temperature	J. Polonsky	481-502 (1964)
The role of metals in some reactions of tryptic hydrolysis	J. Yon	503-517 (1964)

ELECTRONIC ASPECTS OF BIOCHEMISTRY (Contd.)

Title	Authors	Reference
Copper in biological systems	A.S. Brill, R.B. Martin and R.J.P. Williams	519-557 (1964)
Electronic aspects of pharmacology	B. Pullman	559-578 (1964)

ESSAYS IN CO-ORDINATION CHEMISTRY

This is a collection of review articles, in German and English. Edited by W. Schneider, G. Aneregg and R. Gut and dedicated to the 60th birthday of Gerold Schwarzenbach. Published by Birkhauser Verlag, Basel-Stuttgart.

Title	Authors	Reference
Metal amine formation in solution. XII. Complex formation of ethylenediamine with mercury (II) ions	J. Bjerrum and E. Larsen	39-51 (1964)
The use of EDTA and related compounds in gravimetric analysis	R. Přibil	65-73 (1964)
The stabilisation, stereochemistry and reactivity of five-co-ordinate complexes	R.S. Nyholm and M.L. Tobe	112-127 (1964)
Tetrahedral co-ordination in nickel (II) and copper (II) chelates with Schiff bases	L. Sacconi	148-156 (1964)
Biochemistry of the ferrichrome compounds	J.B. Neilands	222-232 (1964)

These are the proceedings of the General Discussions (Symposia) organized by the Faraday Society whose content is directly related to Organic Chemistry. Edited by F.C. Tompkins and published by the Faraday Society, London.

Title	Authors	Reference
The Structure of Electronically Excited Species in the Gas-Phase. A general discussion held at Queens College, University of St. Andrews, Dundee, 2-3 April, 1963.		
Determination of the structures of simple polyatomic molecules and radicals in electronically excited states (Twelfth Spiers Memorial Lecture)	G. Herzberg	35, 7-29 (1963)
Electron-impact spectroscopy	A. Kuppermann and L.M. Raff	35, 30-42 (1963)
Circular dichroism of dissymmetric α,β-unsaturated ketones	R.E. Ballard S.F. Mason and G.W. Vane	35, 43-47 (1963)
Dipole moments and polarizabilities of electronically excited molecules through the Kerr effect	A.D. Buckingham and D.A. Dows	35, 48-57 (1963)
Wave functions of excited states. Allyl cation, radical and anion	J.W. Linnett and O. Sovers	35, 58-70 (1963)
Stereochemistry of hydrocarbon ions. Bridged structure of $C_2H_6^+$	J.C. Lorquet	35, 83-89 (1963)
Electronic absorption spectra of HCO and DCO radicals	J.W.C. Johns S.H. Priddle and D.A. Ramsay	35, 90-104 (1963)
Carbon monoxide flame bands	R.N. Dixon	35, 105-112 (1963)
Spectrum of the C_3 molecule	L. Gausset G. Herzberg A. Lagerqvist and B. Rosen	35, 113-126 (1963)
Rotational analysis of bands of the 3800 Å system of SO_2	A.J. Merer	35, 127-136 (1963)
Absorption spectrum of chlorine dioxide in the vacuum ultra-violet	C.M. Humphries A.D. Walsh and P.A. Warsop	35, 137-143 (1963)
Electronic structure and spectrum of the HCO_2 radical	T.E. Peacock R-UR-Rahman D.H. Sleeman and E.S.G. Tuckley	35, 144-147 (1963)

FARADAY SOCIETY — DISCUSSIONS (Contd.)

Absorption spectra of the hydrides, deuterides and halides of group 5 elements	C.M. Humphries, A.D. Walsh and P.A. Warsop	35,	148-157 (1963)
Electronically excited states of ammonia	A.E. Douglas	35,	158-174 (1963)
The 3820 Å band system of propynal	J.C.D. Brand, J.H. Callomon and J.K.G. Watson	35,	175-183 (1963)
Near ultra-violet spectrum of propenal	J.C.D. Brand and D.G. Williamson	35,	184-191 (1963)
Polarization and assignment of the 3700 Å absorption spectrum of 1,4-diazine vapour	K.K. Innes and L.E. Giddings, Jr.	35,	192-195 (1963)
Spectrum of the tropyl radical	B.A. Thrush and J.J. Zwolenik	35,	196-200 (1963)
Ionization and dissociation energies of the hydrides and fluorides of the first row elements in relation to their electronic structures	W.C. Price, T.R. Passmore and D.M. Roessler	35,	201-211 (1963)

Fundamental Processes in Radiation Chemistry. A general discussion held at the University of Notre Dame, Indiana, 2-4 September, 1963.

(General discussion)		36,	232-325 (1963)
Reactions and Photochemistry of Atoms and Molecules. Introduction to chemical aeronomy	M. Nicolet	37,	7-20 (1964)

FATTY ACIDS

This is a collective volume on the Chemistry, Properties, Production and Uses of Fatty Acids. Part 1 was published in 1960, Part 2 in 1961. Part 3 was completely revised and augmented in 1964. Edited by K.S. Markley and published by Interscience Publishers (a division of John Wiley and Sons, Inc.) New York-London.

Title	Authors	Reference
Biological oxidation of fatty acids	H.R. Mahler	3, 1487-1550 (1964)
Nitrogen derivatives	N.O.V. Sonntag	3, 1551-1715 (1964)
Sulfur derivatives	K.S. Markley	3, 1717-1768 (1964)
Chemical synthesis of fatty acids	K.S. Markley	3, 1769-1965 (1964)
Biogenesis of fatty acids	M. Woodbine	3, 1967-1982 (1964)
Techniques of separation. A. Distillation, salt solubility, low-temperature crystallization	K.S. Markley	3, 1983-2123 (1964)
Techniques of separation. B. Liquid chromatography	H. Schlenk	3, 2125-2247 (1964)
Techniques of separation. C. Gas-liquid chromatography of fatty acids	F.P. Woodford	3, 2249-2282 (1964)
Techniques of separation. D. Countercurrent distribution	C.R. Scholfield	3, 2283-2307 (1964)
Techniques of separation. E. Urea complexes	D. Swern	3, 2309-2358 (1964)

FREE-ELECTRON THEORY OF CONJUGATED MOLECULES

This is a collection of the papers of the Chicago Group, 1949-1961. Edited by J.R. Platt, K. Ruedenberg, C.W. Scherr, N.S. Ham, H. Labhart and W. Lichten and published by John Wiley and Sons, Inc., New York. The articles are referred to by the page numbers in the original publications.

Title	Authors	Reference
Classification of spectra of cata-condensed hydrocarbons	J.R. Platt	J. Chem. Phys. $\underline{17}$ 484-495 (1949)
Free-electron network model for conjugated systems. I. Theory	K. Ruedenberg and C.W. Scherr	J. Chem. Phys. $\underline{21}$ 1565-1581 (1953)
Free-electron network model for conjugated systems. II. Numerical calculations	C.W. Scherr	J. Chem. Phys. $\underline{21}$ 1582-1596 (1953)
Free-electron network model for conjugated systems. III. A demonstration model showing bond order and "free valence" in conjugated hydrocarbons	J.R. Platt	J. Chem. Phys. $\underline{21}$ 1597-1600 (1953)
Free-electron network model for conjugated systems. V. Energies and electron distributions in the FE MO model and in the LCAO MO model	K. Ruedenberg	J. Chem. Phys. $\underline{22}$ 1878-1894 (1954)
Energy levels, atom populations, bond populations in the LCAO MO model and in the FE MO model. A quantitative analysis	N.S. Ham and K. Ruedenberg	J. Chem. Phys. $\underline{29}$
Free-electron model for conjugated systems. VII. A note on the joint condition	K. Ruedenberg	ONR Tech. Report Lab. of Molecular Spectra and Structure Univ. of Chicago, 1955 pp. 230-232
The free-electron theory and the virial theorem	W. Lichten	J. Chem. Phys. $\underline{22}$ 1278-1279 (1954)
The box model and electron densities in conjugated systems	J.R. Platt	J. Chem. Phys. $\underline{22}$ 1448-1455 (1954)
Box model and network model of _sigma_ electron densities in large molecules (the original English version is given here)	J.R. Platt	Calcul des Fonctions d'ondes Moleculaires R. Daudel ed. CNRS, Paris, 1959; pp. 161-178
On the optical properties of interstellar dust	J.R. Platt	Astrophys. J. $\underline{123}$ 486-490 (1956)
Electronic interaction on the free-electron network model for conjugated systems. I. Theory	N.S. Ham and K. Ruedenberg	J. Chem. Phys. $\underline{25}$ 1-13 (1956)

Electronic interaction in the free-electron network model for conjugated systems. II. Spectra of aromatic hydrocarbon	N.S. Ham and K. Ruedenberg	J. Chem. Phys. $\underline{25}$ 13-26 (1956)
FE Theory including an elastic skeleton. I. Spectra and bond lengths in long polyenes	H. Labhart	J. Chem. Phys. $\underline{27}$ 957-962 (1957)
FE Theory including an elastic skeleton. II. Changes of molecular dimensions due to optical excitation	H. Labhart	J. Chem. Phys. $\underline{27}$ 963-965 (1957)
Mobile bond orders in conjugated systems	N.S. Ham and K. Ruedenberg	J. Chem. Phys. $\underline{29}$ 1215-1229 (1958)
Mobile bond orders in the resonance and molecular orbital theories	N.S. Ham	J. Chem. Phys. $\underline{29}$ 1229-1231 (1958)
Theorem on the mobile bond orders of alternant conjugated systems	K. Ruedenberg	J. Chem. Phys. $\underline{29}$ 1232-1233 (1958)
Electronic interaction in the free-electron network model for conjugated systems. III. Spectra of hydrocarbons other than alternants	N.S. Ham	J. Chem. Phys. $\underline{32}$ 1445-1448 (1960)
Quantum mechanics of mobile electrons in conjugated bond systems. III. Topological matrix as generatrix of bond orders	K. Ruedenberg	J. Chem. Phys. $\underline{34}$ 1884-1891 (1961)

FRIEDEL-CRAFTS AND RELATED REACTIONS

This is a collective volume, in English. Edited by G.A. Olah and published by Interscience Publishers Inc., New York.

Title	Authors	Reference
Volume I. General Aspects		
Historical	G.A. Olah and E.A. Dear	1, 1-24 (1963)
Definition and scope	G.A. Olah	1, 25-168 (1963)
Proton acids and Lewis acids	R.J. Gillespie	1, 169-199 (1963)
Catalysts and solvents	G.A. Olah	1, 201-366 (1963)
Lewis acid catalysts in non-aqueous solutions	M. Baaz and V. Gutmann	1, 367-397 (1963)
Coordination compounds of the boron halides	D.R. Martin and J.M. Canon	1, 399-567 (1963)
Coordination compounds of aluminum and gallium halides	N.N. Greenwood and K. Wade	1, 569-622 (1963)
Intermediate complexes	G.A. Olah and M.W. Meyer	1, 623-765 (1963)
Spectroscopic investigations (Vibrational spectra)	D. Cook	1, 767-820 (1963)
Application of isotopic tracers to the study of reactions	R.M. Roberts and G.J. Fonken	1, 821-851 (1963)
Reactivity and selectivity	G.A. Olah	1, 853-935 (1963)
Thermodynamic aspects	D.R. Stull	1, 937-997 (1963)
Stereochemical aspects	H. Hart	1, 999-1015 (1963)
Volume II. Alkylation and Related Reactions. Part I.		
Alkylation of aromatics with alkenes and alkanes	S.H. Patinkin and B.S. Friedman	2, 1-288 (1964)
Alkylation of aromatics with dienes and substituted alkenes	R. Koncos and B.S. Friedman	2, 289-412 (1964)

Alkylation of aromatics with alkynes	V. Franzen	2,	413-416 (1964)
Alkylation of aromatics with haloalkanes	F.A. Drahowzal	2,	417-475 (1964)
Alkylation of aromatics with alcohols and ethers	A. Schriesheim	2,	477-595 (1964)
Alkylation of aromatics with aldehydes and ketones	J.E. Hofmann and A. Schriesheim	2,	597-640 (1964)
Alkylation of aromatics with esters of inorganic acids and alkyl arenesulfonates	F.A. Drahowzal	2,	641-658 (1964)

Volume II. Alkylation and Related Reactions. Part II.

Haloalkylations	G.A. Olah and W.S. Tolgyesi	2,	659-784 (1964)
Cyclialkylation of aromatics	L.R.C. Barclay	2,	785-977 (1964)
Dehydrogenation condensation of aromatics (Scholl and related reactions)	A.T. Balaban and C.D. Nenitzescu	2,	979-1047 (1964)
Isomerization of aromatic hydrocarbons	D.A. McCaulay	2,	1049-1073 (1964)
Alkylation of saturated hydrocarbons	L. Schmerling	2,	1075-1173 (1964)
Alkylation of alkenes with carbonyl compounds (Prins reaction)	C.W. Roberts	2,	1175-1210 (1964)
Isomerization of saturated hydrocarbons	H. Pines and N.E. Hoffman	2,	1211-1252 (1964)
Hydrogen exchange in aromatic compounds	V. Gold	2,	1253-1291 (1964)
Polymerization	D.C. Pepper	2,	1293-1348 (1964)

Volume III. Acylation and Related Reactions. Part I.

Aromatic ketone synthesis	P.H. Gore	3,	1-381 (1964)
Houben-Hoesch and related syntheses	W. Ruske	3,	383-497 (1964)
The Fries reaction	A. Gerecs	3,	499-533 (1964)

FRIEDEL-CRAFTS AND RELATED REACTIONS (Contd.)

Acylation with di- and polycarboxylic acid derivatives	A.G. Peto	3, 535-910 (1964)

Volume III. Acylation and Related Reactions. Part II.

Cycliacylation	S. Sethna	3, 911-1002 (1964)
Mechanism of acylations	F.R. Jensen and G. Goldman	3, 1003-1032 (1964)
Aliphatic acylation	C.D. Nenitzescu and A.T. Balaban	3, 1033-1152 (1964)
Aldehyde syntheses	G.A. Olah and S.J. Kuhn	3, 1153-1256 (1964)
Acid syntheses	G.A. Olah and J.A. Olah	3, 1257-1317 (1964)
Sulfonylation	F.R. Jensen and G. Goldman	3, 1319-1354 (1964)
Sulfonation	K.L. Nelson	3, 1355-1392 (1964)
Nitration	G.A. Olah and S.J. Kuhn	3, 1393-1491 (1964)
Amination	P. Kovacic	3, 1493-1506 (1964)
Perchlorylation	C.E. Inman, R.E. Oesterling and F.L. Scott	3, 1507-1516 (1964)
Halogenation	H.P. Braendlin and E.T. McBee	3, 1517-1593 (1964)

These are the proceedings of the Fourth International Symposium held at Michigan State University, under the auspices of the Analysis Instrumentation Division of the Instrument Society of America, June 17-21, 1963. Edited by L. Fowler and published by Academic Press, New York-London.

Title	Authors	Reference
Trends in gas chromatography (Opening Address)	S. Dal Nogare	1-22 (1963)
Some recent developments in column technology	J.J. Kirkland	77-103 (1963)
Applications of gas chromatography in the petroleum industry	R.L. Martin	127-138 (1963)
Quantitative gas chromatographic analysis of xylene oxidation products	F. Baumann and J.B. Lavigne	139-147 (1963)
Detection devices for gas chromatography	J.C. Sternberg	161-191 (1963)
Some aspects of the quantitative characteristics of catalytic combustion ionization detectors	G. Perkins, Jr. and O.F. Folmer, Jr.	193-202 (1963)
A convenient method for collecting gas chromatography fractions for infrared analysis	L.S. Gray, L.D. Metcalfe and S.W. Leslie	203-212 (1963)

HIGH POLYMERS

This is a series of reviews on the chemistry, physics and technology of high polymers. Edited by H. Mark, P.J. Flory, C.S. Marvel and H.W. Melville and published by Interscience Publishers (a division of John Wiley and Sons), New York-London.

Title	Authors	Reference
Configurational Statistics of Polymeric Chains	M.V. Volkenstein (Editor)	<u>17</u>, x + 562 pp. (1963)
Copolymerization	G.E. Ham (Editor)	
Theory of copolymerization	G.E. Ham	<u>18</u>, 1-65 (1964)
The Q-e scheme	T. Alfrey, Jr. and L.J. Young	<u>18</u>, 67-87 (1964)
Olefin copolymerization with supported metal oxide catalysts	J.P. Hogan	<u>18</u>, 89-113 (1964)
Copolymers of α-olefins	C.A. Lukach and H.M. Spurlin	<u>18</u>, 115-147 (1964)
Block copolymers with ethylene as one of the components	G. Bier and G. Lehmann	<u>18</u>, 149-230 (1964)
Ethylene-propylene copolymers as rubbers	G. Crespi, A. Valvassori and G. Sartori	<u>18</u>, 231-274 (1964)
Other α-olefin copolymers	G. Crespi, A. Valvassori and G. Sartori	<u>18</u>, 275-282 (1964)
Cationic copolymerizations	J.P. Kennedy	<u>18</u>, 283-334 (1964)
Block and graft copolymerizations	A.S. Hoffman and R. Bacskai	<u>18</u>, 335-420 (1964)
Anionic copolymerizations	M. Morton	<u>18</u>, 421-436 (1964)
Copolymerizations involving styrene or substituted styrenes as principal components	L.J. Young	<u>18</u>, 437-524 (1964)
Copolymerizations involving acrylonitrile as a principal component	L.H. Peebles, Jr.	<u>18</u>, 525-586 (1964)
Copolymerizations employing vinyl chloride or vinylidene chloride as principal components	J.F. Gabbett and W. Mayo Smith	<u>18</u>, 587-637 (1964)

Quantitative aspects of high pressure copolymerizations involving ethylene as a principal component	N.L. Zutty and R.D. Burkhart	18,	639-652 (1964)
Copolymerizations involving acrylates and methacrylates as principal components	L.S. Luskin	18,	653-693 (1964)
Copolymerization reativity ratios (Appendix A.)	H. Mark B. Immergut and E.H. Immergut	18,	695-843 (1964)
Tabulation of Q - e values (Appendix B.)	L.J. Young	18,	845-863 (1964)

Crystalline Olefin Polymers II — R.A.V. Raff and K.W. Doak (Editors)

Brittleness	A. Rudin	20,	1-45 (1964)
Stress-cracking	J.B. Howard	20,	47-103 (1964)
Electrical properties of polyolefins	A.J. Curtis	20,	105-130 (1964)
Permeability	V. Stannett and H. Yasuda	20,	131-184 (1964)
Miscellaneous properties and tests	R. Raff E.L. Lyle and R.F. Kratz	20,	185-238 (1964)
Chemical modification	E.M. Fettes	20,	239-300 (1964)
Irradiation - properties changes	V.L. Lanza	20,	301-360 (1964)
Degradation and stabilization	W.L. Hawkins and F.H. Winslow	20,	361-398 (1964)
The processing of crystalline olefin polymers	T.O.J. Kresser	20,	399-550 (1964)
Applications	D.C. Rogers	20,	551-613 (1964)
Statistical summary	E.L. Lyle	20,	615-629 (1964)
Review and preview	H.F. Mark	20,	631-639 (1964)

HIGH PRESSURE PHYSICS AND CHEMISTRY

This is a collection of articles covering most major aspects of this field, including both theoretical and experimental work. Edited by R.S. Bradley and published by Academic Press, London - New York.

Title	Authors	Reference
A brief historical survey of high pressure research in this century	R.S. Bradley	1, 1-9 (1963)
Production and measurement of high pressures	D.C. Munro	1, 11-49 (1963)
Spectroscopic studies of liquids	O.E. Weigang, Jr. and W.W. Robertson	1, 177-206 (1963)
High pressure spectroscopy of solids	S.E. Babb, Jr. and W.W. Robertson	1, 375-409 (1963)
Chemical equilibria in gases	B.F. Dodge	2, 91-130 (1963)
Chemical equilibria in condensed systems	S.D. Hamann	2, 131-162 (1963)
Chemical kinetics	S.D. Hamann	2, 163-207 (1963)
Shock waves	G.E. Duvall and G.R. Fowles	2, 209-291 (1963)
Radiofrequency spectroscopy at high pressures	J.A.S. Smith	2, 293-310 (1963)
Structural determinations by x-rays of systems at high pressure	D.C. Munro	2, 311-323 (1963)
Other miscellaneous effects of pressure	R.S. Bradley	2, 325-337 (1963)

HORMONAL STEROIDS
(Biochemistry, Pharmacology and Therapeutics)

These are the proceedings of the 1st International Congress on Hormonal Steroids. Edited by L. Martini and A. Pecile and published by Academic Press, New York-London.

Title	Authors	Reference
Symposium I. New Steroids with Hormonal-like Activities		
Inverted steroids, their preparation and biological activities	J. Jacques and G. Pincus	1, 3-9 (1964)
Ether derivatives of steroidal hormones	A. Ercoli	1, 11-27 (1964)
New steroids with hormone like activities. C-6 substituted hormone analogs	A. Zaffaroni and A. Bowers	1, 29-38 (1964)
Antiandrogenic, antiestrogenic, and antiovulatory compounds	R.I. Dorfman	1, 39-52 (1964)
The actions of "Retro" progestagens	H.F.L. Scholer	1, 53-65 (1964)
Biological properties of 4-hydroxy-3-keto-Δ^4-steroids	G. Sala	1, 67-77 (1964)
Symposium II. Pathways of Synthesis and Metabolism of Hormonal Steroids		
Some aspects of steroid dynamics	J.F. Tait B. Little S.A.S. Tait et al.	1, 81-102 (1964)
Interaction of thyroid with steroid hormone metabolism	T.F. Gallagher	1, 103-114 (1964)
In vitro conversion of deoxycorticosterone to some acidic metabolites	J.J. Schneider	1, 127-138 (1964)
Symposium III. Control of Synthesis and Release of Steroid Hormones		
Symposium IV. Mechanism of Steroid Action		
The rate of molecular configuration in the interaction of steroid hormones with coenzymes and enzymes	L.L. Engel A.M. Stoffyn and J.F. Scott	1, 291-303 (1964)
Steroid metabolism and steroid action	H.J. Ringold	1, 305-318 (1964)

HORMONAL STEROIDS (Contd.)

Title	Authors	Reference
Steroids, nonsteroids, intermediary metabolism, inflammation and their probable interrelationships	E.M. Glenn	1, 319-352 (1964)

Symposium V. Substances which Affect Synthesis and Action of Steroid Hormones

Nonhormonal inhibitors of adrenocortical steroid biosynthesis	W.W. Tullner	1, 383-398 (1964)
Action of antialdosterone compounds in the laboratory	C.M. Kagawa	1, 445-461 (1964)
The actions of steroid hormones on estradiol-17β in uterine growth and enzymorphology	J.T. Velardo	1, 463-490 (1964)

Symposium VI. Non-Hormonal Activities of Steroids

INFRA-RED SPECTROSCOPY AND MOLECULAR STRUCTURE
(An Outline of the Principles)

This is an edited volume in which the principles and specific aspects of infrared spectroscopy are reviewed by workers active in these particular fields. Edited by M. Davies and published by Elsevier Publishing Company, New York-Amsterdam.

Title	Authors	Reference
The infra-red spectra of simple molecules	W.J. Jones	111-165 (1963)
Characteristic features in the spectra of organic molecules	D. Hadzi	226-269 (1963)
Infra-red spectra of solids: dichroism and polymers	S. Krimm	270-310 (1963)
Quantitative intensity studies and dipole moment derivatives	J. Overend	345-376 (1963)
Hydrogen bonding and solvent effects	H.E. Hallam	405-440 (1963)

The following International Conferences were held under the auspices of the International Union of Pure and Applied Chemistry (IUPAC):

Coordination Chemistry, Stockholm and Uppsala, Sweden, 25-29 June, 1962

Molecular Structure and Spectroscopy, Tokyo, Japan, 10-15 September, 1962

Isotope Mass Effects in Chemistry and Biology, Vienna, Austria, 9-13 December, 1963.

The above publications are available (Butterworths) in book form. The individual papers have also been published in Pure and Applied Chemistry. The specific titles are given on pages 108-116 of this Index.

LESS COMMON MEANS OF SEPARATION

These are the proceedings of the symposium on the Less Common Means of Separation, held in Birmingham, England, 24-26 April, 1963. Sponsored by and published by the Institution of Chemical Engineers, London.

Title	Authors	Reference
Liquid separation using polymer films	J.W. Carter and B. Jagannadhaswamy	35-42 (1964)
The separation of meta- and para-cresols by dissociation extraction	S.R.M. Ellis and J.D. Gibbon	119-123 (1964)

MAGNETIC AND ELECTRIC RESONANCE AND RELAXATION

This is a collection of papers presented at the 11th Colloque Ampère, Eindhoven, July 2-7, 1962. Edited by J. Smidt and published in English, French, and German by Interscience Publishers, Inc. (A division of John Wiley and Sons, Inc.) New York.

Title	Authors	Pages
The dielectric behaviour of oxidized high-pressure polyethylene	C.A.F. Tuijnman	264-271 (1963)
Dielectric saturation in solids containing hydrogen bonds	A. Piekara	297-301 (1963)
Molecular associations of acetic acid in very dilute solutions (in French)	E. Constant and R. De Wavrechin	307-315 (1963)
Solvent effect on intermolecular association of normal saturated alcohols (in French)	R. Liebaert and Y. Leroy	316-320 (1963)
Dipolar Debye absorption and phase change of frozen organic liquids (in French)	M. Freymann, M.R. Freymann and M. Jaffrain	327-337 (1963)
Relationship between dielectric relaxation and viscosity in alcohols	R.G. Bennett and J.H. Calderwood	346-350 (1963)
The relation between dipole orientation and viscous flow	S.H.M. El-Sabeh	351-357 (1963)
Free radicals produced by gamma radiation in desoxyribonucleic acid and derivatives (in French)	A. Van De Vorst and Cl. Williams-Dorlet	398-402 (1963)
C^{13} Hyperfine interactions in semiquinones	M.R. Das and B. Venkataraman	426-430 (1963)
Paramagnetic resonance of hydrazyl-type radicals	K.V. Sane and J.A. Weil	431-438 (1963)
Second-order splitting, anomalous relaxation and sign of the coupling constant in ESR spectra of radical ions	E. de Boer and E.L. Mackor	439-444 (1963)
A proton magnetic resonance investigation of rotational isomerism in ethyl alcohol	R.J. Abraham	589-593 (1963)
Nuclear and electron spin relaxation in solutions of aromatic free radicals	J. Haupt, K. Kramer and W. Müller-Warmuth	709-713 (1963)

MASS SPECTROMETRY

This is a collection of review articles on mass spectrometry of interest to organic chemists. The bibliographies include many references to earlier works. Edited by C.A. McDowell and published by McGraw-Hill Book Company, Inc., New York-London.

Title	Authors	Reference
Mass spectrometry of free radicals	F.P. Lossing	442-505 (1963)
The ionization and dissociation of molecules	C.A. McDowell	506-588 (1963)
Ion-molecule reactions	D.P. Stevenson	589-615 (1963)

This is a collection of review articles, in English, covering the mass spectrometry of organic ions, in detail, with extensive bibliographies to the experimental literature. Edited by F.W. McLafferty and published by Academic Press, New York-London.

Title	Authors	Reference
Quasi-equilibrium theory of mass spectra	H.M. Rosenstock and M. Krauss	2-64 (1963)
Ion-molecule reactions	C.E. Melton	65-115 (1963)
Appearance potential data of organic molecules	M. Krauss and V.H. Dibeler	118-161 (1963)
Negative ion mass spectra	C.E. Melton	163-205 (1963)
Mass spectrometry of organic radicals	A.G. Harrison	207-253 (1963)
Mass spectrometry of ions from electric discharges, flames, and other sources	P.F. Knewstubb	255-307 (1963)
Decompositions and rearrangements of organic ions	F.W. McLafferty	309-342 (1963)
High resolution mass spectrometry	R.A. Saunders and A.E. Williams	343-397 (1963)
Mass spectrometry of long-chain esters	R. Ryhage and E. Stenhagen	399-452 (1963)
Mass spectra of alkylbenzenes	H.M. Grubb and S. Meyerson	453-527 (1963)
Applications to natural products and other problems in organic chemistry	K. Biemann	529-596 (1963)
The molecular structure of petroleum	A. Hood	597-635 (1963)
Mass spectra of terpenes	R.I. Reed	637-699 (1963)

METHODEN DER ORGANISCHEN CHEMIE

(Houben-Weyl) Fourth Edition, edited by E. Muller and published by G. Thieme Verlag, Stuttgart, Germany. This well-known work initiated by T. Weyl and continued by G. Houben (second and third editions, 1921 and 1925) is now in its fourth greatly enlarged edition. The additional volumes which are appearing to complete the predetermined plan of publication greatly enhance the value of this important series.

Title	Authors	Reference
Organic phosphorus compounds. Part 1.	K. Sasse	12/1 1-620 (1963)
Organic phosphorus compounds. Part 2.	K. Sasse	12/2 5-995 (1963)

MOLECULAR ORBITALS IN CHEMISTRY, PHYSICS AND BIOLOGY

This is a collective volume in tribute to R.S. Mulliken. Edited by P.-O. Lowdin and published by Academic Press, New York-London.

Title	Authors	Reference
The structure of methylene and methyl	L. Pauling	207-213 (1964)
Energy calculations for polyatomic carbon molecules	S.J. Strickler and K.S. Pitzer	281-291 (1964)
Rotational structure of the fundamental band v_6 of methyl cyanide	G. Amat and H.H. Nelsen	293-304 (1964)
The resonance formulation of electronically excited π-electron states	E. Heilbronner	329-359 (1964)
π-Electron conjugation in some polyphenyl molecules	T. Nakamura S. Kwun and H. Eyring	421-432 (1964)
Molecular orbital calculations and the aromaticity of some nonbenzenoid aromatic hydrocarbons	T. Nakajima	451-469 (1964)
Application of the molecular orbitals to the study of the base strength	O. Chalvet R. Daudel and F. Peradejordi	475-484 (1964)
Molecular orbital calculations and electrophilic substitution	R.D. Brown	485-511 (1964)
A simple quantum-theoretical interpretation of the chemical reactivity of organic compounds	K. Fukui	513-537 (1964)
π-Molecular orbitals and the processes of life	A. Pullman and B. Pullman	547-571 (1964)

NEW BIOCHEMICAL SEPARATIONS

This is a comprehensive series of articles on new and recent applications of a number of separation methods to biochemical systems. Edited by A.T. James and L.J. Morris and published by D. Van Nostrand Co., Ltd., London-New York.

Title	Authors	Reference	
Methods for the detection and estimation of radioactive compounds separated by gas-liquid chromatography	A.T. James	1-24	(1964)
Gas-liquid chromatographic procedures for the separation, identification and estimation of steroids, alkaloids and sugars	E.C. Horning and W.J.A. Vandenheuvel	25-57	(1964)
Determination of coenzyme A esters by gas-liquid chromatography	M.G. Horning	59-64	(1964)
Gas-liquid chromatography of bile acids	J. Sjovall	65-79	(1964)
Gas-liquid chromatography of the amino acids	A. Karmen and H.A. Saroff	81-92	(1964)
Fractionation of proteins, peptides and amino acids by gel filtration	B. Gelotte	93-109	(1964)
Fractionation of polysaccharides by gel filtration	K. Granath	111-122	(1964)
Characterization of protein hydrolysates and amino acids in biological materials by thin-layer chromatography	M. Brenner A. Niederwieser and G. Pataki	123-156	(1964)
Chromatography of the alkaloids	D. Waldi	157-196	(1964)
Thin-layer chromatography of steroids	R. Tschesche G. Wulff and K.H. Richert	197-245	(1964)
Thin-layer and ion-exchange chromatography of triterpenoid compounds	R. Tschesche I. Duphorn and G. Snatzke	247-259	(1964)
Thin-layer chromatography of bile acids and their derivatives	A.F. Hofmann	261-282	(1964)
Hydroxyapatite as an adsorbent for thin-layer chromatography: separations of lipids and proteins	A.F. Hofmann	283-294	(1964)
Specific separations by chromatography on impregnated adsorbents	L.J. Morris	295-319	(1964)

NEW BIOCHEMICAL SEPARATIONS (Contd.)

Title	Author	Reference
Separation of plant phospholipids and glycolipids	B.W. Nichols	321-337 (1964)
Chromatographic analysis of polar lipids on silicic acid impregnated paper	G.V. Marinetti	339-377 (1964)
Counter-current distribution of lipids	F.D. Collins	379-399 (1964)

NITRO COMPOUNDS

These are the proceedings of the International Symposium held at the Institute of Organic Synthesis, Polish Academy of Sciences, Warsaw, Poland, 18-20 September, 1963. Edited by T. Urbanski and published by Pergamon Press, New York-Oxford.

Title	Authors	Reference
Amphoterism of HNO_3 and the nitronium ion	M.I. Usanovich	20, 1-5 (1964)
Composition of products of direct nitration of high molecular n-alkanes (in German)	F. Asinger	20, 7-14 (1964)
The reaction between n-hexane and dinitrogen tetroxide induced by ionising radiation	J. Falęcki, S. Minc and T. Urbański	20, 15-18 (1964)
Nitration of paraffins. Determination of nitro group position in secondary nitroheptane	W. Biernacki and T. Urbański	20, 19-25 (1964)
Hazards connected with vapour-phase nitration of saturated hydrocarbons	C. Matasă	20, 27-31 (1964)
Nitration of cyclic β-diketenes (in Russian)	E. Gudriniece and G. Vanags	20, 33-36 (1964)
O-Nitration of alcohols	L.T. Eremenko	20, 37-42 (1964)
The substituent effects of positive poles in aromatic nitration	J.H. Ridd	20, 43-51 (1964)

NON-STOICHIOMETRIC COMPOUNDS

This is a collective volume edited by L. Mandelcorn and published by Academic Press, New York - London.

Title	Authors	Reference
Clathrates	H.M. Powell	438-490 (1964)
Organic adducts	L.C. Fetterly	491-567 (1964)
Carbohydrates	F.R. Senti and S.R. Erlander	568-605 (1964)
Physics and chemistry of inclusion compounds	L.A.K. Staveley	606-635 (1964)

PARAMAGNETIC RESONANCE

This is a collection of the papers presented at the First International Conference on Paramagnetic Resonance, held in Jerusalem, July 16-20, 1962. Presented in two volumes, the second is of definite interest to organic chemists. Edited by W. Low and published by Academic Press, New York-London.

Title	Authors	Reference
Paramagnetic resonance in biological materials	D.J.E. Ingram	$\underline{2}$, 809-835 (1963)
Paramagnetic resonance studies of magnetic ions found in nucleic acid pseudocrystals	W.M. Walsh, Jr. L.W. Rupp, Jr. and B.J. Wyluda	$\underline{2}$, 836-854 (1963)
Electron spin resonance of copper phthalocyanine	S.E. Harrison and J.M. Assour	$\underline{2}$, 855-868 (1963)
Trapping in solution of non-equilibrium concentrations of biradicals	R.K. Waring, Jr. and G.J. Sloan	$\underline{2}$, 896-904 (1963)
Low field EPR measurements in DPPH at low temperatures	L. Van Gerven A. Van Itterbeek and L. De Laet	$\underline{2}$, 905-918 (1963)

PEPTIDES

These are the proceedings of the Fifth European Symposium on Peptides, held at Oxford, September, 1962. Edited by G.T. Young and published by Pergamon Press, Inc. (The Macmillan Company) New York.

Title	Authors	Reference
I. Synthesis: Methods of Protection		
Methods of protection: An assessment of the present position	H. Kappeler	3-13 (1963)
Concerning a new type of amino acid protective group (in German)	E. Dane	15 (1963)
On the protection of α-amino and carboxyl groups for peptide synthesis	E. Gazis, B. Bezas, G.C. Stelakatos and L. Zervas	17-21 (1963)
The use of the methylthioethyl group for carboxyl protection in peptide synthesis	H.N. Rydon and J.E. Willett	23-26 (1963)
On cysteine and cystine peptides	L. Zervas, I. Photaki, A. Cosmatos and N. Ghelis	27-35 (1963)
Protection of the thiol group in cysteine with the tertiary butyl residue	A. Chimiak	37 (1963)
Synthesis of α-esters of N-substituted glutamic acid	G.H.L. Nefkens	39-40 (1963)
On the use of the tosyl group for the protection of basic amino-acids	St. Guttmann	41-47 (1963)
Behaviors of protected peptides in the reaction with sodium in liquid ammonia (in German)	S. Bajusz and K. Medzihradszky	49-52 (1963)
II. Synthesis: Methods of Coupling		
Methods of coupling: Progress since 1958	T. Wieland	59-68 (1963)
Use of some new active esters in peptide synthesis	J. Pless	69-72 (1963)
Alkoxylvinyl esters as coupling reagents in peptide synthesis	D. Cohen and H.D. Springall	73-76 (1963)
Side reactions in the azide synthesis of peptides (in German)	E. Schnabel	77-82 (1963)

N-Carboxy-N-trityl- -amino acid anhydrides in peptide synthesis	H. Block and M.E. Cox	83-87 (1963)
Synthesis of proline peptides in clarification of side reactions (in German)	E. Wunsch	89-91 (1963)
Contribution to the reaction mechanism of coupling with dicyclohexylcarbodiimide (in German)	L. Kisfaludy and M. Löw	93-94 (1963)
Influence of substituted benzyloxycarbonyl groups on racemizations occurring in coupling with dicyclohexylcarbodiimide (in German)	L. Kisfaludy	95 (1963)
Gas chromatographic studies of racemization in peptide syntheses (in German)	F. Weygand, A. Prox, L. Schmidhammer and W. König	97-107 (1963)
Detection of diastereoisomers in dipeptides by paper chromatography	E. Taschner, A. Chimiak, J.F. Biernat et al.	109-111 (1963)
Application of the "two spots" method for the detection of racemization during various processes used in the synthesis of peptides	E. Taschner, J.F. Biernat and T. Sokolowska	113-117 (1963)
Further studies of racemization during peptide synthesis	M.W. Williams and G.T. Young	119-121 (1963)
New work on insertion reactions	M. Brenner	123-128 (1963)

III. Synthesis: Special Problems with Uncommon Amino-acids; Abnormal Peptides

Synthesis of peptides with some "unnatural" amino-acids	J. Rudinger	133-164 (1963)
Synthesis of optically active N-methylated amino-acids	P. Quitt	165-169 (1963)
Peptide syntheses with some S-(β-aminoethyl)-L-cystein (in German)	P. Hermann and C.A. Gründig	171-175 (1963)
Alkali instability as a complicating factor in work with 4-oxalysine	G.I. Tesser	177-180 (1963)
Synthesis of peptide derivatives of hydrazinoacetic acids (in German)	H. Niedrich	181-187 (1963)

Synthesis of depsopeptide analogues of ophthalmic acid and glutathione	L.A. Shchukina and A.L. Jouse	189-193 (1963)
Peptolide syntheses (in German)	E. Schröder and K. Lübke	195-206 (1963)
Synthesis of cyclic desipeptides	Yu.A. Ovchinnikov, V.T. Ivanov, A.A. Kiryushkin and M.M. Shemyakin	207-219 (1963)
Tautomerism of hydroxyacylcyclopeptides	V.K. Antonov, A.M. Shkrob and M.M. Shemyakin	221-228 (1963)
A synthetic cyclol tripeptide	D.S. Jones, G.W. Kenner and R.C. Sheppard	229 (1963)
Properties of model compounds related to glycopeptides	V.A. Derevitskaya, L.M. Likhosherstov and N.K. Kochetkov	231-237 (1963)
Preparation of glycosyl esters of amino acids (in German)	L. Kisfaludy and M. Low	239-240 (1963)

IV. Chemical and Physical Properties of Peptides

Dielectric increments and conformations of some simple peptides in aqueous solution	P.M. Hardy, G.W. Kenner and R.C. Sheppard	245-251 (1963)
Non-enzymatic cleavage of peptide chains at the cysteine and serine residues	A. Patchornik and M. Sokolovsky	253-257 (1963)

V. Nomenclature

Report of the committee on nomenclature	Introduced by G.T. Young	261-269 (1963)

PERSPECTIVES IN BIOLOGY

This is a collection of papers dedicated to Bernardo A. Houssay. Edited by C.F. Cori, V.G. Foglia, L.F. Leloir and S. Ochoa and published by Elsevier Publishing Company, Amsterdam-London-New York.

Title	Authors	Reference
Anti-androgenic compounds	R.I. Dorfman	43-55 (1963)
Anti-progestins	G. Pincus and A.P. Merrill	56-61 (1963)
Neurophysin, oxytocin and desaminooxytocin	J.E. Stouffer, D.B. Hope and V. Du Vigneaud	75-80 (1963)

PHYSICAL METHODS IN HETEROCYCLIC CHEMISTRY

This is a collection of review articles concerning physical methods used in the study of heterocyclic chemistry. The volumes are edited by A.R. Katritzky and published by Academic Press, New York-London.

Ionization constants	A. Albert	1,	1-108	(1963)
Heteroaromatic reactivity	J. Ridd	1,	109-160	(1963)
X-ray diffraction studies of heterocyclic compounds	W. Cochran	1,	161-175	(1963)
The solubility of heterocyclic compounds	W. Pfleiderer	1,	177-188	(1963)
Application of dipole moments to heterocyclic systems	S. Walker	1,	189-215	(1963)
Electrochemical properties in solutions	J. Volke	1,	217-323	(1963)
The electronic absorption spectra of heterocyclic compounds	S.F. Mason	2,	1-88	(1963)
Nuclear quadrupole resonance	E.A.C. Lucken	2,	89-102	(1963)
Nuclear magnetic resonance spectra	R.F.M. White	2,	103-159	(1963)
Infrared spectra	A.R. Katritzky and A.P. Ambler	2,	161-360	(1963)

PHYSICAL METHODS IN ORGANIC CHEMISTRY

This is a collective volume, presenting a series of chapters on physical methods in organic chemistry. Edited by J.C.P. Schwarz and published by Oliver and Boyd, Edinburgh-London.

Title	Authors	Reference
Ultraviolet, visible and infrared spectroscopy	P. Bladon and G. Eglinton	22-34 (1964)
Infrared and Raman spectroscopy	G. Eglinton	35-125 (1964)
Ultraviolet and visible spectroscopy	P. Bladon	126-167 (1964)
Magnetic resonance spectroscopy	L.M. Jackman	168-209 (1964)
Optical rotation	J.C.P. Schwarz	210-243 (1964)
The determination of molecular weight	C.T. Greenwood	244-275 (1964)
Diffraction methods	G.A. Sim	276-305 (1964)
Mass spectrometry	R.I. Reed	306-322 (1964)
Dipole moments	B.L. Shaw	323-337 (1964)

PHYSICAL PROPERTIES OF THE STEROID HORMONES

This is a series of comprehensive articles on the physical properties of steroid hormones. Edited by L. L. Engel and published by Pergamon Press (The Macmillan Company) Oxford - New York.

Title	Authors	Reference
Partition coefficients	L. L. Engel and P. Carter	1-36 (1963)
Chromatographic mobilities	R. Neher	37-68 (1963)
Ultraviolet absorption	J. P. Dusza, M. Heller and S. Bernstein	69-287 (1963)
Fluorescence spectra	J. W. Goldzieher	288-320 (1963)
Absorption spectra in concentrated sulfuric acid	L. L. Smith and S. Bernstein	321-448 (1963)

PROCEEDINGS OF THE FIRST INTERNATIONAL PHARMACOLOGICAL MEETING

This meeting was held in Stockholm, Sweden, August 22-25, 1961. The ten volumes of the Proceedings cover the eight main symposia, plenary lectures and abstracts of papers. Discussions of papers are included. The theme of the symposia was " Mode of Action of Drugs ". Selection of topics was made with their interdisciplinary relevance in mind and with emphasis on biochemical pharmacology. The general editing was carried out by B. Uvnas and P. Lindgren acted as assistant editor. The historical background to this first international pharmacological meeting is summarized by C.F. Schmidt in Vol. 1, pp. xii-xvii. The volumes were published by Pergamon Press. Lists of authors, with their addresses are given in the preliminary pages of each volume. Volume 1 was published in 1962 and the succeeding volumes, as noted below. Extensive bibliographies are included. Each volume also contains a general subject index.

Title	Authors	Reference
Plenary Session. Part 1.		1, xi-xxviii (1962)
Pharmacological Control of Release of Hormones Including Antidiabetic Drugs. Part 2.	R. Guillemin and P. Lindgren (Editors)	
Action of neurotropic drugs upon ACTH secretion	S.V. Anichkov A.N. Poskalenko and V.E. Ryzhenkov	1, 1-9 (1962)
Pharmacological control of the secretion of ACTH	P.L. Munson	1, 11-25 (1962)
Mechanisms by which drugs and hormones activate and block release of pituitary gonadotropins	C.H. Sawyer	1, 27-46 (1962)
Pharmacological actions of the hypoglycaemic arylsulphonamides upon the histophysiology and the physiology of the beta cells of the islets of Langerhans of the pancreas	A. Loubatières	1, 47-89 (1962)
Experimental investigations with sulphonylureas	A. Bänder	1, 91-107 (1962)
Pharmacological effects of various chemical substances on the biosyntheses and secretion of thyroid hormone (in French)	R. Michel	1, 109-129 (1962)
Action of some derivatives of aminoethyl-2-benzo-1,4-dioxane on the hormonal equilibria cycle in pseudo pregnancy and pregnancy in the rat (in French)	F. Bovet-Nitti and G. Bignami	1, 131-149 (1962)
Pharmacological control of prolactin secretion and lactation	J. Meites	1, 151-181 (1962)

Selected Additional Papers

Beta-phenethylbiguanide degradation products	A.N. Wick J.L. Faymon and C.J. Stewart	1, 185-188 (1962)

FIRST INTERNATIONAL PHARMACOLOGICAL MEETING (Contd.)

Title	Authors	Vol.	Pages	Year
Effects of sulfonylurea and biguanide compounds on hepatectomized, pancreatectomized and hepato-pancreatectomized dogs	S. Ohashi, M. Tobe	1,	189-190	(1962)
Mode of action of 5-hydroxytryptamine and iproniazid on pregnancy	M. Botros, D. Lindsay, E. Poulson and J.M. Robson	1,	191-194	(1962)
Antigonadotrophic activity of some progestinics given by mouth	A. Ercoli and G. Falconi	1,	195-196	(1962)
Qualitative and quantitative changes brought about in the biological activities of steroids by esterification or etherification	A. Ercoli	1,	197-201	(1962)
The neurogenic influence of dogs and cats on the corticosteroid content in rat plasma. The effect of psychopharmaca and anaesthetics	A. Pekkarinen, E. Tala, N. Niemelä and E. Sotaniemi	1,	203-213	(1962)
The neurogenic influence of dogs on the excretion of corticosteroids in Guinea pigs	N. Niemelä, A. Pekkarinen, E. Sotaniemi and E. Tala	1,	215-228	(1962)
Effects of Drugs on Synthesis and Mobilization of Lipids	E.C. Horning and P. Lindgren (Editors)			
Various mechanisms affecting bile acid metabolism	S. Bergström	2,	1-12	(1963)
Biochemical processes involved in the synthesis, accumulation and release of triglycerides by the liver	M.G. Horning, M. Wakabayashi and H.M. Maling	2,	13-27	(1963)
Drugs as tools in the study of nervous system regulation of release of free fatty acids from adipose tissue	R. Paoletti, R.P. Maickel, R.L. Smith and B.B. Brodie	2,	29-41	(1963)
Transport of fatty acids in the blood: Pathways of transport and the role of catecholamines and the sympathetic nervous system	R.J. Havel	2,	43-65	(1963)
Selected Papers		2,	67-209	(1963)
New aspects of cardiac glycosides	W. Wilbrandt and P. Lindgren (Editors)			
The stereochemistry of the glycosides in relation to biological activity	Ch. Tamm	3,	11-26	(1963)

FIRST INTERNATIONAL PHARMACOLOGICAL MEETING (Contd.)

Title	Author(s)	Vol.	Pages	Year
Possibilities of further developments in the glycoside field by modifying the glycoside structure	K.K. Chen	3,	27-45	(1963)
Metabolism of cardiac glycosides	K. Repke	3,	47-73	(1963)
The effect of digitalis on cardiac metabolism	R.J. Bing	3,	75-85	(1963)
Action of cardiac glycosides on automaticity and contractile activity in single cardiac muscle cells grown in vitro (A slow motion picture)	A. Wollenberger and W. Halle	3,	87-110	(1963)
Cardiac glycosides and ion transport	J.B. Kahn, Jr.	3,	111-135	(1963)
Cardiac glycosides and calcium	A. Farah and P.N. Witt	3,	137-171	(1963)
Cardiac glycosides and actomyosin	P. Waser	3,	173-184	(1963)
Relation of cations to the inotropic and metabolic actions of cardiac glycosides	K.S. Lee	3,	185-201	(1963)
Selected Papers		3,	203-287	(1963)
Drugs and Membranes	C.A.M. Hogben and P. Lindgren (Editors)	4,	1-180	(1963)
Methods for the Study of Pharmacological Effects at Cellular and Subcellular Levels	O.H. Lowry and P. Lindgren (Editors)	5,	5-252	(1963)
Metabolic Factors Controlling Duration of Drug Action	B.B. Brodie, E.G. Erdös, R.P. Maickel and P. Lindgren (Editors)			
Detoxication mechanisms in vivo	R.T. Williams	6,	1-12	(1963)
Oxidation and reduction by microsomal enzymes	J.R. Gillette	6,	13-29	(1963)
Relations between structure, velocity of biological n-hydroxylation and toxicity of aromatic amines	H. Uehleke	6,	31-37	(1963)
Glucuronide conjugation	G.J. Dutton	6,	39-51	(1963)
Sulfate conjugation	J.D. Gregory	6,	53-64	(1963)
Mercapturic acid conjugation	E. Boyland	6,	65-76	(1963)

FIRST INTERNATIONAL PHARMACOLOGICAL MEETING (Contd.)

Title	Author(s)	Vol.	Pages	Year
Mechanisms of enzymatic acylation and peptide bond formation	A. Meister	6,	77-89	(1963)
Comparative studies on sulfanilamide acetylation: an inhibitor in dog liver	K.C. Leibman and A.M. Anaclerio	6,	91-96	(1963)
Demethylation and methylation of drugs and physiologically active compounds	J. Axelrod	6,	97-110	(1963)
Desulfuration	R.D. O'Brien	6,	111-119	(1963)
Enzymatic conversion of cyanide to thiocyanate	B. Sorbo	6,	121-136	(1963)
Esterase action	W. Kalow	6,	137-147	(1963)
Solubilization and isolation of drug-hydrolysing enzymes from microsomes	G. Hollunger and B. Niklasson	6,	149-158	(1963)
Enzymes that inactivate polypeptides	E.G. Erdös	6,	159-178	(1963)
Urinary excretion of drugs	L. Peters	6,	179-191	(1963)
Duration of action of drugs as affected by tissue distribution	T.C. Butler	6,	193-212	(1963)
Drugs as inhibitors of drug metabolism	K.J. Netter	6,	213-233	(1963)
Drugs as activators of drug enzymes	H. Remmer	6,	235-256	(1963)
Physiological impairment of drug metabolism	J.R. Fouts	6,	257-275	(1963)
Species differences and individual variations in drug metabolism	J.J. Burns	6,	277-288	(1963)
Biological inactivation by amine oxidases and time courses of drug action	H. Blaschko	6,	289-298	(1963)
Comparative biochemistry of drug metabolism	B.B. Brodie and R.P. Maickel	6,	299-324	(1963)
Modern Concepts in the Relationship Between Structure and Pharmacological Activity	K.J. Brunings and P. Lindgren (Editors)			
Size and shape of organic molecules and biological activity	A.H. Beckett	7,	5-34	(1963)

Chemical and physical properties and biological activity	A. Burger	7,	35-74 (1963)
An analysis of drug-receptor interactions	B. Belleau	7,	75-99 (1963)
Nature of the cholinergic receptor	P. Waser	7,	101-118 (1963)
Isolation and properties of a drug binding protein from electric tissue of electric eel	S. Ehrenpreis	7,	119-143 (1963)
Protein nature of acetylcholine receptor and the role of energy rich compounds in the realization of acetylcholine action	T.M. Turpajev S.N. Nistratova and T.G. Putintseva	7,	145-155 (1963)
Introduction: Protein and polypeptide structure	A.H. Theorell	7,	157-159 (1963)
Biological specificity in protein-small molecule interactions	D.E. Koshland, Jr.	7,	161-192 (1963)
Inhibitors for liver alcohol dehydrogenase	A.H. Theorell	7,	193-201 (1963)
Structure-activity relationship among physiologically active peptides	R. Schwyzer	7,	203-246 (1963)
Selected Papers		7,	247-382 (1963)
Pharmacological Analysis of Central Nervous Action	W.D.M. Paton and P. Lindgrén (Editors)	8,	1-324 (1962)
Bradykinin and Vaso-Dilating Polypeptides. Part 1.	M.R.E. Silva and U.S. Von Euler (Editors)	9,	1-94 (1962)
Pharmacology of the Lung. Part 2.	D.M. Aviado (Editor)	9,	97-188 (1962)
Abstracts	P. Lindgrén (Editor)	10,	1-175 (1963)

PROGRESS IN BIOPHYSICS AND MOLECULAR BIOLOGY

This is a collective volume, in English, given to review articles in the field of biological structure and function at the macromolecular level and has been active in relation to heredity, protein synthesis and to muscular contraction. The preceding 12 volumes in the series were titled Progress in Biophysics and Biophysical Chemistry. While there is no change in basic policy intended, the new title reflects more accurately the type of articles included. Edited by J.A.V. Butler, H.E. Huxley and R.E. Zirkle and published by Pergamon Press (The Macmillan Company) New York.

Title	Authors	Reference
Radiobiological mechanisms: Comparative distribution and role of ionization, excitation, and energy and charge migration	L.G. Augenstein	13, 1-57 (1963)
Radiation chemistry of aqueous solutions of nucleic acids and nucleoproteins	G. Scholes	13, 59-104 (1963)
X-ray small angle scattering with substances of biological interest in diluted solutions	O. Kratky	13, 107-173 (1963)
Enzymatic synthesis of nucleic acids	M. Grünberg-Manago	13, 175-239 (1963)
Photosynthesis	L.N.M. Duysens	14, 1-104 (1964)
The mechanism of action and the active centre of the alcohol dehydrogenases	J.S. McKinley-McKee	14, 223-262 (1964)
The physical chemistry of phospholipids	D.G. Dervichian	14, 263-342 (1964)

This is a collective volume on the Composition, Structure and Function of Proteins. Second revised and augmented edition. Edited by H. Neurath and published by Academic Press, Inc., New York-London.

Title	Authors	Reference		
Amino acid analysis of peptides and proteins	A. Light and E. L. Smith	1,	1-44	(1963)
Amino acid composition of certain proteins	G. R. Tristram and R. H. Smith	1,	45-51	(1963)
Synthesis and function of peptides of biological interest	K. Hofmann and P. G. Katsoyannis	1,	53-188	(1963)
Chemical aspects of protein synthesis	J. S. Fruton	1,	189-310	(1963)
Concepts and experimental approaches in the determination of the primary structure of proteins	R. E. Canfield and C. B. Anfinsen	1,	311-378	(1963)
Intramolecular bonds in proteins. I. The role of sulfur in proteins	R. Cecil	1,	379-476	(1963)
Intramolecular bonds in proteins. II. Noncovalent bonds	H. A. Scheraga	1,	477-594	(1963)

These are the proceedings of the Third International Symposium held at the Institut fur Organische Chemie der Technischen Hochschule, Stuttgart, September, 1962. Edited by W. Pfleiderer and E.C. Taylor and published by Pergamon Press (The Macmillan Company) New York.

Title	Authors	Reference
The origin of pterin chemistry (in German)	C. Schöpf	3-14 (1964)
Pteridine syntheses with sugars and 2-trifluoromethyl-4-alkyl-5-oxazolones (in German)	F. Weygand, H. Simon, K.D. Keil et al.	15-28 (1964)
A new route to the synthesis of the pyrazine ring (in German)	P. Schmidt, K. Eichenberger and M. Wilhelm	29-35 (1964)
Some aminopteridines (in English)	J. Weinstock and V.D. Wieblhaus	37-46 (1964)
New approaches to the synthesis of 6- and 7-substituted pteridines and their 5-oxides (in English)	I.J. Pachter	47-64 (1964)
Some biologically active pteridines derived from 4-amino-5-nitrosopyrimidines (in English)	T.S. Osdene	65-73 (1964)
Concerning the preparation and bromination of acetonyl-xanthopterin and its homologs (in German)	F. Korte and R. Wallace	75-86 (1964)
Syntheses and structure of pteridine-glycosides and their possible biologic significance (in German)	W. Pfleiderer, R. Lohrmann, F. Reisser and D. Söll	87-110 (1964)
Covalent hydration of pteridines (Hydration of C=N bonds) (in English)	A. Albert	111-128 (1964)
The addition reactions of 2-amino-4-hydroxy-7,8-dihydropteridine and their biochemical significance (in English)	H.C.S. Wood, T. Rowan and A. Stuart	129-141 (1964)
Metal complex formation and nucleophilic additions with 8-alkyl-4-pteridones (in German)	P. Hemmerich	143-167 (1964)
Proton resonance spectra of pteridines. Part 1. (in German)	W. von Philipsborn, H. Stierlin and W. Traber	169-180 (1964)
Synthesis and properties of 5,6- and 5,8-dihydropteridine isomers (in English)	E.C. Taylor, M.J. Thompson and W. Pfleiderer	182-210 (1964)

The alkylation of 2-amino-4-hydroxypteridines (in English)	R.B. Angier	211-218 (1964)
The ease of rearrangement of aminopteridines and aminopyrimidines alkylated on ring nitrogen (in English)	D.J. Brown and J.S. Harper	219-232 (1964)
Separation of synthetic biopterin and isobiopterin (in German)	R. Tschesche B. Hess I. Ziegler and H. Machleidt	233-241 (1964)
Isolation of pteridines by ionic exchange chromatography (in German)	H. Rembold and L. Buschmann	243-254 (1964)
Crithidia assays for unconjugated pteridines (in English)	H.N. Guttman	255-266 (1964)
Considerations of the constitutions of sepia pterin and drosopterins (in German)	M. Viscontini	267-279 (1964)
Recent work on the structure of isosepiapterin and the drosopterins and its relation to pteridene biosynthesis (in English)	H.S. Forrest and S. Nawa	281-293 (1964)
Naturally-occurring tetrahydropterins (in German)	I. Ziegler	295-305 (1964)
Pteridine transformations during the enzymatic conversion of phenylalanine to tyrosine (in English)	S. Kaufman	307-326 (1964)
Further studies of the biogenesis of leucopterins (in German)	H. Simon H. Wacker and J. Walter	327-341 (1964)
The role of 5-methyl tetrahydrofolate in the enzymatic synthesis of methioninemethyl (in English)	J.M. Buchanan A.R. Larrabee S. Rosenthal and R.E. Cathou	343-354 (1964)
Tetrahydrofolic acid — The coenzyme of one-carbon metabolism (in English)	F.M. Huennekens and K.G. Scrimgeour	355-376 (1964)
Pterin phosphate as intermediate in the biosynthesis of folic acid (in German)	L. Jaenicke	377-402 (1964)
Biosynthesis of folic acid and of biopterins	A. Wacker E. Lochmann and S. Kirschfeld	403-408 (1964)
Analogs of folic acid (in German)	A. Wacker E. Lochmann and L. Träger	409-415 (1964)

Folic acid antagonists based on some deazapteridine ring systems (in English)	O.D. Bird V. Oakes K. Undheim and H.N. Rydon	417-426 (1964)
The enzymatic conversion of pteridines into xanthine-8-carboxylic acid (in English)	W.S. McNutt	427-441 (1964)
Enzymic formation of riboflavin (in English)	G.W.E. Plaut	443-463 (1964)
Studies of the metabolism of biopterins and the polarographic characterization of pteridines (in German)	H. Rembold	465-484 (1964)
Tetrahydropteridine and melanophorene differentiation in fish (in German)	I. Ziegler	485-493 (1964)
The role of pteridines in pigmentation (in English)	T. Hama and S. Fukuda	495-505 (1964)
Methods suitable for the elemental microanalysis of the pteridines (in English)	J.E. Fildes	507-509 (1964)
Application of polarography in the studies of pteridines (in German)	J. Komenda	511-520 (1964)

RAPID MIXING AND SAMPLING TECHNIQUES IN BIOCHEMISTRY

This is a collective volume dealing with experimental techniques and their interpretations and uses in biochemistry. Edited by B. Chance, Q.H. Gibson, R.H. Eisenhardt and K.K. Lonberg-Holm and published by Academic Press, New York - London.

Title	Authors	Reference	
The origin of the Hartridge-Roughton rapid reaction velocity method	F.J.W. Roughton	5-13	(1964)
Photochemical activation apparatus using flash tubes	C. Greenwood	157-164	(1964)
Photochemical activation apparatus with optical maser	D. DeVault	165	(1964)
Quenching by squirting into cold imiscible liquids	R.C. Bray	195-203	(1964)
A comparison of the resolution of chemical and optical sampling	T.E. Barman and H. Gutfreund	339-348	(1964)

RODD'S CHEMISTRY OF CARBON COMPOUNDS

This is the revision of the first volume of the well-known series, edited originally by E.H. Rodd. The present editor is S. Coffey and published by Elsevier Publishing Company, Amsterdam-London-New York.

Title	Authors	Reference
The historical development of the structural chemistry of carbon compounds	E.H. Rodd	1, 2-20 (1964)
The classification, nomenclature and literature of carbon compounds	D.F. Styles	1, 21-33 (1964)
The quantitative analysis of carbon compounds and the determination of physical constants and molecular weights	R. Belcher, A.M.G. Macdonald and W.I. Stephen	1, 34-69 (1964)
The physical properties of carbon compounds	H.D. Springall	1, 70-92 (1964)
The crystallography of carbon compounds	A.F. Wells	1, 93-104 (1964)
The absorption of light by carbon compounds	J.E. Page	1, 105-147 (1964)
Isotopically labelled carbon compounds	H.R.V. Arnstein	1, 148-190 (1964)
Modern physico-chemical views on acids and bases and their application to carbon compounds	H.D. Springall	1, 191-200 (1964)
Stereochemistry of carbon compounds	I.G.M. Campbell	1, 201-259 (1964)
Mechanisms of reactions of aliphatic compounds	D.V. Banthorpe	1, 260-305 (1964)
Free radical and homolytic reactions	D.H. Hey and W.A. Waters	1, 306-333 (1964)
Wave mechanics of carbon compounds	W.B. Brown	1, 334-356 (1964)
The saturated or paraffin hydrocarbons. Alkanes	J.D. Downer and K.I. Beynon	1, 357-397 (1964)
Unsaturated acyclic hydrocarbons	R.F. Garwood	1, 398-477 (1964)
Halogen derivatives of the aliphatic hydrocarbons	W.K.R. Musgrave	1, 478-526 (1964)

THE ROYAL INSTITUTE OF CHEMISTRY OF GREAT BRITAIN AND IRELAND
Lectures, Monographs and Reports

This is a collection of review articles published by the Institute

Title	Authors	Reference	
Living molecules (12th Dalton Lecture)	J.N. Daidson	<u>1</u>,	1-17 (1963)
Quantitative evaluation of substituent effects by electronic spectroscopy (Meldola Medal Lecture, 1962)	J.N. Murrell	<u>2</u>,	1-9 (1963)
Quantitative aspects of aromatic substitution (Meldola Medal Lecture, 1962)	R.O.C. Norman	<u>2</u>,	11-32 (1963)
Agriculture and some of its chemical problems (5th P.F. Frankland Memorial Lecture)	E.G. Cox	<u>3</u>,	3-12 (1963)
The petroleum chemicals industry: Present status and future trends	H.M. Stanley	<u>4</u>,	1-25 (1963)
Photosynthesis	J.L. Bailey	<u>5</u>,	1-30 (1963)
Chemical applications of the shock tube (Meldola Medal Lecture, 1961)	J.N. Bradley	<u>6</u>,	1-29 (1963)
Discovery of polythene	R.D. Gibson	<u>1</u>,	1-30 (1964)
Crystal structure studies of aromatic hydrocarbons (Meldola Medal Lecture, 1963)	J. Trotter	<u>2</u>,	1-29 (1964)

THEORY AND STRUCTURE OF COMPLEX COMPOUNDS

These are the proceedings of the Symposium held in Wroclaw, Poland, 15-19 June, 1962. The symposium was organized by the Institute of Physical Chemistry of the Polish Academy of Sciences, the University and Technical University of Wroclaw in collaboration with the Polish Chemical Society. Edited by B. Jeżowska-Trzebiatowska and published by Pergamon Press, Oxford-New York.

Title	Authors	Reference
The role of oxygen in coordination compounds	B. Jeżowska-Trzebiatowska	1-11 (1964)
Chemistry and photochemistry of iron and manganese porphyrins	M. Calvin, P.A. Loach and A. Yamamoto	13-24 (1964)
Coordination polymerization in metal alkoxides	D.C. Bradley	87-93 (1964)
Acid-base complexes and fine structure of the OH stretching vibration band in the hydrogen bonding	H. Ratajczak and L. Sobczyk	229-232 (1964)
Anhydrous uranyl salts in organic solvents	B. Jeżowska-Trzebiatowska, A. Bartecki and B. Kędzia	271-275 (1964)
Structure of uranyl complexes with β-diketones (in Russian, English abstract)	B. Jeżowska-Trzebiatowska and W. Kakołowicz	277-281 (1964)
The magnetic properties of "addition" compounds of chromyl chloride with hydrocarbons	A. Bartecki and S. Wajda	305-309 (1964)
Structure investigation of metal-diphenylcarbazone complexes by infrared spectroscopy	W. Kemula and A. Janowski	321-325 (1964)
Coordination chemistry of o-cyclohexyliminediacetic acid (I) and o-hydroxyanilinediacetic acid (II). (in German)	L. Wolf, W. Matzel and H. Schreiber	349-358 (1964)
Extraction studies on the interaction between $GeCl_4$ molecules and organic diluent	S. Siekierski and R. Olszer	475-478 (1964)
Complexes of chromium (III) with DL-α-alanine	J. Starosta and L. Pajdowski	599-604 (1964)
On thermodynamics of charge-transfer complexes (in German)	G. Junghähnel	697-707 (1964)

THERMAL STABILITY OF POLYMERS

These are the proceedings of the Battelle symposium on Thermal Stability of Polymers sponsored by the Batelle Memorial Institute, Columbus, Ohio, December 5-6, 1963.

Title	Authors	Reference
The mechanism of thermal decomposition for polystyrene	L.A. Wall	A 1
The thermal and photochemical degradation of polymethacrylates	N. Grassie	B 1 - B 20
Thermal degradation of polytetrafluoroethylene in a closed system	H.H.G. Jellinek and H. Kachi	C 1 - C 14
Thermal degradation of polymethylmethacrylate in a closed system	J.E. Clark and H.H.G. Jellinek	D 1 - D 24
Oxidative mechanisms for the degradation of phenol, furan and urea based condensation polymers	R.T. Conley	E 1 - E 68
Mechanisms of thermal degradation of phenolic condensation polymers. II. Thermal stability and degradation schemes of epoxy resins	L.-H. Lee	F 1 - F 23
A method for determining the fundamental and practical behavior of polymers at high temperatures	W.J. Tabor	G 1 - G 22
Thermal decomposition of gamma irradiated fluoropolymers	L.A. Wall and S. Straus	H 1
Thermal stability of chlorinated atactic polypropylene polyvinyl chloride blends	G.A. Grode, W.R. Dunnavant, R.W. Pfeil and J.W. Brasch	I 1 - I 18
Oxidation studies on acrylic polymers	R.T. Conley	J 1 - J 24
The ladder principle — Polymers from vinyl isocyanate	C.G. Overberger, S. Ozaki and H. Mukamal	K 1 - K 2
Degradation of polycarbonates. III. Viscometric study of thermally-induced chain scission	A. Davis and J.H. Golden	L 1 - L 13
Inhibited autoxidation of squalane	J.C.W. Chien	M 1 - M 4
Metal-organic stabilizers and antistabilizers for polyolefin plastics	G.C. Newland and J.W. Tamblyn	N 1 - N 7

THERMAL STABILITY OF POLYMERS (Contd.)

Title	Author(s)	Pages
Some recent advances in the stabilization of polyolefins against thermal oxidation	W.L. Hawkins	O 1 - O 9
New inhibitors of copper-catalyzed oxidation of polypropylene	C.E. Tholstrup	P 1 - P 8
Antioxidant mechanisms in elastomers	J.R. Shelton	Q 1 - Q 2
Thermally stable boron containing polymers	J. Green, M.M. Fein, F.J. Loprest et al.	R 1 - R 18
Synthesis, structure and stability of some polyphosphinates and polyphosphine oxides	E. Steininger and M. Sander	S 1 - S 18
Thermal stability of diolefin cyclopolymers	W.S. Anderson	T 1 - T 10
Mechano-thermal degradation of polyethylene effect on molecular weight distribution	P.M. Kamath, W. Wild and S.G. Howell	U 1 - U 10
Investigation of comonomer distribution in copolymers with thermal methods	K.J. Bombaugh and B.H. Clampitt	V 1 - V 12

These are the proceedings of the symposium, mainly in English, held at the Istituto Superiore di Sanita, Rome, 2-3 May, 1963. Edited by G.B. Marini-Bettolo and published by Elsevier Publishing Company, Amsterdam-London-New York.

Title	Authors	Reference
Development and application of thin-layer chromatography	E. Stahl	1-13 (1964)
Thin-layer chromatography on cellulose plates (in German)	P. Wollenweber	14-31 (1964)
Thin-layer chromatography on loose layers of alumina	L. Lábler	32-44 (1964)
Thin-layer chromatography of natural and synthetic odorants (in French)	E. Demole	45-54 (1964)
Thin-layer electrophoresis	G. Grassini	55-68 (1964)
Mass spectrometry and thin-layer chromatography (in French)	M. Fétizon	69-74 (1964)
Thin-layer chromatography of steroids	R. Neher	75-86 (1964)
Thin-layer chromatography of lipids	F.B. Padley	87-115 (1964)
Chromatography on thin layer of starch with reversed phases	J. Davidek	117-118 (1964)
Centrifugal chromatography. XII. Centrifugal thin-layer chromatography	J. Rosmus, M. Pavlíček and Z. Deyl	119-121 (1964)
Application of circular chromatography on silica gel thin layers. Separation and purification of morphine, papaverine and quinine (in German)	M. von Schantz	122-131 (1964)
Thin-layer chromatography of polynuclear aromatic hydrocarbons (in German)	H.-J. Petrowitz	132-137 (1964)
Thin-layer chromatography of 2,4-dinitrophenyl-hydrazones of aliphatic carbonyl compounds and their quantitative determination	G.M. Nano	138-143 (1964)
Thin-layer chromatography of steroidal bases and holarrhena alkaloids	L. Lábler and V. Cerny	144-148 (1964)
Thin-layer chromatography of alkaloids on magnesia chromatoplates	E. Ragazzi, G. Veronese and C. Giacobazzi	149-154 (1964)

THIN-LAYER CHROMATOGRAPHY (Contd.)

Some applications of thin-layer chromatography for the separation of alkaloids	G. Grandolini C. Galeffi E. Montalvo C.G. Cainovi and G.B. Marini-Bettòlo	155-159 (1964)
Thin-layer chromatography of isomeric oximes. II.	I. Pejković-Tadić M. Hranisavljevic-Jakovljevic and S. Nešić	
Qualitative and quantitative analysis of natural and synthetic corticosteroids by thin-layer chromatography	G. Cavina and C. Vicari	180-194 (1964)
Thin-layer chromatography and the detection of stilboestrol	C. Bonino	195-196 (1964)
Applications of thin-layer chromatography on sephadex to the study of proteins	P. Fasella A. Giartosio and C. Turano	205-211 (1964)
Thin-layer chromatography on silica gel of food colours	M.A. Ciasca and C.G. Casinovi	211-214 (1964)

These are the proceedings of the Symposium on Vacuum Microbalance Techniques, held in Los Angeles, California, October 30, 1962. Edited by K.H. Behrndt and published by Plenum Press.

Title	Authors	Reference
The Cahn recording gram electrobalance	L. Cahn and H.R. Schultz	3, 29-44 (1963)
On some modifications of a torsion microbalance for use in ultrahigh vacuum	H. Mayer, R. Niedermayer, W. Schroen et al.	3, 75-107 (1963)
Application of an automatic recording vacuum microbalance to the study of catalyst surfaces	H.L. Gruber and C.S. Shipley	3, 131-151 (1963)
Microbalance techniques for high-temperature applications. Further developments	N.J. Carrera and R.F. Walker, C.A. Steggerda and W.M. Nalley	3, 153-177 (1963)
Measuring radiation with the Cahn automatic electrobalance	Y. Ohman and B. Rydgren	3, 193-198 (1963)

VITAMINE
(Chemie und Biochemie)

This is a collective volume, in German, of review articles on selected aspects of the chemistry and biochemistry of the vitamins. Edited by J. Fragner and published by VEB Gustav Fischer Verlag, Jena.

Title	Authors	Reference
Nomenclature and classification of the vitamins	J. Fragner	$\underline{1}$, 41-66 (1964)
Relation between chemical constitution and biological activities	J. Fragner	$\underline{1}$, 67-78 (1964)
Biogenesis of the vitamins	J. Kostíř, J. Fragner, K. Slavik and V. Šicho	$\underline{1}$, 79-86 (1964)
Vitamin A group	J. Fragner and K. Hejna	$\underline{1}$, 188-195 (1964)
Chemistry of provitamin	K. Hejna, J. Blattná and J. Fragner	$\underline{1}$, 195-348 (1964)
p-Aminobenzoic acid	J. Fragner, E. Knobloch, J. Jiršak, V. Jelinek, J. Blattna and K. Slavik	$\underline{1}$, 349-382 (1964)
L-Ascorbic acid	Z. Procházka, V. Šanda, V. Jelinek, K. Slavik, K. Ponec and Z. Müller	$\underline{1}$, 383-486 (1964)
Bioflavonoids	J. Fragner, J. Davidek, V. Jelinek, V. Jiraček and K. Ponec	$\underline{1}$, 487-549 (1964)
Biotin	J. Fragner, V. Horák, A. Šimek, V. Jelinek, J. Blattná, K. Slavik and Z. Müller	$\underline{1}$, 552-612 (1964)

VITAMINE (Contd.)

Calciferol		V. Janata J. Fragner E. Knobloch V. Jelinek V. Sicho V. Schreiber J. Blattna and Z. Muller	1, 614-702 (1964)
Choline		J. Kostir J. Fragner E. Knobloch A. Simek V. Jelinek and V. Schreiber	1, 703-737 (1964)
Cortinoids		V. Rabek E. Knobloch A. Simek V. Jelinek K. Slavik V. Schreiber and Z. Muller	1, 740-838 (1964)
Essential fatty acids		J. Fragner J. Zemlicka V. Jelinek J. Blattna V. Sicho and K. Ponec	1, 839-890 (1964)

PART III
MONOGRAPHS ON ORGANIC CHEMISTRY
1963-1964
Pages 221-235

MONOGRAPHS

Title	Authors	Publishers
Modern chemical engineering (2d edition)	A. Acrivos	Chapman and Hall, London (1963)
Instrumentation for process measurement and control	N.A. Anderson	Instruments Publishing Co., Pittsburgh (1964) 79 pp.
Molecular complexes in organic chemistry	L.J. Andrews and R.M. Keefer	Holden-Day, Inc., San Francisco (1964) vii + 196 pp.
Index of biologically active steroids (Vol. II of Steroid Drugs)	N. Applezweig	Holden-Day, Inc., San Francisco (1964) x + 449 pp.
Catalyses and inhibition of chemical reactions	P.G. Ashmore	Butterworths, Inc., Washington, D.C. (1963) 375 pp.
Titrimetric organic analysis. Vol. XV. Part I. Direct methods	M.R.F. Ashworth	Interscience Pub.(John Wiley & Sons, Inc.) (1964) xx + 501 pp.
Applications of neutron diffraction in chemistry	G.E. Bacon	Pergamon Press (The Macmillan Co.) New York (1963) xi + 141 pp.
Mechanochemistry of polymers (Trans. by R.J. Moseley and edited by W.F. Watson) (Rubber and Plastics Research Assoc. of Gt. Britain)	N.K. Baramboim	MacLaren and Sons Ltd., London (1964) 261 pp.
The chemistry of metal complexes	F. Basolo and R.C. Johnson	W.A. Benjamin, Inc., New York (1964) 180 pp.
Synthetic fibers in papermaking	O.A. Battista	Wiley, New York (1964) 340 pp.
Fundamentals of cost engineering in the chemical industry	H.C. Bauman	Chapman and Hall, London (1964) 416 pp.
Physical and chemical methods of separation	E.W. Berg	McGraw-Hill, New York (1963) xiii + 366 pp.
Mass and abundance tables for use in mass spectroscopy	J.H. Beynon and A.E. Williams	Elsevier-Amsterdam (1963) xxi + 570 pp.
Applications of NMR spectroscopy in organic chemistry. Illustrations from the steroid field	N.S. Bhacca and D.H. Williams	Holden-Day, Inc. San Francisco (1964) x + 198 pp.
NMR Spectra catalogue (Vol. II)	N.S. Bhacca D.P. Hillis and L.F. Johnson and E.A. Pier	Varian Associates, Palo Alto, Calif. (1963) 331 pp.

MONOGRAPHS (Contd.)

Title	Author(s)	Publisher
Electron probe microanalysis. Vol. XVII	L.S. Birks	Interscience Pub. (John Wiley & Sons) New York (1964) ix + 253 pp.
Thin-layer chromatography	J.M. Bobbitt	Chapman & Hall, London (1963) 208 pp.
Unsaturated polyesters	H.V. Boenig	American Elsevier Pub. Co., New York (1964) 230 pp.
Radioactivation analysis	H.J.M. Bowen and D. Gibbons	Clarendon Press, Oxford (1963) 295 pp.
Interpretation of mass spectra of organic compounds	H. Budzikiewicz C. Djerassi and D.H. Williams	Holden-Day, Inc. San Francisco (1964) viii + 271 pp.
Structure elucidation of natural products by mass spectrometry (Vol. I. Alkaloids)	H. Budzikiewicz C. Djerassi and D.H. Williams	Holden-Day, Inc., San Francisco (1964) vii + 233 pp.
Structure elucidation of natural products by mass spectrometry (Vol. II. Steroids, Terpenoids, sugars and miscellaneous classes)	H. Budzikiewicz C. Djerassi and D.H. Williams	Holden-Day, Inc., San Francisco (1964) x + 306 pp.
Nucleophilic substitution at a saturated carbon atom. Vol. I.	C.A. Bunton	American Elsevier Pub., Co., New York (1963) 181 pp.
Electronic charges of bonds in organic compounds (Trans. from Russian by J.T. Greaves)	G.V. Bykov	Pergamon Press, Oxford (1964) viii + 191 pp.
An introduction to chemical nomenclature (2d Edition)	R.S. Cahn	Butterworths, London 109 pp.
Medicinal chemistry (Vol. VI)	E.E. Campaigne and W.H. Hartung	Wiley, New York (1963) 356 pp.
Colorimetric determination of elements, principles and methods	G. Charlot	American Elsevier Pub., Co., New York (1964) ix + 449 pp.
Identification of organic compounds	N.D. Cheronis and J.B. Entrikin	Wiley, New York (1963) 477 pp.
Semimicro methods of organic functional group analysis	N.D. Cheronis and T.S. Ma	Wiley, New York (1964) 696 pp.
Polycyclic hydrocarbons (Vols. I and II)	E. Clar	Academic Press, New York (1964) 487 pp. each vol.
Introduction to infrared and Raman spectroscopy	N.B. Colthup L.H. Daly and S.E. Wiberley	Academic Press, New York (1964) 511 pp.

MONOGRAPHS (Contd.)

Chemical bonding	A. L. Companion	McGraw-Hill, New York (1964) 167 pp.
Organic functional group analysis	F. E. Critchfield	The Macmillan Co., New York (1963) 200 pp.
An introduction to practical infra-red spectroscopy (2d Edition)	A. D. Cross	Butterworths, London 86 pp.
Infra-red spectroscopy and molecular structures	M. Davies	American Elsevier Pub., Co., New York (1963)
Naturally occuring oxygen ring compounds	F. M. Dean	Butterworths, London (1963) viii + 670 pp.
Enzymes	M. Dixon and E. C. Webb	Longmans, Green & Co., Ltd. London (1964) xix + 808 pp.
Steroid reactions: An outline for organic chemistry	C. Djerassi	Holden-Day, Inc., San Francisco (1963) 663 pp.
Radiation and radiomimetric chemicals	L. A. Elson	Butterworths, London (1963) 124 pp.
Nuclear reactions (Vol. II)	P. M. Endt and P. B. Smith	Interscience Pub., New York (1963) x + 542 pp.
Modern chemical kinetics	H. Eyring and E. M. Eyring	Reinhold Pub., Corp. New York (1963) ix + 114 pp.
The modern structural theory of organic chemistry (Revised Edition)	W. N. Ferguson	Prentice-Hall, Inc., New Jersey (1963) viii + 600 pp.
The scientific method	L. F. Fieser	Reinhold Pub., Corp. New York (1964) 241 pp.
Current topics in organic chemistry	L. F. Fieser and M. Fieser	Reinhold Pub., Corp. New York (1964) 128 pp.
EDTA Titrations	H. A. Flaschka	Pergamon Press, Oxford (1964) 144 pp.
The chemistry of complex cyanides	M. H. Ford-Smith	British Information Services, New York (1964) 93 pp.
Newer methods in preparative organic chemistry (Vol. II)	W. Foerst	Academic Press, New York (1963) 417 pp.
Newer methods in preparative organic chemistry (Vol. III)	W. Foerst	Academic Press, New York (1964) 490 pp.

MONOGRAPHS (Contd.)

Title	Author(s)	Publisher
Tables for identification of organic compounds	M.F. Frankel and S. Patai	The Chemical Rubber Co., Oxford, Blackwell Scientific Pub., Ltd. (1964) 299 pp.
German-English chemical terminology (4th Revised and enlarged edition)	H. Fromherz and A. King	Weinheim, Bergstr., Ger. (1963) xxi + 588 pp.
The mechanics of aerosols (Trans. from Russian by R.E. Daisley and M. Fuchs)	N.A. Fuchs	Pergamon Press Oxford (1964) xiv + 408 pp.
Polymerization of aldehydes and oxides (Polymer Series. Vol. III)	J. Furukawa and T. Saegusa	Wiley, New York (1963) 482 pp.
The flash of genius	A.B. Garett	D. Van Nostrand Co., Inc. New Jersey (1963) x + 249 pp.
The quantitative analysis of drugs (3rd Edition)	D.C. Garratt	Boots Pure Drug Co., Ltd. Nottingham (1964) 940 pp
High polymers (Vol. XIII. Part I)	N.G. Gaylord	Wiley, New York (1963) 491 pp.
The story of biochemistry	J. Glass	Philosophical Library, Inc., New York (1964) 232 pp.
The biosynthesis of vitamins and related compounds	T.W. Goodwin	Academic Press, New York-London (1963) 366 pp.
Practical chromatographic techniques	A.H. Gordon and J.E. Eastoe	D. Van Nostrand Co., Inc., New Jersey (1964) 200 pp.
Intensity theory for infrared spectra of polyatomic molecules (Trans. by P.P. Sutton)	L.A. Gribov	Consultants Bureau, New York (1964) 113 pp.
Chemical reactions in shock waves	E.F. Greene and J.P. Toennies	Academic Press, New York-London (1964) 352 pp.
X-Ray studies of materials	A. Guinier and D.L. Dexter	Interscience Pub. (Wiley & Sons) New York-London (1963) ix + 156 pp.
Food standards and definitions in the United States	F.L. Gunderson H.W. Gunderson and E.R. Ferguson, Jr.	Academic Press, Inc., New York (1963) x + 269 pp.
An introduction to the chemistry of carbohydrates (2d Edition)	R.D. Guthrie and J. Honeyman	Clarendon Press, Oxford (1964) vi + 144 pp.

MONOGRAPHS (Contd.)

Handbuch der Papier-Chromatographie (Vol. III). Bibliographie 1957-1964 und Anwendungen (Handbook of paper chromatography with bibliography and applications) (in German)	I.M. Hais and K. Macek	VEB Gustav Fischer Verlag-Jena (1963) xix + 700 pp.
The groups of order 2^n (n 6)	M. Hall, Jr. and J.K. Senior	The Macmillan Co., New York (1964) 225 pp.
Statistics in physical science: Estimation, hypothesis testing, and least squares	W.C. Hamilton	The Ronald Press Co., New York (1964) 230 pp.
Die Nucleinsauren: Eine Einfuhrende Darstellung ihrer Chemie, Biochemie und Funktiononen (The nucleic acids. An introduction to their preparation, chemistry, biochemistry and functions) (in German)	E. Harbers	Georg Thieme Verlag Herdweg 63, Stuttgart (1964) xii + 303 pp.
Practical optical crystallography	N.H. Hartshorne and A. Stuart	American Elsevier Pub., Co., New York (1964) 326 pp.
Chemisorption (2d Edition)	D.O. Hayward	Butterworths, London 323 pp.
Clinical chemistry: Principles and techniques	R.J. Henry	Harper & Roe Pub., Inc. New York (1964) xxiv + 1128 pp.
Chemical analysis by flame photometry (2d Revised edition, Vol. XIV) (Trans. by P.T. Gilbert, Jr.)	R. Herrmann and C.T.J. Alkemade	Interscience Pub. (John Wiley & Sons) New York-London (1963) xiv + 244 pp.
Infrared absorption spectra (Index for 1959-1962)	H.M. Hershenson	Academic Press, New York-London (1964) 153 pp.
The shape of carbon compounds	W. Herz	W.A. Benjamin, Inc., New York (1963) 156 pp.
Inorganic chemistry (2d Edition)	R.B. Heslop and P.L. Robinson	Elsevier, Amsterdam (1963)
Indolalkaloide in Tabellen (Tables of indole alkaloids) (in German)	M. Hesse	Strenger Verlag, Heidelberger Platz 3, Berlin (1964) 212 pp.
Divalent carbon (Modern concepts in chemistry)	J. Hine	The Ronald Press Co., New York (1964) vii + 206 pp.
Non-glycolytic pathways of metabolism of glucose (Trans. and revised by O. Touster)	S. Hollmann	Academic Press, Inc., New York (1964) ix + 276 pp.

MONOGRAPHS (Contd.)

Title	Author	Publisher
Identification and analysis of active agents by infrared and chemical methods (Two volumes)	D. Hummel	John Wiley & Sons, London Text Vol. 386 pp. Spectra Vol. 172 pp.
The development of modern chemistry	A.J. Ihde	Harper and Roe, New York (1964) xii + 851 pp.
Molecular structure	K.R. Jennings	Chapman & Hall, Ltd. London (1964) vii + 128 pp.
Theory and structure of complex compounds	B. Jesowska-Trzebiatowski	Pergamon Press (The Macmillan Co.) (1964) 708 pp.
Natural organic macromolecules	B. Jirgensons	Pergamon Press (The Macmillan Co.) (1963) 464 pp.
Preparative inorganic reactions (Vol. I)	W.L. Jolly	Wiley, New York (1964) 264 pp.
Gas phase chromatography (Vol. I) (Trans. by P.H. Scott)	R. Kaiser	Butterworths, London (1963) viii + 199 pp.
Capillary chromatography (Vol. II) (Trans. by P.H. Scott)	R. Kaiser	Butterworths, London (1963) x + 120 pp.
Tables for gas chromatography (Vol. III) (Trans. by P.H. Scott)	R. Kaiser	Butterworths, London (1963) ix + 162 pp.
Bests' safety maintenance directory combined with the manual of modern safety techniques (10th Edition)	C.M. Kellogg	A.M. Best & Co., Inc., New York (1964) 772 pp.
Practical mathematics for chemists	F.H.C. Kelly	Butterworths, Inc. Washington, D.C. (1963) viii + 148 pp.
Index to reviews, symposia volumes and monographs in organic chemistry	N. Kharasch and W. Wolf	Pergamon Press, Oxford-New York (1964) ix + 260 pp.
Spectroscopy and molecular structure	G.W. King	Holt, Rinehart & Winston, Inc New York-London (1964) xiv + 482 pp.
Carbene chemistry (Vol. I) (With contributions by: H.M. Frey, P.P. Gaspar and G.S. Hammond)	W. Kirmse	Academic Press, New York-London (1964) 302 pp.
Chemical kinetics of gas reactions	V.N. Kondrat'ev	Addison-Wesley Pub. Co. Mass. (1964) 812 pp.

MONOGRAPHS (Contd.)

Title	Author(s)	Publisher
Synthetic hetero-chain polyamides (Trans. by N. Kamer) (Israel Program for Scientific Trans.)	V.V. Korshak and T.M. Frunze	Daniel Davey & Co., Inc. New York (1964) x + 564 pp.
Organic name reactions	H. Krauch and W. Kunz	Wiley, New York (1964) 600 pp.
Tetrapyrrole biosynthesis and its regulation	J. Lascelles	W.A. Benjamin, Inc. New York (1964) 132 pp.
Computation of molecular formulas for mass spectrometry	J. Lederberg	Holden-Day, Inc. San Francisco (1964)
Rates and equilibria of organic reactions	J.E. Leffler and E.M. Grunwald	Wiley, New York (1963) 458 pp.
The mitochondrin molecular basis of structure and function	A.L. Lehninger	W.A. Benjamin, Inc. New York (1964) xx + 263 pp.
Compilation of gas chromatographic data	J.S. Lewis	Am. Soc. for Testing and Materials, Philadelphia 625 pp.
Boron hydrides	W.N. Lipscomb	W.A. Benjamin, Inc. New York (1963) ix + 275 pp.
Alicyclic compounds	D. Lloyd	American Elsevier Pub. Co., Inc., New York ix + 171 pp.
Tables of experimental dipole moments	A.L. McClellan	W.H. Freeman & Co., San Francisco (1963) 713 pp.
Mass-spectral correlations	F.W. McLafferty	Am. Chem. Soc., Washington, D.C. (1963)
Photochemistry of proteins and nucleic acids	A.D. McLaren and D. Shugar	Pergamon Press, Oxford (1964) xii + 449 pp.
Thermal degradation of organic polymers	S.L. Madorsky	Interscience Pub., New York-London (1964) xiv + 315 pp.
Fatty acids, their chemistry and physical properties (2d Edition, Part III)	K.S. Markley	Wiley, New York (1964) 944 pp.
Coordination compounds	D.F. Martin and B.B. Martin	McGraw-Hill Book Co., New York 99 pp.
Experimental methods in gas reactions	H. Melville and B.G. Gowenlock	St. Martin's Press, Inc. New York (1964) vii + 446 pp.

MONOGRAPHS (Contd.)

The chemistry of nucleosides and nucleotides	A.M. Michelson	Academic Press, New York-London (1963) 616 pp.
An introduction to polymer chemistry	W.R. Moore	London Univ. Press, London (Aldine Pub. Co.) (1963) 270 pp.
Separation methods in biochemistry	C. Morris and P. Morris	Wiley, New York (1964) 887 pp.
Phthalocyanine compounds	F.H. Moser and A.L. Thomas	Reinhold, New York Chapman & Hall, London (1963) 381 pp.
The theory of the electronic spectra of organic molecules	J.N. Murrell	Methuen & Co., Ltd. London, Wiley, New York (1963) xiv + 328 pp.
Infrared spectra of inorganic and coordination compounds	K. Nakamoto	Wiley, New York (1963) 340 pp.
Vibrational spectra of high polymers	G. Natta	Wiley, New York (1964)
Steroid chromatography	R. Neher	American Elsevier Pub. Co. New York (1963)
Selected works in organic chemistry (Trans. by A. Birron and Z.S. Cole) (Trans. Editor: D.P. Gelfand)	A.N. Nesmeyanov	Pergamon Press, Oxford (1963) xvi + 1172 pp.
New German-English dictionary for chemists	H.H. Neville N.C. Johnston and G.V. Boyd	D. Van Nostrand Co., Inc. New Jersey (1964) xviii + 330 pp.
The collision theory of chemical reactions in liquids	A.M. North	Methuen, London (Wiley, New York) (1964) 153 pp.
Organic semiconductors	Y. Okamoto and W. Brenner	Reinhold Pub. Corp., New York, Chapman & Hall, Ltd., London (1964) vi + 184 pp.
Diene synthesis (Trans. by L. Mandel) (Israel Program for Scientific Trans.)	A.S. Onishchenko	Daniel Davey & Co., Inc. xvi + 685 pp.
Tetracyclic triterpenes (English trans. of Les Triterpenes)	G. Ourisson P. Crabbe and O.R. Rodig	Herman Pub., Ltd. London (1964) 237 pp.
Basic concepts of nuclear chemistry	R.T. Overman	Chapman & Hall, London Reinhold, New York (1963) 128 pp.

MONOGRAPHS (Contd.)

Title	Author(s)	Publisher
Determination of molecular weights and polydispersity of high polymers (Israel Program for Scientific Trans.)	S.R. Pafikov, S.A. Pavlova and I.I. Tverdokhlebova	Daniel Davey & Co., Inc., New York (1964) viii + 357 pp.
Paramagnetic resonance	G.E. Pake	W.A. Benjamin Inc., New York (1962) 220 pp.
The quantum theory of molecular electronic structure	R.G. Parr	W.A. Benjamin, Inc., (1963) xiv + 510 pp.
Steroid analysis by gas liquid chromatography	A.A. Patti and A.A. Stein	Charles C. Thomas Pub. Springfield, Ill. (1964) viii + 95 pp.
The structure of line spectra	L. Pauling and S. Goudsmit	McGraw-Hill, Inc., New York (1963) 263 pp.
Unsaturated nitro compounds (Israel Program for Scientific Trans.)	V.V. Perekalin	Daniel Davey & Co., Inc. New York (1964) ix + 332 pp.
Synthesis of organosilicon monomers	A.D. Petrov, B.F. Mironov, V.A. Ponomarenko and E.A. Chernyshev	Consultants Bureau New York
Organic complexing reagents: Structure behavior and application to inorganic analysis	D.D. Perrin	Interscience Pub. (Wiley & Sons) New York (1964) xii + 365 pp.
Spectra-structure correlation	J.P. Phillips	Academic Press, New York-London (1964) 172 pp.
Radioisotopes and their industrial applications	H. Piraux	S.A. Philips, Paris (1964) 338 pp.
Free-electron theory of conjugated molecules	J.R. Platt et al.	Wiley, New York (1964) 177 pp.
Systematics of the electronic spectra of conjugated molecules	J.R. Platt et al.	Wiley, New York (1964) 378 pp.
Biosynthesis of lipids	G. Popjak and H. Schultz	Pergamon Press (The Macmillan Co.) New York-London (1963) 435 pp.
Organic chemical preparations	F. Popp and H. Schultz	W.B. Saunders Co. (1964) 392 pp.
Progress in reaction kinetics	G. Porter	Pergamon Press (The Macmillan Co.) New York-London (1964) 370 pp.

MONOGRAPHS (Contd.)

Title	Author	Publisher
Chemical infrared spectroscopy (Vol. I)	W.J. Potts	Wiley, New York (1963) 322 pp.
Acylation reactions	P.F.G. Praill	Pergamon Press (The Macmillan Co.) New York-London (1964) 170 pp.
Catalysis and catalysts (Trans. by D. Antin)	M. Prettre	Dover Pub., Inc., New York (1963) vi + 88 pp.
Theoretische Gesichtspunktz in der Organischen Chemie (Theoretical viewpoints in organic chemistry) (in German)	W. Pritzkow	Verlag Theodor Steinkopff Dresden-Leipzig (1963) x + 266 pp.
Spectroscopy and photochemistry	E. Rabinowitch	Pergamon Press (The Macmillan Co.) New York-London (1964) 300 p
Spectroscopy and photochemistry of uranyl compounds	E. Rabinowitch and R.L. Belford	Pergamon Press, Oxford-New York (1964) x + 370 pp.
Crystalline olefin polymers (Vol. XX. Part II) (High Polymers)	R.A.V. Raff and K.W. Doak	Wiley, New York (1964) 675 pp.
Thin-layer chromatography (Trans. by D.D. Libman)	K. Randerath	Verlag Chemie, GmBH Weinheim, Academic Press, New York-London (1963) xiv + 250 pp.
Chemical applications of infrared spectroscopy	C.N.R. Rao	Academic Press, New York (1964) xiii + 683 pp.
Physical chemistry of petroleum solvents	W.W. Reynolds	Chapman & Hall, London (1964) 211 pp.
The biosynthesis of steroids, terpenes and acetogenins	J.H. Richards and J.B. Hendrickson	W.A. Benjamin, Inc. New York (1964) x + 416 pp.
Complexation in analytical chemistry (Vol. XVI)	A. Ringbom	Interscience Pub. (John Wiley & Sons) New York-London (1963) x + 395 pp.
The organic constituents of higher plants	T. Robinson	Burgess, Minneapolis, Minn. (1963) iv + 306 pp.
Morphology of polymers	T.G. Rochow	Wiley, New York (1963) 164 pp.
Chemical bonding and the geometry of molecules	G.E. Ryschkewitsch	Reinhold Pub. Corp., N.Y. (1963) ix + 116 pp.

MONOGRAPHS (Contd.)

Title	Author	Publisher
Ion exchange separations in analytical chemistry	O. Samuelson	Wiley, New York (1963) 474 pp.
Electronic spectra and quantum chemistry	C. Sandorfy	Prentice Hall, Inc. New Jersey (1964) xiii + 385 pp.
Peroxidase	B.C. Saunders	Butterworths, London 271 pp.
Polyurethanes (Vol. XVI) Part II. High Polymers	H.H. Saunders and K.C. Frisch	Wiley, New York (1964) 696 pp.
Modern polarographic methods (Trans. by R.E.W. Maddison)	H. Schmidt and M. von Stackelberg	Academic Press, New York-London (1963) v + 99 pp.
Qualitative organic microanalysis	F.L. Schneider	Academic Press, New York-London (1964) 535 pp.
Interpretation of ultraviolet spectra of natural products	A.I. Scott	Pergamon Press (The Macmillan Co.) (1964) x + 443 pp.
Electronic structure and chemical bonding	D.K. Sebera	Blaisdell (Ginn) New York (1964) ix + 298 pp.
Atomic structure and chemical bonding	F. Seel	Methuen & Co., Ltd. London (1963) 120 pp.
Rheology of polymers	E.T. Severs	Reinhold Pub. Corp. N.Y. (1963) xi + 180 pp.
Chemical constitution and biological activity (3rd Edition)	W.A. Sexton	D. Van Nostrand Co. New Jersey (1963) 503 pp.
Isotopic exchange and the replacement of hydrogen in organic compounds	A.I. Shatenshtein	Consultants Bureau, Inc. New York (1962) 324 pp.
Chemistry of the steroids	C.W. Shoppee	Butterworths, London (1964) viii + 483 pp.
The gas phase oxidation of hydrocarbons (Trans. by M.F. Mullins, Edited by B.P. Mullins)	V. Ya. Shtern	Pergamon Press, Oxford-London (1964) x + 710 pp.
Energetics of propellant chemistry	B. Siegel and L. Schieler	Wiley, New York (1964) xiii + 240 pp.
Guide to gas chromatography literature	A.V. Signeur	Plenum Press, New York (1964) 359 pp.

MONOGRAPHS (Contd.)

Title	Author(s)	Publisher
Spectrometric identification of organic compounds	R.M. Silverstein and G.C. Bassler	Wiley, New York (1963) 177 pp.
Electronic structure properties and the periodic law	H.H. Sisler	Reinhold Pub. Corp. New York, Chapman & Hall, Ltd. London (1963) vii + 120 pp.
A manual of physical methods in organic chemistry	F.L.J. Sixma and H. Wynberg	Wiley, New York (1964) 240 pp.
Principles of magnetic resonance (With examples from solid state physics)	C.P. Slichter	Harper and Roe, New York (1963) 256 pp.
Bridged aromatic compounds	B.H. Smith	Academic Press, New York (1964) 400 pp.
Acrolein	C.W. Smith	Wiley, New York (1962) 273 pp.
Chemical and biological aspects of pyridoxal catalysis	E.E. Snell	Pergamon Press (The Macmillan Co., New York-London (1964) 520 pp.
Commercial methods of analysis (Revised edition)	F.D. Snell and F.M. Biffen	Chemical Pub. Co., Inc. New York (1964) ix + 753 pp.
Hydrogenation in solutions (Trans. from Russian) (Israel Program for Scientific Trans.)	D.V. Sokol'skii	Daniel Davey & Co., Inc. New York (1964) x + 532 pp.
Stereochemistry, mechanism and silicon	L. Sommer	McGraw-Hill Book Co. New York
Free radical reactions in preparative organic chemistry	G. Sosnovsky	Macmillan Co., New York (1964) xvi + 438 pp.
Polyanions and polycations (in French)	P. Souchay	Gauthier-Villars, Paris (1963) 247 pp.
Macromolecular structure of ribonucleic acids (Trans. Editor, J.A. Stekov; P. Gray, Consulting editor)	A.S. Spirin	Reinhold Pub. Corp. New Yor Chapman & Hall, London (1964) x + 210 pp.
Introduction to microwave spectroscopy	T.L. Squires	George Newnes Ltd., London (1963) 140 pp.
An introduction to electron spin resonance	T.L. Squires	Academic Press, New York (1964) 140 pp.

MONOGRAPHS (Contd.)

Title	Author	Publisher
Wood and cellulose science	A.J. Stamm	Ronald Press Co., New York (1964) x + 549 pp.
Organoboron chemistry (Vol. I)	H. Steinberg	Interscience Pub. (John Wiley & Sons) New York (1964) 950 pp.
Oxidation mechanisms in organic chemistry	R. Stewart	W.A. Benjamin, Inc. New York (1963) xi + 179 pp.
Composition tables	G.H. Stout	W.A. Benjamin, Inc. New York 475 pp.
Bailey's industrial oils and fat products (3rd Edition)	D. Swern	Wiley, New York (1964) 1126 pp.
Azeotropy and polyazeotropy (Edited by K. Ridgway)	W. Swietslawski	Pergamon Press (The Macmillan Co.) (1964) 226 pp.
IR Theory and practice of infrared spectroscopy	H.A. Szymanski	Plenum Press, New York (1964) xiv + 375 pp.
Infrared band handbook (Supplements 1 & 2)	H.A. Szymanski	Plenum Press, New York (1963) viii + 484 pp.
Interpreted infrared spectra (Vol. I)	H.A. Szymanski	Plenum Press, New York (1964) 301 pp.
Synthetic methods of organic chemistry (Vol. XV, Includes cumulative index to Vols. XI through XV)	W. Theilheimer	S. Karger, A.G. (1963) 696 pp.
Synthetic methods of organic chemistry (Yearbook, Vol. XVII)	W. Theilheimer	S. Karger, A.G. (1963) 523 pp.
Synthetic methods of organic chemistry (Vol. XVIII)	W. Theilheimer	S. Karger, A.G. (1964) 565 pp.
Solvolysis mechanisms	E.R. Thornton	Ronald Press Co., New York (1964) 200 pp.
Alkylation with olefins	A.V. Topchiev, S.V. Zavgorodnii and V.G. Kryvchkova	American Elsevier Pub. Co. (1964) 335 pp.
Radiolysis of hydrocarbons	A.V. Topchiev	American Elsevier Pub. Co. (1964) 208 pp.
Chemistry and technology of explosives	T. Urbanski	Pergamon Press (The Macmillan Co.) New York-London (1964) 628 pp.

MONOGRAPHS (Contd.)

Title	Author(s)	Publisher
Schmelzpunkt Tabellen organischer Verbindungen (Melting point tables of organic compounds) (in German)	W. Utermark and W. Schicke	Brunswick, Ger., F. Vieweg & Sohn (1963) 715 pp.
Amino-plastics	C.P. Vale and W.G.K. Taylor	Gordon & Breach (1964) 300 pp.
Thermodynamics of irreversible processes	P. Van Rysselberghe	Blaisdell Pub. Co., New York (1964) 165 pp.
The macromolecular chemistry of gelatin	A. Veis	Academic Press, New York (1964) 433 pp.
Vitamins and coenzymes	A.F. Wagner and K. Folkers	Wiley, New York (1964) 532 pp.
Spectroscopy (Vols. I and II)	S. Walker and H. Straw	Macmillan, New York (1963) xix + 267 pp. xxvi + 386 pp.
Formaldehyde (3rd Edition)	Walker	Reinhold Pub. Corp., N.Y. (1964) 736 pp.
Principles and methods of chemical analysis	H.F. Walton	Prentice-Hall, Inc. New Jersey (1964) xvi + 484 pp.
Mechanisms of oxidation of organic compounds	W.A. Waters	Methuen & Co., Ltd. London (1964) 208 pp.
Enzyme and metabolic inhibitors (Vol. I) (General principles of inhibition)	J.L. Webb	Academic Press, New York (1963) 949 pp.
Organic spectral data (Vol. III)	O.H. Wheeler and L. Kaplan	Wiley, New York (1963) 1100 pp.
Thermal methods of analysis (Vol. XIX)	W.W. Wendlandt	Interscience Pub., (John Wiley & Sons) New York-London (1964) x + 424 pp.
Organic chemistry (Two volumes)	F.C. Whitmore	Dover Pub., Inc. New York (1964) 1013 pp.
Solvolysis mechanisms	K.B. Wiberg	Ronald Press Co., New York (1964)
Characterization of organic compounds (2d Edition)	F. Wild	Cambridge Univ. Press (1958) 306 pp.
Organolead chemistry	L.C. Wilemsens	International Lead Zinc Research Org. New York (1964) 11 pp. (Gratis)

MONOGRAPHS (Contd.)

Title	Author	Publisher
Pyrazolones, pyrazolidones and derivatives (Heterocyclic compounds Vol. XX)	R.H. Wiley and P.F. Wiley	John Wiley & Sons, N.Y. (1964) 550 pp.
Liquid fuels	D.A. Williams	Pergamon Press (The Macmillan Co.) New York-London (1964) 95 pp.
Laboratory techniques in organic chemistry (2d Edition)	K.B. Winberg	McGraw-Hill, New York (1960) 262 pp.
Experimental approaches to the development of anti-anginal drugs	M.M. Winbury	Academic Press, New York (1964) 341 pp.
Cellulose plastics: Cellulose acetate, cellulose ethers, regenerated celluloses and cellulose nitrate	V.E. Yarsley	Gordon & Breach, New York (1964) 224 pp.
Infrared spectroscopy of high polymers	R. Zbinden	Academic Press, New York (1964) xii + 264 pp.
Organic polarographic analysis (Vol. XII)	P. Zuman	Pergamon Press (The Macmillan Co.) New York-London (1964) x + 313 pp.
Twenty-five year cumulative index (Journal of Chemical Education) (Vols. 1-25)		Chemical Education Pub., Co.
Ten year cumulative index (Journal of Chemical Education) (Vols. 26-35)		Chemical Education Pub., Co.
Safety measures in chemical laboratories (3rd Edition)		H.M.S.O., London (1964) iv + 36 pp.

INDEXES

Author index 239-267

Subject index 269-321

Addresses of Publishers 323-326

ABDERHALDEN, R. 146
ABDULNUR, S. 166
ABEL, E.W. 117
ABEL, K. 24
ABELSON, P.H. 132
ABERNATHY, J.L. 73
ABRAHAM, R.J. 183
ABRAHAMSSON, S. 107, 138
ABRAMOVITCH, R.A. 13, 49
ABU-ISA, I. 94
ACHESON, R.M. 12
ACHTERBERG, A. 59
ACKER, L. 143
ACRIVOS, A. 221
ADAMS, R.M. 157
ADDISON, C.C. 14
ADJANGBA, M.S. 44
ADLER, G. 132
AEBI, H. 114
AFANAS'EV, M.I. 128
AGAR, J.N. 39
AIRAPETYANTS, A.V. 91
AKERFELDT, S. 136, 137
AKHREM, A.A. 128, 130
AKHTAR, M. 18
AKHTAR, S. 49
ALBERT, A. 195, 205
ALBERTSSON, P-A. 132
ALEXANDER, P. 119
ALFREY, Jr., T. 177
ALIMARIN, I.P. 111
ALKEMADE, C.T.J. 225
ALLEN, F.W. 36
ALLEN, J.A. 53
ALLEN, M.J. 164
ALLEN, P.E. 84
ALLEN, P.E.M. 160
ALLESTON, D.L. (see D. Seyferth et al), 42
ALLINGER, N.L. 71
ALM, T. 137
ALPATOVA, N.M. 130
ALTSZULER, N. 121
AMAT, G. 187
AMBA-RAO, C.L. 93
AMBE, L. 122
AMBLER, A.P. 195
AMBROZ, J. 84

ANACLERIO, A.M. 201
ANDERER, F.A. 20
ANDERSON, F.R. 83
ANDERSON, H.C. 95
ANDERSON, J.S. 101, 150
ANDERSON, N.A. 221
ANDERSON, W.S. 213
ANDRAC, M. 35, 43
ANDRAKO, J. 78
ANDRASCHECK, H.J. 29
ANDRE, C. 43
ANDREWS, L.J. 221
ANDREWS, R.D. 93
ANDRIANOV, K.A. 128
ANEREGG, G. 167
ANET, E.F.L.J. 3
ANFINSEN, C.B. 20, 204
ANGELINI, P. 25
ANGIER, R.B. 206
ANICHKOV, S.V. 198
ANSELL, M.F. 118
ANSON, M.L. 20
ANTONINI, E. 20
ANTONOV, V.K. 114, 193
APARAJITHAN, K. 141
APPLEZWEIG, N. 221
ARBUSOW, B.A. 115
ARBUZOV, Yu.A. 131
ARCHER, A. 152
ARCHER, S. 100
ARENS, J.F. 143
ARIENS, E.J. 80
ARMAND, Y. 35
ARMAREGO, W.L.F. 12
ARMENTOR, J. 65
ARMSTRONG, G.T. 138
ARMSTRONG, J.G. 7
ARNAUD, A. 35
ARNAUD, P. 60
ARNDT, Chr. 69
ARNETT, E.M. 105
ARNOLD, J.R. 21
ARNSTEIN, H.R.V. 209
AROESTE, H. 5
ASHBY, W.D. 63
ASHMORE, P.G. 221
ASHWORTH, M.R.F. 221
ASINGER, F. 30, 189

ASSELINEAU, C. 35
ASSELINEAU, J. 35
ASSOUR, J.M. 190
ASTON, J.G. 150
ATEN, Jr., A.H.W. 14
ATTRILL, J.B. 23
AUBRUN, P. 35
AUGENSTEIN, L.G. 165, 203
AUGSTKALNS, V.A. 93
AUTIAN, J. 79
AUVRAY, J. 76
AVERY, J. 165
AVIADO, D.M. 78, 202
AVRAMENKO, L.I. 17
AVRON, M. 124
AXELROD, J. 201
AYREY, G. 48
AZIZOV, U. 87

BAAZ, M. 173
BABB, Jr., S.E. 179
BABCOCK, J.C. 153
BACCAREDDA, M. 87, 89
BACHMANN, O. 29
BACK, N. 49
BACON, G.E. 221
BACSKAI, R. 177
BADGER, G.M. 12
BADDILEY, J. 65
BAER, E. 107
BAHNER, F. 146
BAICH, A. 155
BAIER, E. 27
BAILEY, J.L. 210
BAILLIE, L.A. 21
BAIR, H.E. 94
BAITINGER, W.F. 141
BAJUSZ, S. 191
BAKER, S.C. 40
BAKER, W.O. 92
BALABAN, A.T. 140, 174, 175
BALANDIN, A.A. 158
BALDESCHWIELER, J.D. 47
BALIAH, V. 140, 141
BALL, E.G. 122
BALLARD, R.E. 168
BALLESTER, M. 141

BALOGH, V. 92
BALTAZZI, E. 48
BALTSCHEFFSKY, H. 136
BAMFORD, C.H. 92
BANBROOK, A.K. 90
BANDER, A. 198
BANDERET, A. 87
BANGERT, R. 68
BANGHAM, A.D. 15
BANKS, C.V. 123
BANKS, W. 2
BANTHORPE, D.V. 209
BAPSERES, P. 59
BARAMBOIM, N.K. 221
BARASSIN, J. 35
BARBOUR, A.K. 11
BARCLAY, L.R.C. 174
BARKALOV, I.M. 88
BARKER, H. 126
BARKER, I.R.L. 55
BARKER, J.A. 38
BARMAN, T.E. 208
BARRER, R.M. 65
BARRETT, G.C. 79
BARRIOL, J. 75
BARRY, R.D. 49
BARTECKI, B. 211
BARTOK, W. 71
BARTON, D.H.R. 101, 109, 114
BARU, V.G. 129
BASERGA, A. 22
BASOLO, F. 135, 221
BASSHAM, J.A. 10
BASSIGNANA, P. 59, 141
BASSLER, G.C. 232
BATTISTA, O.A. 221
BATTERSBY, A.R. 101, 109
BATZER, H. 54
BAUER, S.H. 133
BAUM, E. 50
BAUMAN, H.C. 221
BAUMANN, F. 176
BAUMGARTEN, E. 33
BAUMGARTNER, W.E. 22
BAUMLER, J. 57
BAWN, C.E.H. 86
BAX, D. 145
BAYER, E. 24, 31, 56

BAYSAL, B. 89
BEATTIE, I.R. 117
BEBUCH, H. 107
BECHTLER, G. 29
BECK, E. 64
BECK, J.C. 122
BECK, R. 97
BECKETT, A.H. 201
BECKHAM, L.M. 156
BEETS, M.G.J. 59
BEGUIN, Cl. 46
BEHRNDT, K.H. 216
BEKASOVA, N.I. 131
BEKE, D. 12
BELCHER, R. 1, 209
BELEN'KII, L.I. 131, 141
BELF, L.J. 11
BELFORD, R.L. 230
BELL, J.A. 105
BELLEAU, B. 202
BELOV, B.I. 127
BELOV, V.N. 127, 131
BEL'SKII, I.F. 128
BELYAKOVA, L.D. 25
BENARD, J. 151
BENASSI, C.A. 8
BENDER, A.E. 53
BENDER, M.A. 97
BENDER, M.L. 149
BENEDIKT, G. (see R. Koster et al), 31
BENES, M. 91
BENLIAN, D. 59
BENNETT, J.C. 36
BENNETT, R.G. 183
BENOIT, H. 86, 92
BENSASSON, R. 88
BENSON, A.A. 15
BENSON, S.W. 17
BENT, H.A. 72
BENTLEY, F.F. 64
BENTLEY, K.W. 65
BENTLEY, R. 156
BEREZOVSKII, V.M. 128
BERG, E.W. 221
BERG, R. 41
BERGEL, F. 109
BERGELSON, L.D. 31, 115, 140

BERGSON, G. 137
BERGSTROM, S. 199
BERLINER, E. 105
BERNFELD, P. 155
BERNHAUER, K. 10, 31
BERNSTEIN, S. 197
BERONIUS, P. 138
BEROZA, M. 132, 151
BERSOHN, R. 165
BERRY, A.J. 101
BERRY, J.P. 83
BERRY, M. 117
BERTELE, E. 32
BESNAINOU, S. 77
BESTIAN, H. 27, 29, 30
BESTMANN, H.J. 115
BETTELHEIM, F.A. 76
BEYERMANN, K. 146
BEYNON, J.H. 221
BEYNON, K.I. 029
BEZAS, B. 191
BEZIAT, Y. 44
BHACCA, N.S. 221
BIAIS, J. 76
BIALY, G. 121
BIANCHI, E. 87
BIANCHI, U. 87
BIEDERMANN, G. 137
BIEL, J.H. 152
BIELANSKI, A. 158
BIEMANN, K. 36, 114, 185
BIER, G. 177
BIERI, J.G. 107
BIERNACKI, W. 189
BIERNAT, J.F. 192
BIFFEM, F.M. 232
BIGELEISEN, J. 75, 112
BIGG, P.H. 126
BIGNAMI, G. 198
BIGORGNE, M. 59
BILES, J.A. 78
BINDERNAGEL, K.O. 154
BING, R.J. 200
BIRCH, A.J. 111
BIRCH, G.G. 2
BIRD, O.D. 207
BIRKS, L.S. 222
BIRLEY, A.W. 52

BISERTE, G. 45
BISHOP, C.T. 3
BISHOP, J.S. (see R.C. DeBodo et al), 121
BJERRUM, J. 167
BJORNERSTEDT, R. 37
BLACK, W.P. (see J.F. Tait et al), 180
BLADON, P. 196
BLAEDEL, W.J. 1
BLAGOEV, B. 43
BLAIR-WEST, J.R. 121
BLANTON, Jr., C.D. 79
BLASCHKE, G. 31
BLASCHKO, H. 201
BLASIUS, E. 62
BLATTNA, J. 217, 218
BLATZ, P.E. 91
BLAU, M. 97
BLEI, I. 133
BLOCH, K. 133
BLOCK, H. 192
BLOM, L. 144
BOATMAN, J.C. 150
BOBBITT, J.M. 222
BOBINSKI, J. 74
BODANSKY, O. 157
BODILY, D.M. 94
BOEHM, H-P. 33
BOENIG, H.V. 222
BOES, J. 144
BOGDANOVA, A.V. 130
BOGDANOV, G.N. 127
BOGDANOV, R.V. 130
BOGORAD, L. 155
BOHLMANN, F. 69
BOISSONNAS, R.A. 16
BOLTZ, D.F. 123
BOMAN, H.G. 137
BOMBAUGH, K.J. 213
BOND, G.C. 4
BONDI, A. 139
BONINO, C. 215
BONNER, J. 68, 155
BONNETT, R. 48
BOOR, Jr., J. 81
BOOS, H. 32

BOOTH, G. 14
BORAK, E. 75
BORCIC, S. 113, 114
BOREK, E. 134
BORESKOV, G.K. 4
BORGARDT, F.G. 48
BORKOVEC, A.B. 151
BORNOWSKI, H. 69
BOS, H. 145
BOSSERT, W.H. 122
BOSSHARD, E. 57
BOSTICK, E.E. 81
BOSTSARRON, A-M. (see J. Flahaut et al), 150
BOTROS, M. 199
BOTTEI, R.S. 48
BOTTER, R. (see H. Hering et al), 112
BOTTY, M.C. 83
BOULANGER, P. 45
BOUQUET, G. 59
BOURAT, G. 84
BOUSQUET, W.F. 21
BOVET, D. 109
BOVET-NITTI, F. 198
BOVEY, F.A. 70
BOWEN, E.J. 17
BOWEN, H.J.M. 222
BOWERS, A. 180
BOY, Jr., R.E. 94
BOYD, B.R. 64
BOYD, G.V. 228
BOYER, P.D. 36
BOYLAND, E. 165, 200
BOYS, F.L. 64
BRAAMS, R. 6
BRADLEY, D.C. 211
BRADLEY, D.F. 165
BRADLEY, J.N. 210
BRADLEY, R.S. 179
BRADY, R.O. 36
BRAENDLIN, H.P. 11, 175
BRAND, J.C.D. 19, 169
BRADFORD, E.B. 83
BRANDENBERGER, H. 57
BRASCH, J.W. 212
BRATERMAN, P.S. 6

BRATOZ, S. 77
BRAUN, D. 50, 85, 92
BRAUN, T. 139
BRAUN, W. 28
BRAUNITZER, G. 20
BRAY, R.C. 208
BRAZHNIKOVA, H.G. 40
BRECK, D.W. 74
BREDERECK, H. 30
BREITENBACH, J.W. 90
BRENDLE, M. 87
BRENNER, M. 188, 192
BRENNER, N. 64
BRENNER, W. 228
BRESKY, D.R. 64
BRESLER, S.E. 70
BRESSAN, G. 84
BRETHERICK, L. 55
BRIEGLEB, G. 32
BRILL, A.S. 167
BRIMACOMBE, J.S. 3, 8
BRINI, M. 85
BROCHMANN-HANSSEN, E. 78
BROCKMAN, R.W. 2
BROCKMANN, H. 68
BRODIE, B.B. 199, 200, 201
BROOKS, C.J.W. 25
BROT, C. 76
BROWN, D.A. 126
BROWN, D.J. 206
BROWN, D.M. 16
BROWN, D.W. 90
BROWN, E.J. 115
BROWN, H.C. 19, 100
BROWN, Jr., J.F. 81
BROWN, R.D. 187
BROWN, W.B. 209
BROWNING, B.L. 163
BROWNING, D.R. 126
BROWNLEE, G. 80
BRUBAKER, C.H., Jr. 123
BRUDERRECK, H. 25
BRUNER, F.A. 25
BRUNINGS, K.J. 201
BRYCE SMITH, D. 43
BUC, H. 34
BUCH, H.A. 73

BUCHANAN, J.M. 206
BUCHANAN, M.A. 163
BUCHER, T. 108
BUCHI, G. 114
BUCHWALD, H.D. 133
BUCKINGHAM, A.D. 168
BUCKINGHAM, D.A. 159
BUCKWALTER, F.H. 78
BUCOURT, R. 46
BUDNIKOV, G.K. 131
BUDY, A.M. 142
BUDZIKIEWICZ, H. 141, 222
BUEHLER, C.A. 48
BULKA, E. 13
BULLOCK, R.M. 53
BU'LOCK, J.D. 104
BUNNETT, J.F. 38
BUNTON, C.A. 222
BURAWOY, A. 140
BURGADA, R. 34, 44
BURGER, A. 202
BURKE, Jr., J.A. 149
BURKE, R.F. 11
BURKHART, R.D. 178
BURNS, J.J. 201
BURR, J.G. 113
BURSTALL, M.L. 160
BURWELL, Jr., R.L. 38
BUSCH, D.H. 123, 149
BUSCH, H. 2, 37, 157
BUSCHMANN, L. 206
BUSEV, A.I. 139
BUTENANDT, A. 98
BUTLER, J.A.V. 203
BUTLER, T.C. 201
BUTTA, E. 89
BUTTERFIELD, D.E. 54
BUTTERY, R.G. 25
BUXTON, M.W. 11
BUYSKE, D.A. 22
BYKOV, G.V. 222
BYWATER, S. 160

CAHN, L. 216
CAHN, R.S. 73, 101, 117, 222
CAINOVI, C.G. 215
CAIRNS, J. 65

CALDERWOOD, J.H. 183
CALDWELL, D.J. 39
CALVIN, G.J. 23
CALVIN, M. 4, 23, 41, 211
CAMBEY, L.A. 1
CAMPAIGNE, E.E. 95, 222
CAMPBELL, I.G. 14
CAMPBELL, I.G.M. 209
CAMPBELL, P.N. 53
CAMPBELL, W.J. 63
CAPON, B. 117
CAMPOS, F.P. 150
CANFIELD, R.E. 204
CANON, J.M. 173
CANONNE, P. 145
CAPRAIO, V. 40
CAPUTO, A. 20
CAROL, J. 78
CARR, D.R. 22
CARRERA, N.J. 216
CARRINGTON, A. 48, 117
CARROLL, B. 133
CARRUTHERS, W. 104
CARTER, F.L. 49
CARTER, J.W. 182
CARTER, P. 197
CARTON, G.P. 25
CASINOVI, C.G. 62, 215
CASPAR, D.L.D. 20
CASSA, F.F. 20
CASSIDY, H.G. 72
CASTAING, R. 143
CASTELLI, R. 86
CASTILLE, Y. 92
CATHOU, R.E. 206
CAUNT, A.D. 84
CAVA, M.P. 140
CAVALIERI, L.F. 103
CAVALLITO, C.J. 152
CAVINA, G. 215
CECIL, R. 204
CEDER, O. 138
CEDWALL, J. 137
CERRAI, E. 62
CHALVET, O. 166, 187
CHAMBERLIN, M. 41
CHAMPETIER, G. 59
CHANCE, B. 108

CHAPEVILLE, F. 41
CHAPIRO, A. 86, 90, 92
CHAPMAN, O.L. 17, 116
CHAPPEL, C.I. 102
CHAPPELEAR, D.C. 93
CHARLES, S.W. 110
CHARLESBY, A. 90
CHARLESTON, D.C. 97
CHARLOT, G. 222
CHATTORAJ, S.C. 24
CHAUDRON, T. 59
CHAUVIERE, G. 43
CHAZERAIN, J. 34
CHEESEMAN, G.W.H. 12
CHELDELIN, V.H. 155
CHEN, E. 73
CHEN, K.K. 200
CHERNICK, C.L. 123
CHERNYSHEV, E.A. 229
CHERONIS, N.D. 222
CHERRY, C. 65
CHERTKOV, Ya. B. 162
CHEVERDINA, N.I. 44 (see K.A. Kochechkov et al), 84,
CHICHESTER, C.O. 11, 155
CHIDESTER, G.H. 163
CHIEN, J.C.W. 212
CHIEN, W.K.W. 90
CHIH-YUAN, LI 128
CHILD, Jr., W.C. 118
CHILDRESS, S.J. 79
CHILOV, E. 45
CHILTZ, G. 48
CHIMIAK, A. 191, 192
CHINOPOROS, E. 47
CHISTYAKOV, I.G. 128
CHIU, J. 26
CHOPPIN, G.R. 73
CHREMOS, G.N. 57
CHRISTENSON, L.D. 151
CHRISTIAN, J.E. 21, 78
CHRISTMANN, A. 29
CHUMAEVSKII, N.A. 129
CIARDELLI, F. 84
CIASCA, M.A. 215
CIEPLINSKI, E.W. 145
CIER, A. 166
CIFONELLI, J.A. 21

CERNY, V. 214
CLAMPITT, B.H. 94, 213
CLAR, E. 58, 222
CLARK, H.C. 11
CLARK, J. 118
CLARK, J.E. 212
CLARK, J.R. 125
CLARK, S.J. 63
CLARK, V.M. 33, 101
CLARKE, A.E. 125
CLAUSS, K. 27, 30
CLEMENT, G-M. 60
CLERMONT, Y. 122
CLEVELAND, F.F. 63
COATS, A.W. 23
COCHRAN, W. 195
COE, K.L. 132
COELHO, R. 75
COFFEY, G.L. 21
COFFEY, S. 209
COGHLAN, J.P. 121
COGNION, J-M. 60
COGROSSI, C. 59, 141
COHEN, D. 191
COHEN, H.M. (see D. Seyferth et al), 42
COHEN, M.D. 116
COHEN, S.G. 105
COHN, M. 37
COHN, W.E. 41, 103
COLBURN, C.B. 11
COLCLOUGH, R.O. 86
COLEMAN, B.D. 86
COLEMAN, M.H. 15
COLLINS, C.J. 19, 113
COLLINS, F.D. 189
COLLMAN, J.P. 149
COLONGE, J. 44
COLTHUP, N.B. 222
COMMONER, B. 23
COMPANION, A.L. 223
CONDIT, P.C. 84
CONIA, J-M. 123
CONLEY, R.T. 212
CONNOR, J.A. 14
CONNORS, K.A. 79
CONSTANT, E. 76, 183
CONWAY, T.F. 64

COOK, A.H. 111
COOK, D. 173
COOK, J. 104
COOK, P.W. 24
COOKSON, R.C. 116
COOPER, J. 78
COOPER, W. 85, 160
COOVER, Jr., H.W. 92
COPE, A.C. 100
COPIUS-PEEREBOOM, J.W. 145
COPP, D.H. 122
COPP, F.C. 9
CORDIER, P. 35, 86
CORLESS, J.T. 113
CORI, C.F. 194
COSMATOS, A. 191
COTE, W.A. 83
COTTON, F.A. 74, 102
COUPEK, J. 91
COURTOT, P. 34
COUSINEAU, G.H. 114
COWLING, E.B. 153
COX, E.G. 210
COX, J.D. 112, 140
COX, Jr., J.R. 49
COX, M.E. 192
COX, R.H. 94
CRABBE, P. 228
CRAIG, D.P. 159
CRAIG, L.C. 133
CRAIG, P.N. 152
CRAIG, R.D. 56
CRAM, D.J. 110
CRANE, F.L. 107
CRANE, R.K. 21
CRANSTON, J.A. 101
CRAWFORD, V.A. 94
CREAN, G.P. 142
CREECH, B.G. (see E.C. Horning et al), 121
CRESPI, G. 177
CRESPI, H.L. 114
CRICK, F.H.C. 132
CRITCHFIELD, F.E. 223
CROMBIE, L. 68
CROSS, A.D. 223
CSAPILLA, J. 57
CUDDIHY, E. 87

CUNDALL, R.B. 106, 160
CUNNINGHAM, B.B. 37
CURIEN, H. 76
CURTIS, A.J. 178
CUSANO, C.M. 88
CVETANOVIC, R.J. 17, 106
CZVIKOVSZKY, T. 90

DAENIKER, H.U. (see R.B. Woodward et al), 140
DAFLER, J.R. 73
DAHMEN, E.A.M.F. 145
DAIDSON, J.N. 210
DAKSHINAMURTI, K. 142
DALL'ASTA, G. 33
DAL NOGARE, S. 176
DALY, J.W. 29
DALY, L.H. 222
DALZIEL, A. 22
DANE, E. 191
DANUSSO, F. 92
DAS, M.R. 183
DAUBEN, W.G. 116
DAUDEL, R. 42, 166, 187
DAUTREVAUX, M. 45
DAVID, C. 90
DAVID, R.C. 135
DAVIDEK, J. 214, 217
DAVIDSON, J.N. 103
DAVIES, M. 181, 223
DAVIES, R.C. 6
DAVIS, A. 212
DAVIS, B.A. 49
DAVIS, H.M. 126
DAVYDOV, B.E. 91
DAWANS, F. 45, 91
DAWBER, J.G. 23
DAY, J.H. 47
DEAN, F.M. 223
DEAR, E.A. 173
DE BODO, R.C. 121
deBOER, E. 183
DECAREAU, R.V. 11
DECORA, A.W. 141
DEFFET, L. 57
de GROOT, M.S. 77
De HAAS, G.H. 15

243

DeKOCK, R.J. 145
De LAET, L. 190
de la LLOSA, P. 45
DELEST, P. 59
DE LIGNY, C.L. 145
DELL, F. 56
DELPUECH, J.J. 46
DEL RE, G. 166
DELUZARCHE, A. (see F. Schue et al), 85
DE MAEYER, L. 38, 106
de MAYO, P. 116
DEMOLE, E. 214
De MORE, W.B. 17
DEN BOER, F.C. 145
DENO, N.C. 105, 135
DENTON, D.A. 121
DEREVITSKAYA, V.A. 193
DERGE, K. 50
DERVICHIAN, D.G. 203
DESSAUER, R. 17
DESSY, R.E. 43, 71
DETERMANN, H. 28
DETTLI, L. 40
DeVAULT, D. 208
DEWAR, J.S. 31
DeWAVRECHIN, R. 183
DeWOLFE, R.H. 71
DEXTER, D.L. 119, 224
DEYL, Z. 62, 214
DIBELER, V.H. 185
DIEHL, E. 33
DIEHL, P. 76
DIEHR, H.J. 33
DIEKMANN, H. 10
DIJKSTRA, R. 145
DILL, E.H. 93
DIMICK, K.P. 26
DIMONIE, M. 89
DINER, S. 166
DINGMAN, W. 133
DIPPY, J.F.J. 140
DIRKSEN, H.W. 31
DIXON, M. 223
DIXON, R.N. 126, 168
DIXON-LEWIS, G. 117
DJERASSI, C. 101, 109, 114, 222, 223
DOAK, K.W. 178, 230

DOBO, J. 90
DOBRATZ, I.W. 26
DODGE, B.F. 179
DOLE, M. 94
DOLIN, P.I. 119
DOLPHIN, A. 111
DOMANGE, L. 150
DOMINGUEZ, X.A. 72
DONALD, H.J. 94
DORFMAN, R.I. 121, 180, 194
DORING, C.E. 154
DOSE, K. 146
DOUGLAS, A.E. 169
DOURGARVAN, S.G. 89
DOUZOU, P. 6, 77, 92, 166
DOWNER, J.D. 209
DOWS, D.A. 168
DOYLE, F.P. 9
DRAGEL, D.T. 64
DRAGO, R.S. 102
DRAHOWZAL, F.A. 174
DRAKE, G.W. 73
DRAWERT, F. 29, 50
DREW, R.C. 119
DREYER, W.J. 36
DRILLIEN, G. 45
DRINKARD, W.C. 149
DROUHET, E. 40
DRYER, H.T. 64
DRYSDALE, G.R. 21
DUANE, W.C. 124
DUBACH, U.C. 67
DuBEAU, N.P. 132
DUBININ, M.M. 158
DUBOIS, J.E. 44
DUCHEFDELAVILLE, G. 151
DUCHESNE, J. 5, 165
DUCHEYLARD, G. (see H. Hering et al), 112
DUCHON d'ENGENIERES, M. 35, 42
DUDIN, A. 178
DUFFIN, G.F. 13
DUHAMEL, L. 34
DUNBAR, D. 65
DUNIGAN, E.P. 88
DUNITZ, J.D. 32, 56
DUNN, A. (see R.C. DeBodo et al), 121

DUNNAVANT, W.R. 212
DUPHORN, I. 188
DURAND, M.H. (see J.E. Dubois et al), 44
DURBECK, H.W. 144
DUSZA, J.P. 197
DUTCHER, J.D. 2
DUTTON, G.J. 200
DUTTON, H.J. 21, 107
DUVALL, G.E. 179
duVIGNEAUD, V. 133, 194
DUYSENS, L.N.M. 203
DWORKIN, A. 88
DWYER, F.P. 159
DYKE, Jr., F.H. 63
DYNE, P.J. 38
DYRSSEN, D. 136

EADS, E.A. 73
EASTHAM, A.M. 160
EASTOE, J.E. 224
EAVES, D.E. 85
EBERSON, L. 136
EBSWORTH, E.A.V. 14
ECKHARD, S. 143
ECKSTEIN, Z. 12
EDELHOCH, H. 165
EDELSON, M.R. 82
EDSALL, J.T. 20
EDVARSON, K. 37
EDWARDS, M.B. 87
EFFENBERGER, F. 30
EGAMI, F. 103
EGLINTON, G. 16, 196
EHRENBERG, A. 6, 166
EHRENPREIS, S. 202
EHRENSON, S. 105
EHRLICH, G. 4
EICHENBERGER, K. 205
EICHHORN, G.L. 149
EICHLER, S. 31
EIDINOFF, M.L. 22
EIDUS, Ya. T. 128, 131
EIERMANN, K. 94
EIGEN, M. 30, 96, 106, 108
EISCHENS, R.P. 134
EISENBERG, H. 20
EISENBRAND, J. 143

EISENHARDT, R.H. 208
ELEY, D.D. 4, 6, 160, 165
ELIAN, M. (see I. Necsoiu et al), 140
ELIASON, R.R. 90
EL KHADEM, H. 2
ELLENBOGEN, L. 142
ELLIOT, D.F. 23
ELLIOTT, G.R.B. 150
ELLIS, G.P. 102
ELLIS, S.R.M. 182
EL-SABEH, S.H.M. 183
ELSINGER, F. (see E. Bertele et al), 32
ELSON, L.A. 223
ELVING, P.J. 111
EMELEUS, H.J. 14
EMERSON, G. 71
EMERSON, G.F. 16
EMMERSON, J.L. 79
EMMONS, W.D. 161
ENDE, H.A. 87
ENDT, P.M. 223
ENGEL, L.L. 180, 197
ENIKOLOPYAN, N.S. 88
ENTRIKIN, J.B. 222
ERATH, E.H. 83
ERB-DEBRUYNE, F. 35
ERCOLI, A. 180, 199
ERDOS, E.G. 200
ERDTMAN, H. 109
EREMENKO, L.T. 189
EREMENKO, T.V. 128
ERHARDT, R. 93
ERIKSEN, S.P. 79
ERLANDER, S.R. 190
ERSHOV, V.V. 127
ERUSALIMSKII, B.L. 129
ERWIN, J. 133
ESAYAN, G.T. 129
ESCHENMOSER, A. 32 (see E. Bertele et al), 111
ESHELMAN, H.C. 65
ETIENNE, Y. 89, 161
ETTRE, L.S. 145
EUSTON, C.B. 64
EVANS, R.S. 56
EVANS, T.C. 119
EYRING, E.M. 135, 223

EYRING, H. 38, 39, 72, 187, 223

FABIAN, J. 31
FAHEY, R.C. 31
FAIGENBAUM, H.M. 47
FALCONI, G. 199
FALECKI, J. 189
FALES, H-M. 156
FALK, J.E. 159
FANELLI, A. 20
FANNING, J.C. 49
FANTA, P.E. 161
FARAH, A. 200
FARBER, E. 157
FARBEROV, M.L. 130
FARINA, M. 84
FARMER, R.H. 52
FARRAR, W.V. 101
FASELLA, P. 215
FAUVARQUE, J.F. 43
FAYMON, J.L. 198
FEDOROVA, A.V. 129
FEDOROVA, N. (see Gy. Hardy et al), 89
FEDOSEEV, A.D. 128
FEIGL, F. 1
FEIN, M.M. 213
FELDMAN, E.G. 78
FELIX, A.M. 166
FELNER, I. (see E. Bertele et al), 32
FELLOUS, R. (see J.E. Dubois et al), 44
FENNELL, T.R.F.W. 144
FENNEMA, O. 11
FERGUSON, E.E. 77
FERGUSON, G. 19
FERGUSON, J. 54
FERGUSON, K.A. 121
FERGUSON, W.N. 223
FERGUSON, Jr., E.R. 224
FERNANDEZ-ALONSO, J.I. 5
FERRARO, J.R. 63
FERRIER, J. 84
FERRIER, R.J. 55
FERRY, J.D. 70
FETIZON, M. 214
FETTERLY, L.C. 190

FETTERS, L.J. 81
FETTES, E.M. 178
FETTIS, G.C. 106
FEUERBACHER, D.G. 73
FEUGEAS, C. 43
FICINI, J. 43
FIEDLER, H. (see E. Bayer et al), 31
FIELD, B.D. 118
FIELDS, D.L. 161
FIESER, L.F. 72, 223
FIESER, M. 223
FIGGIS, B.N. 102
FILDES, J.E. 207
FILLER, R. 47
FINCH, C.A. 53
FINKEL, A.J. 114
FINKEL'SHTEIN, A.I. 131
FINLEY, K.T. 49
FIOSHIN, M. Ya. 127
FIRSCHING, F.H. 139
FISCHER, E. 71
FISCHER, E.O. 27
FISCHER, H.G. 133
FISCHER, N. 161
FISCHER, W. 33
FISCHER-HJALMARS, I. 137, 141
FISH, A. 118
FLAHAUT, J. 150
FLASCHKA, H.A. 223
FLEISCHMANN, M. 164
FLOCK, F.H. 30
FLODIN, P. 137
FLORIA, J.A. 26
FLORINI, J. 22
FLORY, P.J. 177
FLOSS, H-G. 28
FLOWERS, M.C. 118
FLUCK, E. 14, 28, 50, 51
FLUGEL, R. (see O. Glemser et al), 29
FLYGARE, W.H. 132
FOERST, W. 27, 223
FOGLIA, V.G. 194
FOLKERS, K. 234
FOLMER, Jr., O.F. 176
FONER, S.N. 133
FONKEN, G.J. 173
FONTAINE, G. 43

FONTANA, C.M. 160
FORD-SMITH, M.H. 223
FORCHIELLI, E. 121
FORREST, H.S. 206
FORRETTE, J.E. 63
FORSS, D.A. 25
FORSTER, E.O. 164
FORSTER, T. 110
FORT, Jr., R.C. 49
FOTHERBY, K. 142
FOURNIER, M. 59
FOUTS, J.R. 201
FOWDEN, L. 36
FOWKES, F.M. 151
FOWLER, L. 176
FOWLES, G.R. 179
FOWLES, G.W.A. 102
FOX, T.G. 86
FOYE, W.O. 78
FRAENKEL-CONRAT, H. 72
FRAGNER, J. 217, 218
FRAMPTON, V. 49
FRANCK, B. 31
FRANCOIS, B. 86
FRANCOIS, R. 60
FRANCQ, J.C. 92
FRANKEL, M.F. 224
FRANKLIN, J.L. 72
FRANTA, E. 86
FRANZEN, V. 174
FRAUENDORF, H. 146
FREE, A.H. 8
FREELAND, L.T. 26
FREEMAN, N.K. 107
FREEMAN, N.L. 21
FREERKSEN, E. 40
FREHSE, H. 143
FREIDLIN, L. Kh. 130
FREIFELDER, M. 4
FREIS, E.D. 152
FREMER, K.E. 163
FRENKEL, S. Ya. 127
FREON, P. 44
FRESENIUS, W. 143
FREUNDLICH, M.M. 133
FREY, H.M. 106, 116
FREYMANN, M. 183
FREYMANN, M.R. 183

FRICK, G. 136
FRIEDENBERG, R. 49
FRIEDLANDER, G. 37
FRIEDLANDER, H.Z. 90, 91, 92
FRIEDMAN, B.S. 173
FRIEDMAN, H.L. 95
FRIEDMAN, Jr., H.G. 73
FRIEDKIN, M. 7
FRIEND, D.G. 152
FRIEND, J.A. 164
FRIES, B.A. 84
FRISCH, K.C. 231
FRITZ, G. 69
FRITZ, H.P. 16
FROBERG, C-E. 138
FROMHERZ, H. 224
FROSINI, V. 87
FROVOOST, F. 90
FRUNZE, T.M. 227
FRUTON, J.S. 204
FRY, A. 113
FUCHS, N.A. 224
FUCHS, W. 50
FUHRMAN, F.A. 133
FUKUDA, S. 207
FUKUI, K. 187
FUOSS, R.M. 81
FURUKAWA, J. 85, 224
FUSARI, S.A. 21
FUSON, R.C. 16
FYVIE, A.C. 54

GABBETT, J.F. 177
GABBIANI, G. 122
GAJDOS, A. 41
GAJDOS-TOROK, M. 41
GALEFFI, C. 215
GALLAGHER, T.F. 180
GALLARD, J. 91
GALLOT, Y. 86
GAMICHE, J.M. 124
GANDINO, M. 59
GANTMAKHER, A.R. 85
GARCIA-FERNANDEZ, H. 42
GARDNER, W.L. 26
GARETT, A.B. 224
GARRATT, D.C. 224

GARRETT, H.E. 53
GARROD, L.P. 152
GARVAN, F.L. 159
GARWOOD, R.F. 209
GASCOIGNE, J.A. 53
GAUDEMAR, F. 43
GAUDEMAR, M. 43, 44
GAUMANN, A. 58
GAUMANN, T. 58
GAUMETON, A. 34
GAUSSET, L. 168
GAUTIER, J.A. 34, 42, 50
GAWEHN, K. 96
GAYLORD, N.G. 224
GAZIS, E. 191
GEACINTOV, C. 95
GEHMAN, W.G. 71
GEISELER, G. 154
GEISS, F. 145
GEISSLER, A.W. 96
GEISSMAN, T.A. 155
GELEJI, F. 90
GELOTTE, B. 188
GERECS, A. 174
GERGELY, J. 166
GERMAIN, J-E. 60
GERRARD, W. 75
GETOFF, N. 99
GHELIS, N. 191
GIACOBAZZI, C. 214
GIARTOSIO, A. 215
GIBB, Jr., T.R.P. 150
GIBBON, J.D. 182
GIBBONS, D. 222
GIBBS, F.W. 126
GIBBS, M. 124
GIBSON, D.M. 107
GIBSON, Q.H. 106, 208
GIBSON, R.D. 210
GIDDINGS, J.C. 1, 24
GIDDINGS, Jr., L.E. 169
GIERSCHNER, K. 50
GILHAM, J.K. 94
GILL, D. 84
GILLARD, R.D. 23
GILLESPIE, R.J. 173
GILLESPIE, T. 83
GILLETTE, J.R. 67, 200

GILLIER, H. 46
GILLIS, J. 73
GILMAN, H. 16, 42
GINNCHI, G. 40
GINSBURG, V. 10
GIULOTTO, L. 76
GIUSTI, P. 89
GJUROVIC-DELETIS, O. 113
GLASS, J. 224
GLAUDEMANS, C.P.J. 82
GLAZKI, A.J. 21
GLEITSMAN, G.B. 28
GLEMSER, O. 29
GLENN, E.M. 121, 181
GLICKSMAN, M. 11
GLOCKNER, P. 31
GLUCKMAN, M.I. 79
GOAMAN, L.C.G. 140
GODFREY, L.E.A. 29
GODING, J.R. (see J.R. Blair-West et al), 121
GODULA, E.A. 153
GOHRE, H. (see H. Mayer et al), 216
GOL'BERT, K.A. 128
GOLD, V. 19, 112, 113, 174
GOLDANSKII, V.I. 88
GOLDBERG, E.P. 87
GOLDBERG, I.H. 104
GOLDBY, S.D. 149
GOLDEN, J.H. 212
GOL'DFARB, Ya. L. 141
GOLDFINGER, P. 48
GOLDHAMER, D. 72
GOLDIN, A. 7
GOLDMAN, G. 175
GOLDZIEHER, J.W. 197
GOLITZ, D. 56
GOLLNICK, K. 116
GOLOVNYA, R.V. 131
GOMPPER, R. 12, 32
GOOD, N.E. 124
GOODMAN, C.H.L. 52
GOODMAN, M. 166
GOODWIN, H.A. 159
GOODWIN, T.W. 224
GORDON, A.H. 224
GORDON, M. 152

GORDON, S. 22
GORDY, W. 119
GORE, P.H. 174
GORING, D.A.I. 82
GOSSELCK, J. 30
GOSSNER, K. 38
GOTZE, K. 99
GOUDSMIT, S. 229
GOULD, R.F. 149
GOURLEY, D.R.H. 67
GOUTAREL, R. 45
GOUVERNEUR, P. 144
GOWENLOCK, B.G. 227
GRABAR, D.G. 83, 88
GRANATH, K. 188
GRANDOLINI, G. 215
GRANT, D.F. 140
GRANT, D.M. 39
GRANT, D.W. 25
GRASSELLI, J.G. 64
GRASSIE, N. 212
GRASSINI, G. 214
GRAUE, G. 143
GRAY, H.B. 73
GRAY, L.S. 176
GRAY, P. 117
GRAYSON, M. 115
GREEN, B.N. 56
GREEN, D.E. 107
GREEN, H. 139
GREEN, J. 61, 213
GREEN, L.E. 25
GREEN, M. 27
GREEN, N. 151
GREENBERG, D.M. 10
GREENBLATT, R.B. 122
GREENE, E.F. 224
GREENWOOD, C. 208
GREENWOOD, G.T. 2, 196
GREENWOOD, N.N. 14, 53, 108, 173
GREGORY, J.D. 200
GREKOV, A.P. 131
GREWER, T. 94
GRIBI, H.P. (see E. Bertele et al), 32
GRIBOV, L.A. 224
GRIFFIN, Jr., R.W. 47

GRISEBACH, H. 126
GRISWOLD, A.A. 116
GRITZKY, R. (see H.W. Wanzlick et al), 32
GROB, E.C. 57
GRODE, G.A. 212
GRONOWITZ, S. 12
GROOTENHUIS, P. 52
GROSS, B. (see M. Andrac et al), 43
GROSS, J.I. 21
GROSS, P.R. 114
GROVE, J.F. 117
GRUBER, H.L. 216
GRUBB, H.M. 185
GRUNBERG-MANAGO, M. 203
GRUNDIG, C.A. 192
GRUNDMANN, Ch. 28
GRUNDON, M.F. 104
GRUNWALD, E.M. 227
GRYAZNOV, V.M. 127
GSCHWEND, H. (see E. Bertele et al), 32
GUAITA, M. 85
GUARINO, A.J. 155
GUDRINIECE, E. 189
GUILLEMIN, R. 122, 198
GUINIER, A. 224
GUINN, V.P. 22
GUITTARD, M. (see J. Flahaut et al), 150
GUNDERMANN, K.D. 30
GUNDERSON, F.L. 224
GUNDERSON, H.W. 224
GUNN, E.L. 26
GUNNING, H.E. 17, 101
GUNSTONE, F.D. 54
GUNTHER, D. 29
GUPTA, P.C. 143
GURD, F.R.N. 108
GURPIDE, E. 121
GUSEINOV, I.I. 127
GUT, J. 12
GUT, M. 21, 121, (see J.F. Tait et al), 180
GUT, R. 167
GUTFREUND, H. 208
GUTHRIE, R.D. 224
GUTMANN, F. 164

GUTMANN, V. 99, 173
GUTOWSKY, H.S. 110
GUTSULYAK, B.M. 127
GUTTMAN, H.N. 206
GUTTMANN, St. 191
GUYER, P. 57
GUYOT, A. 85
GWYNN, G.W. 21

HAAF, W.R. 93
HAAHTI, E.O.A. 121
HAAS, W. 143
HACKER, H. (see Th. Kauffmann et al), 32
HADDOW, A. 2
HADZI, D. 181
HAFNER, K. 27, 31
HAFNER, K.H. 27
HAGE, S.M. (see Th. Kauffmann et al), 32
HAGEMEYER, Jr., H.J. 87
HAIDUC, I. 128
HAIS, I.M. 225
HAJDU, P. 146
HAJEK, M. 30
HALA'SZ, I. 24
HALCOUR, K. 30
HALE, D. 65
HALEVI, E.A. 105, 113
HALL, E.T. 126
HALL, L.D. 3
HALL, S.A. 151
HALL, Jr., M. 225
HALLAM, H.E. 181
HALLE, W. 200
HALY, A.R. 94
HAM, G.E. 177
HAM, N.S. 171, 172
HAMA, T. 207
HAMANN, S.D. 39, 179
HAMELIN, R. 43
HAMER, F.M. 161
HAMER, J. 49
HAMILTON, L.D. 97
HAMILTON, W.C. 225
HAMILTON, Jr., J.M. 11
HAMILTON-MILLER, J.M.T. 80

HAMMACK, T.J. 93
HAMMES, G.G. 38
HAMMOND, G.S. 17, 133
HAMMOUDA, Y. 45
HAMRIK, O.H. 84
HANAHAN, D.J. 36
HANCOCK, R.L. 122
HANDLOSER, J.S. 22
HANLEY, H.J.M. 74
HANSEL, W. 30
HANSEN, B. 136
HANSEN, J. (see Th. Kauffmann et al), 32
HANSS, M. 92
HAOLI, C. 66
HARBERS, E. 225
HARDELIUS, L.O. 138
HARDWICK, R.B. 53
HARDY, C.J. 118
HARDY, Gy. 89
HARDY, P.M. 193
HARLAN, J.W. 21
HARPER, J.S. 206
HARPER, N.J. 9
HARPER, P.V. 97
HARRAH, L. 65
HARRINGTON, F.D. 64
HARRIS, C.M. 159
HARRIS, H. 103
HARRIS, J.F. 163
HARRIS, L.S. 152
HARRIS, R.S. 142
HARRIS, J.F., Jr. 100
HARRISON, A.G. 185
HARRISON, S.E. 190
HART, E.J. 134
HART, G.M. 81, 87
HART, H. 173
HARTEL, K. 30
HARTLEY, G.S. 54
HARTLEY, S.B. 117
HARTMANN, H. 5, 26
HARTSHORNE, N.H. 225
HARTUNG, H.W. 78
HARTUNG, W.H. 95, 222
HARTZELL, Jr., R.W. 114
HARWOOD, J.H. 52

HASLAM, E. 104
HASSELL, W.F. 141
HASSID, W.Z. 2
HATAKEYAMA, S. 90
HATEFI, Y. 10
HATZ, R. 51
HAUPT, J. 183
HAUSER, M. 47
HAUSSER, K.H. 76
HAUSSLER, A. 146
HAVEL, R.J. 199
HAWARD, R.N. 54
HAWKES, S.J. 24
HAWKING, F. 7
HAWKINS, K.H. 21
HAWKINS, W.L. 178, 213
HAWTHORNE, J.N. 15, 142
HAWTHORNE, M.F. 14
HAWORTH, R.D. 104
HAYANO, M. 21
HAYASHI, C. 93
HAYASHI, K. 88
HAYES, S. 43, 46
HAYNES, L.J. 2
HAYWARD, D.O. 225
HEACOCK, R.L. 133
HEALY, T.W. 125
HECK, W. 33
HECKER, E. 98
HEIDEPRIEM, H. (see H.W. Wanzlick et al), 32
HEILBRONNER, E. 69, 187
HEIMPEL, A.M. 151
HEINMETS, F. 166
HEINZE, H. (see F. Hubenett et al), 30
HEINZELMAN, R.V. 67
HEISE, J.J. 65
HEJNA, K. 217
HELCKE, T.R. 54
HELDT, W.Z. 149
HELLER, C.G. 122
HELLER, M. 197
HELLMUTH, E. 94
HEMMERICH, P. 205
HEMPEL, R. 66
HENBEST, H.B. 101
HENDERSON, J.F. 85

HENDERSON, M.E.K. 111
HENDRICKS, S.B. 10, 132
HENDRICKSON, J.B. 230
HENKLER, H. (see Th. Kauffmann et al), 32
HENNEBERG, D. 144
HENNEQUIN, F. 44
HENNIG, J. 95
HENRIKSEN, T. 119
HENRY, R.J. 225
HERBST, H. 67
HERING, H. 112
HERLINGER, H. 27
HERMANN, P. 192
HERMANS, J.J. 81, 87
HERMANS, Jr., J.J. 82
HERMANS, P.H. 66
HERNDON, W.C. 73
HERRMANN, R. 225
HERSCH, P. 1
HERSCHBERG, I.S. 144
HERSHENSON, H.M. 225
HERZ, J. 91
HERZ, W. 225
HERZBERG, G. 168
HERZOG, G. 28
HERZOG, L.F. 1
HESLOP, R.B. 225
HESS, B. 206
HESSE, M. 225
HEUSLER, K. 32
HEWLETT, P.S. 54
HEY, D.H. 209
HEYNA, J. 27
HICKS, G.P. 1
HIGASHIMURA, T. (see S. Okamura et al), 88, 86
HIGHLEY, D.R. 142
HIGINBOTHAM, W.A. 22
HILDEBRAND, J.H. 75
HILLER, G. 28
HILLER, G.H. 50
HILLIS, D.P. 221
HILSCHMANN, N. 20
HILSE, K. 20
HIMMELBLAU, D.M. 49
HINE, J. 225
HINSCHBERGER, A. 85

HIRAOKA, M. 89
HIRS, C.H.W. 37
HIRSCH, W. 73
HISCOCKS, E.S. 126
HOARE, D.E. 18
HOBSON, Jr., M.C. 4
HOCH, G.E. 123
HOCHSTRASSER, R.M. 119
HOCK, C.W. 83
HOCK, K-L. (see E. Bayer et al), 31
HODGINS, J.W. 82
HODNETT, E.M. 113
HOFMANN, A. 30
HOFMANN, A.F. 188
HOFMANN, Hd. (see F. Hubenett et al), 30
HOFMANN, J.E. 174
HOFMANN, K. 108, 204
HOFMANN, U. 69
HOFFMAN, A.S. 84, 177
HOFFMAN, C.J. 48, 117
HOFFMAN, J. 114
HOFFMAN, N.E. 174
HOFFMANN, E.G. 44
HOFFMANN, H. 27, 33
HOFFMANN, R. 120
HOFFMANN, T.A. 5
HOFHEINZ, W. 126
HOGAN, J.P. 177
HOGANSON, E. 116
HOGBEN, C.A.M. 200
HOGFELDT, E. 137
HOL, P.A.H.M. 145
HOLCOMB, D.N. 39
HOLDEN, H.W. 94
HOLLEY, Jr., C.E. 112
HOLLMANN, S. 225
HOLLUNGER, G. 201
HOLMAN, R.T. 107
HOLMES, R.R. 71
HOLMES, W.N. 122
HOLMES, W.S. 117
HOLMSTEDT, B. 25
HOLOUBEK, K. 30
HONEYCUTT, Jr., E.H. 93
HONEYMAN, J. 224
HONIG, B. 165

HOOD, A. 185
HOOGENBOOM, B.E. 74
HOPE, D.B. 194
HOPFF, H. 57
HOPPE, J.O. 95
HOPPE, W. 144
HORAK, V. 217
HORECKER, B.L. 134
HORLEIN, G. 30
HORNER, L. 29, 115
HORNING, E.C. 25, 26, 36, 107, 121, 156, 188, 199
HORNING, M.G. 25, 188, 199
HORSFIELD, A. 56
HORTON, D. 2
HOUWINK, D.R. 54
HOW, M.J. 3
HOWARD, G.E. 53
HOWARD, J.B. 178
HOWELL, S.G. 213
HOWLAND, E.H. 52
HRANISAVLJEVIC-JAKOVLJEVIC, M. 215
HSIA CHEN, C.S. 83, 88
HSU, J.M. 142
HUANG, R.Y-M. 82
HUBENETT, F. 30
HUBER, J.F.K. 145
HUBER, M. 151
HUBER, P. 66
HUCKEL, W. 42, 144
HUDSON, F.L. 54
HUDSON, R.F. 14, 27, 115
HUENNEKENS, F.M. 206
HUFFMAN, C.W. 48
HUGGINS, D. 149
HUGGINS, M.L. 86
HUGHES, B.G. 150
HUGHES, E.M. 52
HUGHES, R.E. 38
HUGHES, R.H. 94
HUGHES, S.R.C. 140
HUGHES, W.L. 22
HUHEEY, J.E. 71
HUISGEN, R. 29, 30
HUISMAN, T.H.J. 8
HULETT, J.R. 118
HUMES, E.S. 94

HUMMEL, D. 28, 226
HUMMEL, J.P. 36
HUMPHRIES, C.M. 159, 168
HUMPHRIES, S.G. 155
HUNG, N.N. 34
HUNGER, A. (see R.B. Woodward et al), 140
HUNIG, S. 32
HURLEY, G.F. 81
HUTCHINSON, D.W. 33
HUTCHINSON, F. 37
HUTCHISON, Jr., C.A. 123
HUTSON, D.H. 2
HUXLEY, H.E. 203
HUYBRECHTS, G. 48
HUYSER, E.S. 100
HUYSKENS, P. 75
HUYTEN, F.H. 145
HYMAN, H.H. 134

IHDE, A.J. 226
ILLUMINATI, G. 13
ILVESPAA, A.O. 57
ILYINA, D.E. 89
IMAI, Y. 87
IMMERGUT, B. 178
IMMERGUT, E.H. 163, 178
ING, H.R. 67, 108
INGOLD, C.K. 123
INGRAM, D.J.E. 190
INMAN, C.E. 175
INNES, K.K. 169
INOUE, S. 85
IRELAND, R.E. 123
IRGOLIC, K. 143
ISAGULYANTS, V.I. 129
ISENBERG, I. 165
ISENBERG, I.H. 163
ISHIMORI, M. 85
ISLER, O. 16
ISSLEIB, K. 115
IVANOFF, D. 43
IVANOV, V.T. 193
IVANOVA, N.M. 162
IWAKURA, Y. 87
IZAWA, S. 124

JACKMAN, L.M. 196
JACKSON, H. 67
JACOBSON, M. 132, 151
JACQUE, L. 59
JACQUES, J. 180
JACQUES, J.K. (see S.B. Hartley et al), 117
JAENICKE, L. 36, 68, 206
JAFFE, H.H. 13
JAFFRAIN, M. 183
JAGANNADHASWAMY, B. 182
JAGGER, J. 6
JAKABHAZY, S.Z. 88
JAMES, A.T. 23, 188
JANATA, V. 218
JANES, W.H. 86
JANIK, J.A. 76
JANOT, M-M. 109
JANOWSKI, A. 211
JANSEN, C.J. 7
JANSSEN, M.J. 141
JARDETZKY, O. 6
JAROVITZKY, P.A. 84
JARVIS, D. 133
JAVOREL, J. 75
JAY, P. 60
JEAN, L. 149
JEANNIN, Y. 151
JEANLOZ, R.W. 10
JEFFAY, H. 21, 22
JEFFREY, G.A. 2
JEGER, O. 116
JEHLE, H. 166
JELINEK, V. 217, 218
JELLINEK, H.H.G. 212
JENCKS, W.P. 105, 135
JENDRYCHOWSKA-BONAMOUR, A-M. 90
JENKINS, L.T. 94
JENNEWEIN, C-M. 144
JENNINGS, K.R. 226
JENSEN, F.R. 175
JENSEN, H. 27
JENSEN, K.A. 13
JENSEN, M.L. 113
JENSEN, R. 64
JENSEN, W. 163
JENTZSCH, D. 145
JENTZSCH, J. 28

JERZY GLASS, G.B. 8
JESOWSKA-TRZEBIATOWSKI, B. 226
JETSCHMANN, K. 41
JEZOWSKA-TRZEBIATOWSKA, B, 158, 211
JICHA, D.C. 149
JIRGENSONS, B. 226
JIRSAK, J. 217
JOB, C. 87
JOHNS, J.W.C. 168
JOHNSON, A.W. 111
JOHNSON, I.S. 7
JOHNSON, L.F. 221
JOHNSON, M.J. 54
JOHNSON, P. 53
JOHNSON, R.C. 221
JOHNSON, W.C. 126
JOHNSON, W.S. 111
JOHNSTON, N.C. 228
JOLIOT, P. 41
JOLIVET, C. 43
JOLLES, P. 30, 45
JOLLY, W.L. 226
JONASSEN, H.B. 49
JONES, D.M. 3
JONES, D.S. 193
JONES, G.D. 160
JONES, I.C. 122
JONES, J.K.N. 82
JONES, M.E. 132
JONES, M.M. 74, 149
JONES, P.R. 48
JONES, R.A.Y. 52
JONES, W.J. 181
JORDAN, J. 56
JORGENSEN, Chr. K. 5
JOSIEN, M-L. 77
JOSSEM, E.L. 63
JOUSE, A.L. 193
JOZEFOWICZ, E. 158
JOZEFOWICZ, J. 86
JUCKER, E. 29, 67, 109
JUJII, H. 85
JUKES, T.H. 23
JULIA, M. 123
JUNGAS, R.L. 122
JUNGHAHNEL, G. 211
JUNKMANN, K. 146

JURKOWITZ, L. (see H.G. Williams-Ashman et al), 122
JUSTER, N.J. 72
JUTISZ, M. 45

KABANOV, V.A. 88, 89
KABARA, J.J. 21
KACHI, H. 212
KADESCH, R.G. 107
KAESZ, H.D. 71
KAGAWA, C.M. 181
KAHN, Jr., J.B. 200
KAISER, R. 145, 226
KAKOLOWICZ, W. 211
KALNITSKY, G. 36
KALOW, W. 201
KALVODA, J. 32
KAMATH, P.M. 213
KAMEN, M.D. 22, 72, 132
KAMLET, M.J. 53
KANA'AN, A.S. 14
KAPLAN, H.S. 134
KAPLAN, L. 234
KAPLAN, N.D. 7
KAPPELER, H. 191
KARASZ, F.E. 94
KARGIN, V.A. 88, 89, 92
KARLSON, P. 28
KARMEN, A. 24, 107, 133, 188
KARNOJITZKI, V.J. 60
KASHA, M. 119
KATES, M. 15
KATO, M. 49
KATRITZKY, A.R. 12, 52, 195
KATSOYANNIS, P.G. 204
KATZ, J.J. 114
KAUFMAN, S. 206
KAUFMANN, H. 56
KAUFFMAN, G.B. 72
KAUFFMANN, Th. 32
KAUSS, J.M. 64
KAWASAKI, C. 142
KAYSER, D. (see O. Warburg et al), 96
KAZBEKOV, E.N. 70
KE, B. 93
KEARNS, D.R. 6
KEATS, A.S. 152

KEDZIA, B. 211
KEEFER, R.M. 221
KEIL, K.D. 205
KEITH, H.D. 83
KELLOGG, C.M. 226
KELLY, F.H.C. 226
KELLY, R. 22
KEMMNER, G. 145
KEMP, P. 15
KEMULA, W. 1, 211
KENDALL, B.R.F. 1
KENDREW, K.C. 132
KENNEDY, J.P. 70, 85, 177
KENNER, G.W. 193
KERLER, W. 28
KERN, W. 50, 84, 85, 92
KERSAINT, G. 42
KESSICK, M.A. 113
KESSLER, Yu. M. 130
KESTNER, N.R. 166
KETALAAR, J.A.A. 75
KEULEMANS, A.I.M. 145
KEVAN, L. 17
KEYWORTH, D.A. 139
KHALETSKII, A.M. 129
KHANANASHVILI, L.M. 128
KHARASCH, N. 226
KHATOON, T. 67
KHOMYAKOV, K.P. 131
KIEFFER, W.F. 71
KIENITZ, H. 143
KIEWLICZ, W. 42
KIHARA, T. 5
KIHLBERG, R. 137
KILIAN, H.G. 94
KILLMANN, E. 70
KIMURA, F. 90
KING, A. 224
KING, G.W. 226
KING, K.W. 153
KIRBY, A.J. 33
KIRBY, K.S. 103
KIREEV, V.A. 131
KIRKLAND, J.J. 176
KIRMSE, W. 104, 226
KIRRMANN, A. 43, 46
KIRSCHFELD, S. 206
KIRYUSHKIN, A.A. 193

KISELEV, A.V. 25
KISFALUDY, L. 192, 193
KISIELESKI, W.E. 22
KISS, J. 66
KIVELSON, D. 38
KJAER, A. 110
KJELLSTROM, S. 137
KLABUNOVSKII, E.I. 158
KLEB, K.G. 32
KLEIN, P.D. 113
KLEINE-WEISCHEDE, K. (see O. Glemser et al), 29
KLEMA, F. 50
KLEMM, W. 29
KLENK, E. 107
KNAPP, J.Z. 24
KNAPPE, W. 95
KNEWSTUBB, P.F. 185
KNOBLER, C.M. 38
KNOBLOCH, E. 217, 218
KNOTHE, H. 40
KNOX, J.H. 24, 106
KNOX, K.L. 25
KNOX, K.W. 40
KNOX, R. 80
KNOX, R.S. 119
KNUDSEN, E.T. 40
KNUNYANTS, I.L. 128, 131
KNUPPEN, R. 146
KOBAYASHI, E. 88
KOCHECHKOV, K.A. 84
KOCHETKOV, N.K. 13, 193
KONDRAT'EV, V.N. 226
KOEBEL, K. 66
KOGAN, L.M. 130
KOIDE, S. 5
KOLB, B. 145
KOLESNIKOVA, R.V. 17
KOLINSKY, M. 91
KOLOSOV, M.N. 108
KOLTHOFF, I.M. 139
KOMENDA, J. 207
KONCOS, R. 173
KONIG, C. 27
KONIG, W. 28, 192
KONOPIK, N. 99
KOOYMAN, E.C. 110, 123
KORSHAK, V.V. 88, 91, 227
KORTE, F. 141, 144, 205

KORTEN, E. 50
KORTUM, G. 28
KOSEL, Ch. (see Th. Kauffmann et al), 32
KOSHLAND, Jr., D.E. 41, 133
KOSOWER, E.M. 77
KOSSEL, H. 146
KOSSLER, I. 89
KOST, A.N. 130
KOSTER, R. 31
KOSTIR, J. 217, 218
KOTANI, M. 5
KOTCHECHKOV, K.A. 44
KOVACIC, P. 175
KOVACS, A.J. 70
KOVACS, G. (see Gy. Hardy et al), 89
KOVALEVA, N.V. 25
KOVALEVA, V.P. 89
KOWLASKY, A. 37
KOZLOV, V.V. 127
KRAMER, H. 30
KRAMER, K. 183
KRATKY, O. 203
KRATZ, R.F. 178
KRAUCH, H. 227
KRAUS, M.H. 144
KRAUSE, A. 99
KRAUSE, R.A. 149
KRAUSS, M. 185
KREEVOY, M.M. 45, 74
KREMERS, R.E. 163
KRENTSEL, B.A. 84, 89, 91
KRESGE, A.J. 112
KRESSER, T.O.J. 178
KRESZE, G. 144
KREUDER, M. (see K. Hafner et al), 27
KREUTZBERGER, A. 69
KREUGER, P.J. 63
KREUZBICHLER, L. 29
KRIEG, D.R. 103
KRIMEN, L.I. 48
KRIMM, H. 28
KRIMM, S. 181
KRIPPAHL, G. 41
KRITCHEVSKY, D. 15, 114
KROGMANN, D.W. 123, 124
KROHNKE, F. 28

KRUGER-THIEMER, E. 40
KRUSE, W. 106
KRYSZEWSKI, M. 91
KRYVCHKOVA, V.G. 233
KUBLIK, Z. 1
KUBOTA, H. 85
KUCHEROV, V.F. 129
KUDRYAVTSEV, Yu. P. 128
KUEMMERLE, H.P. 40
KUHN, H.J. 66
KUHN, S.J. 175
KUHN, W. 66, 99
KUIVILA, H.G. 16
KUKHARSKAYA, E.V. 128
KULESHOVA, N.D. 131
KUNTZ, Jr., I.D. 41
KUNZ, W. 67, 227
KUPCHAN, S.M. 78
KUPPER, W. 143
KUPPERMANN, A. 168
KURASHEV, M.V. 131
KURATA, M. 70
KURCZEWSKA, H. 91
KUROSAKI, T. 87
KURZER, F. 29
KUSCHMIERS, R. 154
KUSS, E. 57
KUSTIN, K. 38
KUTZBACH, C. 68
KUZNETSOVA, A.I. 128
KWIATEK, J. 150
KWUN, S. 187
KYOGOKU, Y. 6

LABARRE, J-F. 34
LABHART, H. 171, 172
LABLER, L. 214
LaBRECQUE, G.C. 151
LaBROSSE, E.H. 22
LACROIX, M. 165
LADIK, J. 5, 166
LAGERQVIST, A. 168
LAGOWSKI, J.J. 12, 14
LAL, J. 160
La MER, V.K. 125
LAMLA, E. 96
LANDAIS, J. 34

LANDAU, R. 59
LANDO, J.B. 88
LANGER, Jr., A.W. 70
LANGER, G. 57
LANOS, F. 60
LANTERMAN, E. 63
LANZA, V.L. 178
LAPIDOT, A. 52
LARBIG, W. (see R. Koster et al), 31
LARRABEE, A.R. 206
LARSEN, E. 167
LASCELLES, J. 227
LaSELL, J.H. 64
LATHROP, K.A. 97
LAUTOUT-MAGAT, M. 86
LAUE, P. 154
LAVER, W.M. 113
LAVIGNE, J.B. 176
LAVOREL, J. 41
LAVROVA, M.F. 40
LAWESSON, S-O. 31, 136
LAWS, E.Q. 23
LAWSON, E.J. 153
LAYER, R.W. 48
LAZAROW, A. 121
LAZER, L. 22
LEA, C.H. 53
LEACH, S. 77
LEAL, G. 72
LEARMONTH, G.S. 62
Le BAIL, H. 112
Le BEL, P. 87
Le BEL, R.G. 82
Le BRIS, A. 35
LEBRUN, A. 76
LECOMTE, J. 6
LEDEN, I. 112
LEDER, P. 134
LEDERBERG, J. 227
LEDERER, E. 32, 42, 66, 110
LEDERER, M. 61
LEE, K.S. 200
LEE, L-H. 212
LEEMANN, H.G. 56, 67
LEERMAKERS, P.A. 74
LEES, W.A. 140
LEETE, E. 155

LEFEBVRE, G. 35
LEFEBVRE, R. 77
LEFFLER, J.E. 227
LEFTIN, H.P. 4
LEGRAND, A.P. 76
LEGRAND, D.G. 93
LeGUILLANTON, G. 44
LEHMAN, I.R. 103
LEHMAN, R.L. 165
LEHMANN, A. 41
LEHMANN, G. 177
LEHMANN-HORCHER, M. (see H.W. Wanzlich et al), 32
LEHMKUHL, H. 30
LEHN, J.M. 141
LEHNINGER, A.L. 227
LEHRLE, R.S. 90
LEIBMAN, K.C. 201
LEICESTER, H.M. 71
LEISTEN, J.A. 73
LELOIR, L.F. 194
LEMANCEAU, B. 76
LEMIEUX, R.U. 36
LEMONS, J.F. 150
LENZ, G. (see O.L. Chapman et al), 116
LEONARDO, G.L. 73
LERNER, L.J. 122
LEROY, Y. 183
LESLIE, R.B. 6, 165
LESLIE, S.W. 64, 176
LETORT, M. 59
LEUSSING, D.L. 150
LEVINE, A.K. 23
LEVINE, J. 79
LEVINSKAS, G.J. 157
LEWIS, A.F. 94
LEWIS, J. 102
LEWIS, J.R. 54
LEWIS, J.S. 227
LI, N.C. 149
LIAO, S. 122
LIBBY, W.F. 17, 132
LIBOWITZ, G.G. 150
LICHTIN, N.N. 105
LICHTEN, W. 171
LICHTENTHALER, F.W. 31
LIDE, Jr., D.R. 38

LIEBAERT, R. 183
LIEBERMAN, S. 121
LIEHR, A.D. 5, 102
LIFSON, S. 165
LIGHT, A. 204
LIGHTBODY, J.J. 124
LIKHOSHERSTOV, L.M. 193
LIM, D. 86, 91
LINDEN, H.R. 153
LINDENBERG, A.B. 76
LINDGREN, P. 198, 199, 200, 201, 202
LINDENMEYER, P.H. 81
LINDERSTROM-LANG, C.U. 112
LINDGREN, F.T. 107
LINDSAY, D. 199
LINDSEY, A.S. 125
LINDSEY, E.E. 93
LINEBACK, D.R. 36
LIN'KOVA, M.G. 131
LINNETT, J.W. 23, 168
LIPMANN, F. 103
LIPPERT, E. 77
LIPSCOMB, W.N. 227
LISTER, J.H. 125
LITTLE, B. 180
LITTLE, W.F. 135
LITTLEWOOD, A.B. 24
LIU, W-K. 66
LIVINGSTONE, S.E. 159
LLOYD, D. 126, 227
LLOYD JONES, H. 13
LOACH, P.A. 211
LOCHMANN, E. 206
LOCKE, E.G. 163
LOCKYER, R. 1
LODER, J.W. 125
LOFGREN, N. 137
LOFTIFIELD, R.B. 21
LOGAN, N. 14
LOHRMANN, R. 205
LOMAKINA, N.M. 40
LONG, F.A. 19
LONGCHAMBON, L. 98
LONGIAVE, C. 86
LONGUET-HIGGINS, H.C. 75, 165
LOPEZ-DELGADO, R. 77

LOPREST, F.J. 213
LORAINE, J.A. 142
LORENZ, S. (see O. Warburg et al), 96
LORQUET, J.C. 168
LOSSING, F.P. 184
LOTZSCH, K. (see Th. Kauffmann et al), 32
LOUBATIERES, A. 198
LOUDON, J.D. 118
LOURY, M. 60
LOVELOCK, J.E. 24
LOW, M. 192, 193
LOW, W. 190
LOWDIN, P-O. 166, 187
LOWE, B.M. 112
LOWE, H.J. 156
LOWRY, O.H. 200
LOWS, A.E. 21
LUBKE, K. 66, 193
LUCAS, R.A. 102
LUCIER, J.J. 64
LUCK, H. 56
LUCK, J.M. 36
LUCK, W. 32
LUCKEN, E.A.C. 195
LUKACH, C.A. 177
LUKINA, M. Yu. 129
LUM, A. (see J. Green et al), 213
LUMBROSO, H. 76, 161
LUND, P.K. 64
LUNDBERG, W.O. 107
LUNIN, A.F. 90
LUSHBAUGH, C.C. 22
LUSKIN, L.S. 178
LUSSAN, C. 76
LUTTRINGHAUS, A. 31, 32
LUTZ, J.D. 97
LUUKKAINEN, T. 121
LUZZATI, V. 91, 103
LYLE, E.L. 178
LYLE, G.G. 73
LYLE, R.E. 73

MA, T.S. 222
MAASS, G. 106
MACALUSO, A. 49

MacDONALD, A.M.G. 209
MacDONALD, P.C. 121
MACEK, K. 225
MacFARLANE, M.G. 15
MACH, K. 89
MACHLEIDT, H. 206
MACHUS, F.F. (see I.M. Paushkin et al), 92
MACINTRYE, W.M. 74
MACKINNEY, G. 36
MACKLE, H. 140
MACKOR, E.L. 113, 183
MacLEAN, C. 113
MacRAE, A.V. 132
MADER, W.J. 79
MADOR, I.L. 150
MADORSKY, S.L. 227
MAERCKER, A. 100
MAGAT, E.E. 87
MAGAT, M. 84, 88
MAGEE, J.L. 119
MAGEE, M.D. 141
MAHESH, V.B. 122
MAHLER, H.R. 170
MAHLER, J. 71
MAICKEL, R.P. 199, 200, 201
MAIER, L. 102
MAIER, R. 1
MAIER, W. 77
MAILLARD, A. (see F. Schue et al), 85
MAIRANOVSKII, S.G. 129
MAITTE, P. 43
MAJER, J.R. 17
MAJZOUB, M. (see H. Hering et al), 112
MALEKNIA, N. 45
MALING, H.M. 199
MALING, J.E. 165
MALINOWSKI, S. 42
MALKIN, T. 107
MAMUZINC, R.I. 141
MANDEL, P. 104
MANDELCORN, L. 190
MANDELKERN, L. 39
MANDELS, M. 153
MANECKE, G. 92, 96
MANGINI, A. 110
MANGOLD, H.K. 1

MANNELLA, G.G. 47
MANTSAVINOS, R. 155
MARCHAL, J. 85
MARCHESSAULT, R.M. 82
MARCHI, R.P. 72
MARCINKIEWICZ, S. 61
MARCOT, B. 59
MARCUS, R.A. 38
MARCUS, Y. 47
MARGERUM, D.W. 123
MARGOLIS, L. Ya. 4
MARGRAVE, J.L. 14
MARINI-BETTOLO, G.B. 214, 215
MARINETTI, G.V. 189
MARION, L. 109
MARK, H. 134, 177, 178
MARK, H.F. 57, 178
MARK, Jr., H.B. 1
MARKHAM, R. 103
MARKI, F. 66
MARKLEY, K.S. 170, 227
MARKOVITZ, A. 21
MARKS, H. 81
MAROTZ, R. 143
MARS, P. 4
MARSH, Jr., H.V. 124
MARSHALL, D.J. 1
MARSHALL, Jr., E.K. 7
MARSHALL, G.R. 62
MARTELL, A.E. 149
MARTENS, G. (see G. Chiltz et al), 48
MARTHE, J-P. 45
MARTI, E. 66
MARTIN, B. 113
MARTIN, B.B. 227
MARTIN, D.F. 150, 227
MARTIN, D.R. 173
MARTIN, G. 59
MARTIN, H. 94
MARTIN, R.B. 167
MARTIN, R.J.L. 164
MARTIN, R.L. 176
MARTIN, W.B. 100
MARTIN-SMITH, M. 67
MARTINI, L. 180
MARVEL, C.S. 177
MARX, R. 88

MARXER, A. 57
MASON, R. 165
MASON, S.F. 54, 104, 117, 168, 195
MASSIN, M. 76
MATASA, C. 60, 189
MATHES, W. 27
MATHIES, J.C. 64
MATHIESON, A.R. 160
MATSUOKA, S. 93
MATTHEY, R. 66
MATTOCK, G. 1
MATTOON, J.R. 155
MATZEL, W. 211
MAVEL, G. 35, 76
MAYER, E.A. 144
MAYER, H. 216
MAYER, R. 28, 31
MAYERHOF, W.E. 37
MAYES, N. (see J. Green et al), 213
MAYNTZ, R. 144
MAYO SMITH, W. 177
MAZZANTI, G. 33
MEAD, C.A. 119
MEBANE, A.D. 68
MEDVEDEV, S.S. 85
MEDZIHRADSZKY, K. 191
MEHLITZ, A. 50
MEHTA, M.D. 9
MEIBOOM, S. 38
MEISLICH, H. 72
MEISTER, A. 201
MEITES, J. 198
MELANDER, L. 136
MELIA, T.P. 126
MELLER, A. 82
MELLETT, L.B. 67
MELLON, Jr., E.K. 14
MELLOR, D.P. 159
MELLOR, J.M. 118
MELTON, C.E. 185
MELVILLE, H. 227
MELVILLE, H.W. 81, 177
MENTZNER, C. 98
MERER, A.J. 168
MERIWETHER, H.T. 21
MERMOUD, Ch. 90
MERRIFIELD, R.E. 120

MERRILL, A.P. 194
MERRITT, Jr., C. 25
METCALFE, L.D. 176
MEYER, A.S. 24
MEYER, B. 49 (see O. Glemser et al), 29
MEYER, E.F. (see E. Bertele et al), 32
MEYER, H. 137
MEYER, M.W. 173
MEYERSON, S. 1, 185
MICHALSKI, J. 42
MICHEL, R. 198
MICHELSON, A.M. 228
MICOVIC, V.M. 141
MIETZSCH, F. 30
MIGINIAC, L. (see M. Andrac et al), 43
MIGINIAC-GROIZELEAU, L. 44
MIGINIAC, P. (see M. Andrac et al), 43
MIHAILOVIC, M. Lj. 141
MILBURN, R.M. 150
MILES, R.B. 95
MILLAUER, H. (see F. Weygand et al), 205
MILLEN, D.J. 53
MILLER, G.H. 90
MILLER, I.K. 87
MILLER, M. 139
MILLER, S.A. 52
MILLER, S.I. 73
MILLER, W.I. 121
MILLIKEN, R.S. 75
MINAMI, S. 93
MINAS, Th. 50
MINC, S. 189
MINCKLER, Jr., L.S. 85
MIOCQUE, M. 34, 35, 42
MIRKHAIDAROVA, Ch.Kh. 162
MIRONOV, B.F. 229
MIRONOV, G.A. 131
MIRONOV, G.S. 130
MISKIDZH'YAN, S.T. 130
MISLIN, H. 66
MISTRY, S.P. 142
MITACEK, Jr., P. 150
MITCHELL, Jr., J. 26
MIYA, T.S. 79
MOAGANIN, J. 87

MODELL, W. 132
MOELWYN-HUGHES, E.A. 101
MOHRMANN, S. (see H.W. Wanzlick et al), 32
MOLDAVE, K. 155
MOLDAVER, B.L. 129
MOLE, M.F. 117
MOLLER, K.E. 30
MOLNARFI, S. 44
MON, T.R. 25
MONTALVO, E. 215
MOODY, G.J. 3
MOONEY, E.F. 24
MOORE, C.D. 53
MOORE, D. 21
MOORE, J.A. 161
MOORE, N.P.W. 52
MOORE, R.S. 93
MOORE, S. 41
MOORE, W.R. 228
MOORE, Jr., L.D. 83
MORAWETZ, H. 81, 88
MORGAN, K.Z. 132
MORGAN, P.W. 90
MORGENSTERN, J. 31
MORIAMEZ, M. 76
MOROSOFF, N. 88
MORRIS, C. 228
MORRIS, J. 90
MORRIS, L.J. 188
MORRIS, M.L. 149
MORRIS, P. 228
MORTON, J.R. 49
MORTON, M. 81, 177
MOSCHETTO, Y. 45
MOSER, F.H. 73, 228
MOSES, L.E. 134
MOSHER, H.S. 133
MOTHES, K. 28, 29
MOUSSERON, M. 44, 115
MOYE, C.J. 125
MRAK, E.M. 11
MUKAMAL, H. 212
MULLER, E. 186
MULLER, F.H. 94
MULLER, G. 99
MULLER, H. (see Th. Kauffmann et al), 32
MULLER, J-Cl. 46

MULLER, N. 63
MULLER, O. 10, 31
MULLER, Z. 217, 218
MULLER-SCHIEDMAYER, G. 29
MULLER-WARMUTH, W. 183
MULLHOFER, G. 113
MUNAVALLI, S. 45, 118
MUNIER, R.L. 45
MUNNICH, K.O. 96
MUNRO, D.C. 179
MUNISON, P.L. 198
MUR, V.I. 130
MURALT, A.v. 66
MURIN, A.N. 130
MURRAY, W.J. 139
MURRELL, J.N. 210, 228
MURTY, T.S.S.R. 141
MUSAJO, L. 8, 109
MUSGRAVE, W.K.R. 209
MUSSO, H. 30
MUSTAFA, A. 17
MUSZIK, J. 77
MUXFELDT, H. 68
MYAMLIN, V.A. 127
MYSZKOWSKI, J. 60

McBEE, E.T. 11, 175
McCANN, S.M. 122
McCARLEY, R.E. 150
McCLELLAN, A.L. 227
McCLELLAND, B.J. 49
McCLURE, D.S. 119
McCLURE, J.H. 26
McCOLLUM, J.D. 1
McCOUBREY, J.C. (see S.B Hartley et al), 117
McCRAE, W. 16
McDOWELL, C.A. 184
McELROY, W.D. 10
McFADDEN, W.H. 25
McGARRY, E.E. 122
McGLASHAN, M.L. 112
McHALE, D. 61
McKAY, H.A.C. 55
McKENDRY, M.A. 21
McKINLEY-McKEE, J.S. 203
McLAFFERTY, F.W. 151, 185, 227

McLAREN, A.D. 227
McLAREN, L. 24
McLEAN, F.C. 142
McLENNAN, Jr., J.A. 5
McMAHON, F.G. 152
McNESBY, J.R. 18
McNUTT, W.S. 207
McOMIE, J.F.W. 16

NACHMANSOHN, D. 41
NAGAKURA, S. 77, 110
NAIR, V. 22
NAKAI, N. 113
NAKAJIMA, T. 187
NAKAMOTO, K. 228
NAKAMURA, T. 187
NAKAYAMA, T.O.M. 155
NALLEY, W.M. 216
NAMETKINE, N.S. 89
NANCE, W.E. 132
NANO, G.M. 214
NAPOLY, C. 59
NASINI, A.G. 85
NATTA, G. 33, 59, 66, 84, 86, 98, 228
NAWA, S. 206
NAYAK, R. 122
NAYLER, J.H.C. 9
NAZARENKO, I.I. 129
NECHTSCHEIN, M. 91
NECSOIU, I. 140
NEEB, R. 69
NEFEDOV, B.K. 128
NEFEDOV, V.D. 129
NEFKENS, G.H.L. 191
NEHER, R. 61, 197, 214, 228
NEIDHARDT, F.C. 104
NEILANDS, J.B. 167
NEILL, K.G. 125
NEILL, W.K. 94
NEIMAN, M.B. 129
NEL'KENBAUM, Ya. I. 162
NELSON, H.H. 187
NELSON, E. 40, 78
NELSON, F.A. 134
NELSON, K.L. 175
NELSON, W.O. 122

NENITZESCU, C.D. 174, 175
 (see I. Necsoiu et al), 140
NESIC, S. 215
NESMEYANOV, An.N. 128, 228
NESYNOV, E.P. 131
NETTER, K.J. 201
NETUPSKAYA, S.V. 162
NEUFELD, E.F. 2
NEUKOM, H. 56
NEUMANN, W.P. 27
NEURATH, H. 204
NEUSS, N. 7
NEUWIRTH, W. 28
NEVILLE, H.H. 228
NEWHAM, J. 47
NEWLAND, G.C. 212
NEYROLLES, J. 44
NICHOL, C.A. 142
NICHOLAS, H.J. 155
NICHOLS, A.V. 107
NICHOLS, B.W. 189
NICHOLSON, M. 164
NICKLESS, G. 62
NICOL, J.A.C. 65
NICOLAU, C.S. 6
NICOLET, M. 169
NIEDENZU, K. 30, 151
NIEDERMAYER, R. 216
NIEDERWIESER, A. 188
NIEDRICH, H. 192
NIELSEN, A.H. 38
NIEMANN, A. 146
NIEMANN, C. 133
NIEMELA, N. 199
NIERMANN, H. 29
NIESSEN, H. 143
NIKLASSON, B. 201
NIKITIN, Yu.S. 25
NIKOLAEV, A. Ya. 128
NIKOLAEV, L.A. 130
NIKOLAEV, N.I. 130
NIKOLETIC, M. 114
NIRENBERG, M. 134
NISHI, H.M. 88
NISHIJIMA, Y. 70
NISMAN, B. 104
NISTRATOVA, S.N. 202
NITSCHMANN, H. 56

NITZSCHKE, M. 28
NOBLE, F.W. 24
NOBLE, Jr., P. 48
NOBLES, W.L. 79
NOEL, Y. 45
NOFRE, C. 166
NOISILLIER, G. 60
NORD, F.F. 10, 155
NORIN, T. 137
NORMAN, R.O.C. 12, 210
NORMANT, H. 43
NORTH, A.M. 86, 228
NORTHCOTE, D.H. 36, 111
NOYES, R.M. 71, 106
NOYES, Jr., W.A. 17, 115
NURNBERG, H.W. 144
NYHAN, W.L. 157
NYITRAI, K. 89
NYHOLM, R.S. 159, 167
NYSTROM, R.F. 21

OAKES, V. 207
OAKEY, R.E. 62
OAKS, D.M. 26
OBER, R.E. 21
OBOLSENTSEV, R.D. 162
O'BRIEN, C. 48
O'BRIEN, R.D. 201
OCHOA, S. 66, 194
ODOR, L. 90
OEHME, F. 50
OESTERHELT, G. 145
OESTERLING, R.E. 175
OGREN, W.L. 124
OGRYZLO, E.A. 72
OGUCHI, T. 5
O'HARE, P.A.G. 140
OHASHI, S. 199
OHMAN, Y. 216
OKABE, H. 18
OKADA, T. 90
OKAMOTO, Y. 228
OKAMURA, S. 56, 86, 88
OLAH, G.A. 173, 174, 175
OLAH, J.A. 175
OLLIS, W.D. 140
OLSZER, R. 211

O'NEILL, L.A. 52
ONISHCHENKO, A.S. 228
O'REILLY, J.M. 94
ORGEL, L.E. 65, 165
ORLOVA, E.M. 130
ORMEROD, M.G. 119
OROSHNIK, W. 68
ORTLIEB, C. 85
OSBOND, J.M. 142
OSBORN, M.J. 134
OSDENE, T.S. 205
OSHIMA, K. 89
OSIPOV, O.R. 130
OSTER, G. 70
OTTERBACH, D. (see E. Bayer et al), 31
OURISSON, G. 101, 228
OVCHINNIKOV, Yu. A. 193
OVERBERGER, C.G. 70, 81, 84, 212
OVEREND, J. 181
OVEREND, W.G. 52
OVERMAN, R.T. 228
OWEN, G.D.T. 85
OZAKI, S. 212

PACHTER, I.J. 205
PADDOCK, N.L. 118
PADIEU, P. 45
PADLEY, F.B. 214
PAFIKOV, S.R. 229
PAGE, I.H. 152
PAGE, J.E. 209
PAJDOWSKI, L. 211
PAKE, G.E. 229
PALEEV, O.A. 84
PALEEVA, I.E. 44
PALLAUD, R. 59
PALMA, F.E. 63
PALMER, M.H. 52, 118
PAOLETTI, R. 15, 199
PAOLO, G. 84
PANKOW, B. (see H.W. Wanzlick et al), 32
PAPA, L.J. 1
PAPARIELLO, G.J. 79
PAPISSOV, I.M. 88
PAQUOT, Ch. 45

PARELLO, J. 46
PARHAM, W. 100
PARKE, W.C. 166
PARKS, Jr., R.E. 157
PARIS, J.P. 17
PARKER, C.A. 18
PARKER, R.E. 11
PARR, R.G. 229
PARROD, J. 86, 91
PARTON, H.N. 164
PASCARU, I. 140
PASCAUD, M. 15
PASQUALINI, J.R. 34
PASQUON, I. 86
PASSAGLIA, E. 82
PASSMORE, T.R. 169
PASTUSKA, G. 50
PATAI, S. 224
PATALAKH, I.I. (see I.M. Paushkin et al), 92
PATAKI, G. 188
PATANELLI, D.J. 122
PATAT, F. 30, 57, 70
PATCHETT, A.A. 67
PATCHORNIK, A. 193
PATINKIN, S.H. 173
PATON, W.D.M. 202
PATRICK, C.R. 84
PATRIE, M. 150
PATTI, A.A. 229
PAULING, L. 187, 229
PAULIK, F. 43, 71
PAUSHKIN, I.M. 90, 92
PAVLICEK, M. 62, 214
PAVLOVA, S.A. 229
PAVLOVSKAYA, T.E. 131
PASYNSKII, A.G. 131
PAYNE, Jr., D.A. 73
PEACHEY, J.E. 53
PEACOCK, T.E. 168
PEARSON, G.S. 18
PECILE, A. 180
PECK, V.G. 83
PEDERSEN, C. 13
PEEBLES, F.N. 93
PEEBLES, Jr., L.H. 177
PEETS, E. 22
PEGONE, G. 85

PEJKOVIC-TADIC, I. 215
PEKKARINEN, A. 199
PELLETIER, S.W. 66
PEN'KOVSKII, V.V. 131
PENNEY, W.H. 64
PENSET-HARSTROM, I. 75
PEPPER, D.C. 174
PERADEJORDI, F. 166, 187
PEREKALIN, V.V. 229
PEREZ, A. 84
PERI, J.B. 38
PERKAMPUS, H.H. 33
PERKINS, A.J. 63
PERKINS, Jr., G. 176
PERRIN, D.D. 118, 229
PERRY, J.A. 64
PESARO, M. (see E. Bertele et al), 32
PESKA, J. 91
PETERLIN, A. 56
PETERS, D. 141
PETERS, J. 64
PETERS, J.M. 142
PETERS, L. 201
PETERSEN, S. 27
PETERSON, R.E. 22
PETO, A.G. 175
PETROV, A.A. 129
PETROV, A.D. 229
PETROV, I.N. 164
PETROVA, R.S. 25
PETROW, V. 52
PETROWITZ, H-J. 50, 214
PETTIT, R. 16, 71
PEUSCHEL, G. (see O. Glemser et al), 29
PFEIL, R.W. 212
PFLEIDERER, W. 31, 195, 205
PHILIPPOFF, W. 92
PHILLIPS, G.O. 2
PHILLIPS, J.G. 122
PHILLIPS, J.N. 159
PHILLIPS, J.P. 229
PHOTAKI, I. 191
PICKTHALL, J. 54
PIEKARA, A. 183
PIER, E.A. 221
PIERRE, J.L. 60

PIERSON, A.K. 152
PIGMAN, W. 41
PIHL, A. 119
PILCHER, G. 117
PILPEL, N. 47
PILYUGIN, G.T. 127
PIMENTEL, G.C. 110
PINCUS, G. 121, 153, 180, 194
PINDER, A.R. 49
PINES, H. 4, 174
PINNER, S.H. 160
PINO, P. 84
PIRAUX, H. 229
PIRT, S.J. 55
PISANO, J.J. 156
PITTS, Jr., J.N. 17
PITZER, K.S. 187
PLASKETT, L.G. 62
PLAT, M. 46
PLATAU, G.O. 132
PLATE, N.A. 89
PLATT, J.R. 171, 229
PLAUT, G.W.E. 207
PLESKOV, Yu.V. 127
PLESCH, P.H. 160
PLESS, J. 191
PLOSS, G. (see K. Hafner et al), 27
PLOTNIKOVA, G.I. 130
POCHI, P.E. 121
POHL, H.A. 165
POILBLANC, R. 59
POISSON, J. 35
POLAK, L.S. 92
POLANSKY, M.M. 142
POLLARD, E.C. 6
POLONOVSKI, J. 41
POLONSKY, J. 166
POMEY, G. 42
PONEC, K. 217, 218
PONOMARENKO, V.A. 229
POPJAK, G. 21, 229
POPOV, U.A. 91
PORATH, J. 108, 132
PORTER, G. 77, 106, 116, 229
PORTER, G.B. 72
POSKALENKO, A.N. 198
POST, J. 114
POTTS, W.J. 230

POULIGUEN, F. 60
POULSON, E. 199
POULSON, R.E. 25
POWELL, H.M. 190
POWERS, W.H. 123
POWRIE, W.D. 11
POZDEEV, N.M. 162
POZDEEV, V.V. 128
PRAILL, P.F.G. 230
PRANDOS, J.W. 93
PRAT, M.R. 77
PRATESI, P. 109
PRATT, R. 78
PREETZ, W. 62
PRELOG, V. 108, 109, 111, 114
PRESCOTT, D.M. 103
PRESTON, R.D. 65
PRETTRE, M. 230
PREVOST, C. (see M. Andrac et al), 43
PREZIOSI, P. 40
PRIBIL, R. 167
PRIBYL, E.J. 95
PRICE, F.P. 83
PRICE, W.C. 169
PRIDDLE, S.H. 168
PRIGOGINE, I. 5
PRILEZHAEVA, E.N. 128
PRINCIPE, A.H. 64
PRINS, W. 136
PRINZ, E. 27
PRIOLA, A. 85
PRITCHARD, G.O. 90
PRITZKOW, W. 230
PROCHAZKA, Z. 217
PROKHOROVA, A.A. 131
PROUT, C.K. 126
PROX, A. 28, 192
PRUTHI, J.S. 11
PTAK, M. 77, 92
PULLMAN, A. 77, 165, 187
PULLMAN, B. 165, 167, 187
PURCELL, K.F. 102
PURNELL, J.H. 65
PUTINTSEVA, T.G. 202
PUZITSKII, K.V. 131
PYATNITSKII, I.V. 127
PYRKOV, L.M. 127

QUANE, D. 48
QUINKERT, G. 116
QUITT, P. 192

RAAB, G. (see D. Seyferth et al), 42
RAAEN, V.F. 113
RABEK, V. 218
RABINOVITCH, B.S. 18, 118
RABINOWITCH, E. 230
RACHLIN, A.I. (see B.P. Vaterlaus et al), 66
RACKER, E. 41
RACZY, L. 76
RADDA, G.K. 12
RAFF, L.M. 168
RAFF, R.A.V. 178, 230
RAGAZZI, E. 214
RAHMAN, R-UR, 168
RAINS, H.G. 54
RAMALINGAM, K.V. 82
RAMEL, A. 66
RAMIREZ, F. 115
RAMIREZ, V.D. 122
RAMSAY, D.A. 168
RAMSAY, O.B. 49
RANBY, B. 137
RANDALL, E.W. 47
RANDERATH, K. 230
RAO, C.N.R. 230
RAO, Y.S. 49
RAOUL, Y. 41
RAPHAEL, R.A. 16
RAPP, A. 50
RAPSON, W.H. 82
RATAJCZAK, H. 211
RATOVSKAYA, A.A. 162
RATTEE, I.D. 57
RATZSCH, M. 154
RAUCH, E. (see Th. Kauffmann et al), 32
RAUSCH, M.D. 149
RAVID, B. 113
READ, B.E. 93
REASONER, J. (see O.L. Chapman et al), 116
REDFERN, J.P. 23
REED, R.I. 16, 185, 196
REED, W.L. 48

REES, C.W. 13
REESE, E.T. 153
REICH, E. 104
REICHENBERG, D. 65
REICHHERZER, R. 99
REID, P.E. 82
REILLEY, C.N. 1
REINER, J.M. 22
REINERT, K. (see R. Koster et al), 31
REINMOLLER, M. 84
REIO, L. 137
REISS-HUSSON, F. 91
REISSER, F. 205
REITHEL, F.J. 20
REMBAUM, A. 87
REMBOLD, H. 96, 206, 207
REMMER, H. 201
REMPP, P. 86, 91
RENAUD, P. 46
REPKE, K. 200
REUTCHNICK, R. 40
REUTOV, O.A. 43, 110
REYNOLDS, C.A. 123
REYNOLDS, D.D. 161
REYNOLDS, T.M. 11
REYNOLDS, W.W. 230
RHOADS, S.J. 141
RHODES, W. 120
RICHARDS, F.M. 20, 36
RICHARDS, G.N. 90
RICHARDS, J.H. 230
RICHARDS, P. 97
RICHARDSON, M.J. 83
RICHERT, K.H. 188
RIDD, J. 195
RIDD, J.H. 189
RIEKE, J.K. 81, 87
RIERA, J. 141
RIESS, G. 87
RIJDERS, G.W.A. 145
RINGBOM, A. 111, 230
RINGOLD, H.J. 180
RINGSDORF, H. 89
RIONDEL, A. (see J.F. Tait et al), 180
RISI, S. 146
RITCHIE, C.D. 105

RITCHIE, E. 125
RITTENBERG, D. 75
RITTER, F.J. 145
ROBB, J.C. 90
ROBERTS, C.W. 174
ROBERTS, E. 142
ROBERTS, J.C. 53
ROBERTS, J.D. 27
ROBERTS, R.M. 173
ROBERTSON, J.M. 19, 101, 115
ROBERTSON, W.W. 179
ROBINS, D.C. 80
ROBINSON, B. 48, 52
ROBINSON, C.V. 22
ROBINSON, G.W. 39
ROBINSON, L. 141
ROBINSON, M. 83
ROBINSON, P.L. 225
ROBINSON, R. 52, 109
ROBINSON, T. 230
ROBSON, J.M. 199
ROCHOW, E.G. 42
ROCHOW, T.G. 83, 230
RODD, E.H. 209
RODE, V.V. 129
RODEGKER, W. 22
RODEL, E. 145
RODIG, O.R. 228
RODIONOV, A.N. (see K.A. Kochechkov et al), 84
RODRIGO, R. 111
ROESSLER, D.M. 169
ROGERS, D.C. 178
ROGOVIN, Z.A. 131
ROGOWSKI, F. 69
RONDEAU, R. (see D. Dunbar et al), 65
ROSE, F.L. 54
ROSE, J.B. 160
ROSEN, B. 168
ROSEN, F. 142
ROSEN, S.M. 134
ROSENBERG, B.H. 103
ROSENBLUM, C. 21, 139
ROSENSTEIN, I.D. 91
ROSENSTEIN, R.D. 2
ROSENSTOCK, H.M. 185
ROSENTHAL, S. 206

ROSMUS, J. 62, 214
ROSOWSKY, A. 161
ROSS, S.D. 105
ROSSI, C. 87
ROSSINI, F.D. 112, 137
ROTH, J-P. 91
ROTH, L.J. 22
ROTH, W.R. 27
ROTHBLAT, G.H. 114
ROTHCHILD, S. 21
ROTHFIELD, L. 134
ROTERMUND, G. (see R. Koster et al), 31
ROUGHTON, F.J.W. 208
ROUSSET, A. 77
ROWAN, T. 205
ROWLAND, F.J. 33
RUBY, R.H. 41
RUDINGER, J. 111, 192
RUDLOFF, V. 20
RUEDENBERG, K. 171, 172
RUKHADZE, E.G. 129
RUNDLE, R.E. 135
RUNNSTROM-REIO, V. 136
RUPE, B.D. 21
RUPP, Jr. L.W. 190
RUSKE, W. 174
RUSSEL, J.H. 125
RUZICKA, L. 66, 109
RYABCHIKOV, D.I. 129
RYBICKA, S. 52
RYDGREN, B. 216
RYDON, H.N. 207
RYHAGE, R. 185
RYSCHKE-WITSCH, G.E. 230
RYZHENKOV, V.E. 198
RZESZOTARSKA, B. (see E. Taschner et al), 192

SACCONI, L. 167
SADRON, C. 6, 28, 92, 98
SAEGUSA, T. 85, 224
SAEMAN, J.F. 163
SAGER, W.F. 105
SAGITULLIN, R.S. 130
SAKURADA, I. 90
SALA, G. 180

SALEM, L. 166
SALINGER, R.M. 135
SALTIEL, J. 135
SALYERS, A. 166
SALZER, F. 144
SAMUEL, D. 52, 114
SAMUELSON, O. 138, 231
SAND, H. 32
SANDA, V. 217
SANDER, M. 213
SANDERSON, R.T. 73
SANDORFY, C. 231
SANDSTROM, J. 136, 141
SANGER, F. 101
SANE, K.V. 183
SANNER, T. 119
SANTOS-RUIZ, A. 35
SANYER, N. 163
SAPPOK, R. 33
SARETT, L.H. 67
SARGESON, A.M. 159
SARKANEN, K.V. 163
SAROFF, H.A. 188
SARTORI, G. 177
SARTORI, M.F. 47
SARRAZIN, G. 45
SASAGURI, K. 93
SASSE, K. 32, 186
SASSE, W.H.F. 12
SATCHELL, D.P.N. 117
SATHIANANDAN, K. 63
SATYAMURTHY, N.S. 150
SAUKKONEN, J.J. 62
SAUNDERS, B.C. 231
SAUNDERS, F.J. 122
SAUNDERS, F.L. 87
SAUNDERS, H.H. 231
SAUNDERS, R.A. 185
SAUS, A. (see F. Asinger et al), 30
SAVOST'YANOVA, M.V. 129
SAWYER, C.H. 198
SAYRE, E.V. 37
SCANLON, W.W. 133
SCHAEFFER, R. 135
SCHAFER, K.L. 69
SCHAFER, W. 30, 104
SCHAFFEL, G.S. 59

SCHAFFNER, K. 116
SCHALEGER, L.L. 19
SCHANKER, L.S. 9
SCHEFFER, K. 50
SCHEFFOLD, R. (see E. Bertele et al), 32
SCHENCK, G.O. 116
SCHENK, G. (see E. Bayer et al), 31
SCHENKER, E. 67
SCHENKER, K. (see R.B. Woodward et al), 140
SCHERAGA, H.A. 204
SCHERER, O. 30
SCHERR, C.W. 171
SCHETTLER, P.D. 24
SCHETTY, G. 57
SCHICH, R.L. 22
SCHICKE, W. 234
SCHIELER, L. 231
SCHIFF, L.I. 37
SCHIFFERS, A. 50
SCHILDKNECHT, H. 30, 56
SCHILL, G. 32
SCHILLER, A.M. 81
SCHIMMELSCHMIDT, K. 27
SCHINDLER, A. 84
SCHJEIDE, O.A. 107
SCHLAGER, C.A. 121
SCHLENK, H. 170
SCHLIEBENER, C. 70
SCHLINGLOFF, G. 31
SCHLOSSER, M. 31, 32
SCHMAUCH, L.J. 25
SCHMERLING, L. 174
SCHMID, H. 99
SCHMID, K. 58
SCHMIDHAMMER, L. 28, 192
SCHMIDT, C.F. 198
SCHMIDT, H. 231
SCHMIDT, M. 99
SCHMIDT, P. 205
SCHMIDT, U. 32
SCHMIDT-BLEEK, F. 33
SCHMITZ, E. 12, 31
SCHMUTZLER, R. 149
SCHNABEL, E. 191
SCHNECKO, H. 84
SCHNEIDER, F.L. 231

SCHNEIDER, J.J. 180
SCHNEIDER, W. 25, 44, 167
SCHNELL, H. 28
SCHNEPP, O. 38
SCHOEDLER, Cl. 46
SCHOENECK, W. (see Th. Kauffmann et al), 32
SCHOFFA, G. 6
SCHOLFIELD, C.R. 107, 170
SCHOLER, H.F.L. 180
SCHOLES, G. 203
SCHOLTEN, J.J.F. 4
SCHONIGER, W. 143
SCHOPF, C. 205
SCHORN, P.J. 145
SCHOTLAND, R.S. 95
SCHRAGE, A. 160
SCHRAUZER, G.N. 31
SCHREIBER, H. 211
SCHREIBER, T.P. 63
SCHREIBER, V. 218
SCHRIESHIEM, A. 71, 174
SCHRODER, E. 66, 193
SCHRODER, J. (see O. Glemser et al), 29
SCHROEDER, W.A. 36
SCHROEN, W. 216
SCHROETER, L.C. 78
SCHUBERT, W.J. 155
SCHUDEL, P. 16
SCHUE, F. 85
SCHUELER, F.W. 152, 153
SCHUERCH, C. 82, 163
SCHULENBERG, J.W. 100
SCHULKEN, Jr., R.M. 94
SCHULL, W.J. 97
SCHULTE, K.E. 67
SCHULTZ, H. 229
SCHULTZ, H.R. 216
SCHULY, H. 27
SCHULZ, G. (see K. Hafner et al), 27
SCHULZ, G.V. 70
SCHULZ, J. (see Th. Kauffmann et al), 32
SCHULZ, R.C. 32
SCHUTTE, H.R. 28, 29, 154
SCHWAB, G. 143
SCHWAB, G-M. 38

SCHWABE, K. 99
SCHWARTZ, G. 166
SCHWARTZ, M.A. 78
SCHWARZ, H. 96
SCHWARZ, J.C.P. 196
SCHWARZENBACH, G. 57
SCHWEBKE, G.L. 16
SCHWENKER, Jr., R.F. 93
SCHWEIGER, H-G. 96
SCHWEIZER, E.E. 100
SCHWYZER, R. 36, 108, 202
SCIUCHETTI, L.A. 78
SCOTT, A.F. 135
SCOTT, A.I. 231
SCOTT, F.L. 175
SCOTT, J.F. 180
SCOTT, R.P.W. 24, 56
SCRIMGEOUR, K.G. 206
SCROCCO, E. 5
SEARCY, A.W. 135
SEARLES, Jr., S. 161
SEBBAN-DANON, J. 86
SEBERA, D.K. 231
SECOR, R.M. 47
SEEL, F. 56, 69, 231
SEGALOFF, A. 153
SEGRE, E. 37
SELBY, K. 153
SELIGER, H.H. 10
SELYE, H. 122
SEMENIDO, G.E. 89
SEMPLE, R.E. 82
SENAN, K.C. 113
SENATORE, P. 76
SENIOR, J.K. 225
SENTI, F.R. 190
SEREBRYAKOV, E.P. 129
SETHNA, S. 175
SETSER, D.W. 18
SEVERS, E.T. 231
SEXTON, W.A. 231
SEYDEL, J. 40
SEYFERTH, D. 42
SEYLER, J.K. 150
SHAMMA, M. 48
SHAPIRO, I. 139
SHARKOV, V.I. 28

SHARPE, A.G. 11, 14
SHATENSHTEIN, A.I. 19, 231
SHAW, B.L. 196
SHAW, N. 111
SHAW, R.R. 132
SHCHERBAKOVA, K.D. 25
SHCHUKINA, L.A. 193
SHEEHAN, J.C. 108, 152
SHELTON, J.R. 213
SHEMYAKIN, M.M. 31, 108, 114, 115, 140, 190
SHEPPARD, H. 22
SHEPPARD, R.C. 193
SHEVCHENKO, L.L. 127
SHIMANOUCHI, T. 6, 110
SHINER, Jr., V.J. 113
SHIPLEY, C.S. 216
SHKOLINA, M.A. 88
SHKROB, A.M. 193
SHOEMAKE, G.R. 24
SHOKINA, V.V. 128
SHOPPEE, C.W. 231
SHORT, R.V. 122
SHOSTAKOVSKII, M.F. 128, 130
SHRINER, R.L. 47, 48
SHTERN, V. Ya. 231
SHUBIN, V.N. 119
SHUGAR, D. 227
SHUIKIN, N.I. 128
SHULMAN, A. 159
SHUSTER, L. 37
SICHO, V. 217, 218
SIDHU, S.S. 150
SIDOROVA, L.G. 84
SIEDLE, A.R. 157
SIEFKEN, W. 27
SIEGEL, A. 99
SIEGEL, B. 71, 231
SIEGEL, E. 32
SIEGEL, S. 71
SIEGELMAN, H.W. 10
SIEKIERSKI, S. 211
SIERILA, P. 163
SIGNEUR, A.V. 231
SIGODINA, A.B. 130
SIGWALT, P. 45
SILVA, M.R.E. 202

SILVER, B. 52
SILVERMAN, D.A. (see H.G. Williams-Ashman et al), 122
SILVERSTEIN, R.M. 232
SIM, G.A. 196
SIMBORG, D.W. 113
SIMEK, A. 217, 218
SIMON, H. 113, 205, 206
SIMON, R.W. 59
SIMON, W. 32
SIMONIS, A.M. 80
SIMONS, J.P. 17
SIMONSEN, D.G. 142
SIMMONDS, A.B. 9
SIMPSON, W.T. 119
SINANOGLU, O. 5, 38, 120, 166
SINN, H. 30
SINN, V. 86
SISLER, H.H. 232
SIXMA, F.L.J. 232
SJOBERG, K. 136
SJOVALL, J. 156, 188
SKELLON, J.H. 53
SKELLY, W.G. 48
SKINNER, H.A. 39, 112, 117
SKODA, J. 103
SKORKO, M. 91
SKOULIOS, A.E. 86
SKVARCHENKO, V.R. 129
SKVORTSOVA, N.I. 127, 131
SLADE, Jr., P.E. 94
SLADKOV, A.M. 91, 128
SLADKOVA, T.A. 130
SLAVIK, K. 217, 218
SLIAM, E. (see I. Necsoiu et al), 140
SLICHTER, C.P. 232
SLOAN, G.J. 190
SLEEMAN, D.H. 168
SLUSARCHYK, W.A. 48
SMALLER, B. 6
SMETS, G. 81, 92, 160
SMID, J. 106
SMIDT, J. 183
SMITH, B.H. 232
SMITH, C.N. 151
SMITH, C.W. 232

SMITH, D.R. 38
SMITH, E.L. 204
SMITH, G.F. 12, 139
SMITH, G.G. 140
SMITH, H.A. 49
SMITH, J.A.S. 179
SMITH, J.T. 80
SMITH, J.W. 76
SMITH, L.L. 197
SMITH, P.B. 223
SMITH, R.H. 20, 204
SMITH, R.J. 64
SMITH, R.L. 199
SMITHEN, E.E. 13
SMYTH, D.G. 23
SNATZKE, G. 188
SNAVELY, M.K. 64
SNELL, E.E. 36, 232
SNELL, F.D. 232
SNELL, J.F. 21, 22
SNYDER, L.R. 1
SOBCZYK, L. 211
SOBEL, J. (see Th. Kauffmann et al), 32
SOBOTKA, H. 8
SOBUE, H. 85, 89
SOGOLOVA, T.I. 84
SOINE, T.O. 79
SOKALSKI, Z. 158
SOKOLINSKAYA, T.A. (see I.M. Paushkin et al), 92
SOKOL'SKII, D.V. 232
SOKOLOV, N.D. 128
SOKOLV, S.D. 13, 131
SOKOLOVSKY, M. 193
SOKOLOWSKA, T. (see E. Taschner et al), 192
SOLL, D. 205
SOLOMON, D.H. 125
SOLOMON, O.F. 89
SOLOVEICHIK, S. 73
SOLTYS, E. 42
SOMMER, J.M. 46
SOMMER, L. 232
SOMOGYI, A. 90
SONDHEIMER, F. 111
SONNTAG, N.O.V. 170
SORBO, B. 201

SORKINA, T.I. 46
SOSIN, S.L. 91
SOSNOVSKY, G. 31, 136, 232
SOTANIEMI, E. 199
SOUCHAY, P. 108, 232
SOULADZE, G.T. 86
SOULAS, R. 89, 161
SOUTIF, M. 77
SOUTY, N. 76
SOVERS, O. 168
SPAFFORD, N. 21
SPARAPANY, J.J. 113
SPAUDE, S. (see Th. Kauffmann et al), 32
SPEDDING, H. 2
SPENCER, G. 52
SPENSER, I.D. 13
SPIEGELBERG, H. (see B.P. Vaterlaus et al), 66
SPIESS, B. (see F. Weygand et al), 205
SPINDEL, W. 114
SPINKS, A. 102
SPIRIDONOVA, N.V. 131
SPIRIN, A.S. 103, 232
SPONA, J. 99
SPORN, M.B. 133
SPRAGUE, J.M. 152
SPRINGALL, H.D. 191, 209
SPURLIN, H.M. 177
SQUIRES, T.L. 232
SRINIVASAN, P.R. 134
SRINIVASAN, R. 17
STACEY, F.W. 100
STACEY, M. 3, 8, 11, 53
STAEHELIN, M. 103
STAHL, E. 33
STALLBERG-STENHAGEN, S. 107
STAMM, A.J. 233
STANBURY, J.B. 121
STANG, Jr., L.G. 97
STANLEY, H.M. 210
STANNETT, V. 87, 178
STAPLE, E. 155
STARBUCK, W.C. 37
STARKWEATHER, Jr., H.W. 81
STAROSTA, J. 211

STATTON, W.O. 83
STAUDE, E. 29
STAVELEY, L.A.K. 151, 190
STAVERMAN, A.J. 70
STECHER, O. 29
STECKI, J. 5
STEELE, D. 117
STEELE, R. 121
STEELE, W.J. 2
STEELINK, C. 72
STEELMAN, S. 67
STEGGERDA, C.A. 216
STEIN, A.A. 229
STEIN, R.S. 92, 93
STEINBERG, D. 21
STEINBERG, H. 233
STEINER, R.F. 165
STEININGER, E. 213
STEKOL, J.A. 152
STELAKATOS, G.C. 191
STENHAGEN, E. 107, 144, 185
STENLAKE, J.B. 102
STEPHEN, W.I. 1, 209
STERN, M.J. 112, 113
STERNBERG, J.C. 25, 176
STERNHELL, S. 125
STEVENS, A. 36
STEVENSON, D.P. 184
STEVENSON, P.E. 73
STEWART, C.J. 198
STEWART, C.P. 8
STEWART, F.H.C. 49
STEWART, G.F. 11
STEWART, R. 233
STHAL, E. 214
STICH, K. 56, 67
STIERLIN, H. 205
STJERNSTROM, N.E. 136
STOCK, L.M. 19
STOCKEL, E. 154
STOCKMAYER, W.H. 70
STOFFYN, A.M. 180
STOLKA, M. 89
STONE, B.A. 125
STONE, F.G.A. 16
STONE, F.S. 54
STONE, J.A. 38

STORCK, W. 92
STORK, G. 114
STORVICK, C.A. 142
STOUFFER, J.E. 194
STOUT, G.H. 233
STRANDBERG, B. 136
STRAUB, F.B. 10
STRAUS, S. 212
STRAUSS, J.S. 121
STRAUSZ, O.P. 17
STRAW, H. 234
STREITWEISER, Jr., A. 105
STRENG, A.G. 48
STRICKLAND, K.P. 155
STRICKLER, S.J. 187
STROHMEIER, W. 33
STUART, A. 205, 225
STUART, H.A. 70
STULL, D.R. 173
STUNKEL, D. (see H. Mayer et al), 216
STUPEL, H. 54
STURM, E. (see K. Hafner et al), 27
STYLES, D.F. 209
SUGDEN, T.M. 65
SUGRUE, M.F. 80
SULLIVAN, D.M. 93
SULLY, B.D. 54
SUND, H. 10
SUNDEN, N. 137
SUNKO, D.E. 113 114
SUPINA, W.R. 156
SUPPAN, P. 116
SURTEES, J.R. 125
SUSA, E. 86
SUSOR, W.A. 124
SUTHERLAND, G. 110
SUTTON, J. 112
SUZUOKI, K. 86
SWARTZ, C.J. 78
SWEELEY, C.C. 24, 156
SWEELEY, G.C. 107
SWERN, D. 170, 233
SWETS, J.A. 133
SWIETSLAWSKI, W. 233
SWIFT, S.M. 25
SYKES, G. 53

SYMONS, M.C.R. 19
SYMONS, N.K.J. 83
SYMONS, R.H. 125
SZCZEPANIK, P.A. 113
SZENT-GYORGYI, A. 36
SZMUSZKOVICZ, J. 16, 67
SZWARC, M. 81, 106, 137
SZYMANSKI, A. 91
SZYMANSKI, H.A. 156, 233

TABATA, Y. 89
TABOR, W.J. 212
TACUSSEL, J. 45
TADMOR, J. 26, 62
TAFT, R.W. 105
TAIT, J.F. 180
TAIT, S.A.S. 180
TAITS, S.Z. 141, 162
TAKAHAHSHI, K. 103
TAKAYANAGI, M. 93
TAKEDA, M. 88
TALA, E. 199
TALALAEVA, T.V. (see K.A. Kochechkov et al), 84
TAMBLYN, J.W. 212
TAMM, Ch. 199
TANG, W.K. 94
TANNER, D. 87
TAPLIN, G.V. 97
TASCHNER, E. 192
TATIBOUET, F. 44
TATLOW, J.C. 11, 65
TAYLOR, E.C. 16, 205
TAYLOR, H.S. 23
TAYLOR, J.H. 22
TAYLOR, L.S. 97
TAYLOR, R. 140
TAYLOR, W.G.K. 234
TCHOUBAR, B. 43, 46
TEE, P.A.H. 71
TEGNER, C. 102
TEGGINS, J.E. 150
TELFORD, J. 152
TENNANT, G. 118
TERANISHI, R. 25
TEREKHOVA, S.F. (see I.M. Barkalov et al), 88

TERENIN, A. 4 18
TERENT'EV, A.P. 129
TESSER, G.I. 192
TEYSSIE, P. 45, 91
THATCHER, J.W. 63
THEILHEIMER, W. 233
THIELE, O.W. 146
THIERFELDER, W. 154
THIREAU, A-M. 45
THEORELL, A.H. 202
THO, P.Q. 85
THOLSTRUP, C.E. 213
THOMAS, A.L. 73, 228
THOMAS, L. 122
THOMAS, M. 67
THOMAS, R.M. 85
THOMPSON, G. 101
THOMPSON, H.W. 20, 110
THOMPSON, M.C. 149
THOMPSON, M.J. 205
THOMPSON, N.S. 163
THOMPSON, R.L. 7
THOMPSON, Jr., G.A. 36
THOMSON, C. 38
THOMSON, W.H.S. 8
THORNTON, E.R. 233
THRUSH, B.A. 169
TIERS, G.V.D. 70
TIETZE, E. 27
TILL, J.E. 37
TIMELL, T.E. 3, 82
TINOCO, Jr., I. 39, 120
TIPSON, R.S. 2
TISCHER, T.N. 150
TISELIUS, A. 132
TISHLER, M. 152
TITOV, A.I. 140
TITOV, Yu. A. 130
TOBE, M. 199
TOBE, M.L. 167
TOBIN, M.C. 63
TOBOLSKY, A.V. 23
TODD, A. 110
TOEPEL, T.H. 60
TOENNIES, J.P. 224
TOEPFER, E.W. 142
TOLBERT, B.M. 21, 22

TOLG, G. 143
TOLGYESI, W.S. 174
TOLGYESSY, J. 139
TOMIKAWA, K. (see S. Okamura et al), 88
TOMILOV, A.P. 127
TOMILOV, S.M. 130
TOMPKINS, F.C. 168
TONGE, B.L. 71
TOPCHIEV, A.V. 84, 89, 91, 131, 233
TOPSOM, R.D. 125
TORGOV, I.V. 46, 109
TOROPOVA, M.A. 129
TORP, B.A. 150
TOSCH, W.C. 63
TRABER, W. 205
TRAGER, L. 206
TRAMS, E.G. 36
TRAYNARD, Ph. 91
TRAYNHAM, J.G. 72
TREHARNE, R.W. 65
TREICHEL, P.M. 16
TREKOVAL, J. 86
TRELOAR, F.E. 160
TRENDELENBERG, F. 67
TRESZCZANOWICZ, E. 158
TRIEM, H. (see F. Asinger et al), 30
TRIPPETT, S. 115, 117
TRISTRAM, G.R. 20, 204
TROFIMOVA, G.M. (see I.M. Barkalov et al) 88
TROSSARELLI, L. 85
TROTMAN-DICKENSON, A.F. 101
TROTTER, J. 210
TROZZOLO, A.M. 77
TRZEBIATOWSKI, W. 158
TSCHESCHE, R. 188, 206
TSUBOI, M. 6
TSURUTA, T. 85
TUAN, D.F.T. 38
TUCHWEBER, B. 122
TUCK, L.D. 79
TUCKER, S.W. 141
TUCKLEY, E.S.G. 168
TUIJNMAN, C.A.F. 183

TULLAR, B.F. 152
TULLNER, W.W. 181
TUNITSKII, N.N. 130
TURANO, C. 215
TURLEY, S.G. 81
TURNBULL, J.H. 66
TURNER, A.B. 118
TURNER, G.B. 53
TURPAJEV, T.M. 202
TURRO, N.J. 133
TVERDOKHLEBOVA, I.I. 229
TYLER, Jr., V.E. 78

UBBELOHDE, A.R.J.P. 5, 75
UCHIDA, T. 103
UEBERSFELD, J. 76
UEHLEKE, H. 200
UEMURA, S. 93
UGI, I. 13
ULRICH, W. 64
UMA, M. 140
UNDERWOOD, A.L. 1
UNDHEIM, K. 207
UNGER, I. 115
UNO, K. 87
URBANSKI, T. 12, 189
USANOVICH, M.I. 189
USMANOV, Kh.U. 87
USOL'TSEVA, V.A. 128
UTERMARK, W. 234
UVNAS, B. 198
UYEO, S. 111

VAFINA, E.V. 162
VAKULA, V.L. 130
VALE, C.P. 234
VALENTINE, L. 52
VALLINO, M. 43
VALVASSORI, A. 177
VANAGS, G. 189
vanBEYLEN, M. 160
VAN DEENEN, L.L.M. 15
VANDENBERG, E.J. 81
VANDEN HEUVEL, W.J.A. 25, 26, 36, 156, 188 (see E.C. Horning et al), 121

VANDERCHMITT, A. (see H. Hering et al), 112
VANDERHOFF, J.W. 83
VANDERSLICE, T.A. 133
VAN DE VORST, A. 183
VANDE WIELE, R.L. 121
van DUIJN, J. 145
VAN ITTERBEEK, A. 190
VAN der KERK, G.J.M. 59
van der WAALS, J.H. 77
VANDER WAL, R.J. 15
VAN DORMAEL, A. 60
van DUUREN, B.L. 47
VANE, G.W. 168
VAN KAMPEN, E.J. 145
VANGHEMEN, M. 29
VAN HERPEN, G. 6
VAN MEURS, N. 144
VANNERBERG, N-G. 138
van ROSSUM, J.M. 80
van RYSSELBERGHE, P. 234
van WOERDEN, H.F. 48
VARAGNAT, J. 44
VARGA, J. 89
VARTANYAN, S.A. 130
VASIL'EV, G.S. 127
VASYUNINA, N.A. 158
VATERLAUS, B.P. 66
VAUGHAN, G. 85
VAUGHAN, L.G. 42
VAUPOTIC, J. 22
VDOVINE, V.M. 89
VEIBEL, S. 144
VEIS, A. 234
VEKSLER, V.I. 131
VELARDO, J.T. 181
VELDE, D. (see O. Glemser et al), 29
VENANZI, L.M. 32
VENDITTI, J.M. 7
VENKATARAMAN, B. 183
VENNESLAND, B. 124
VERBEKE, G. (see G. Chiltz et al), 48
VERDIER, E.T. 58
VERESHCHAGIN, A.G. 131
VERDUYN, G. 90

VERKADE, P.E. 44
VERONESE, G. 214
VESELY, K. 84
VESLEY, G.F. 74
VICARI, C. 215
VICEK, A.A. 102
VIEL, C. 34, 45, 60
VIGDERGAUZ, M.S. 128
VIGNERON, M. 35
VILESSOV, F. 18
VIRNIK, A.D. 131
VISCONTINI, M. 206
VISHNYAKOVA, T.P. 92
VIZGERT, R.V. 127
VLADIMIROVA, M.V. 130
VOELTER, W. (see E. Bayer et al), 31
VOGEL, E. 27
VOGEL, H. 146
VOGT, Jr., L.H. 47
VOITENKO, R.M. 91
VOLKE, J. 195
VOLKENSTEIN, M.V. 177
VOLMAN, D.H. 17
VOLOD'KIN, A.A. 127
von ARX, E. 61
von E. DOERING, W. 27
von EULER, U.S. 202
von FRANKENBERG, C.A. 38
von PHILIPSBORN, W. 205
von R. SCHLEYER, P. 141
von SCHANTZ, M. 214
von STACKELBERG, M. 231
von SEEMANN, C. 102
VOPEL, K.H. (see K. Hafner et al), 27
VOSKRESENSKII, V.A. 130
VOSS, W. 77
VOYUTSKII, S.S. 130
VUL'FSON, N.S. 128
VYSTRCIL, A. 141

WACKER, A. 103, 206
WACKER, H. 206
WADDINGTON, G. 137
WADE, K. 173
WADSO, I. 112

WAGNER, A.F. 234
WAGNER, C.D. 22
WAGNER, F. 10, 31
WAINERDI, R.E. 132
WAJDA, S. 211
WAKABAYASHI, M. 199
WALDI, D. 188
WALDMANN-MEYER, H. 61
WALDRON, J.D. 56
WALKER, B. 24
WALKER, R.F. 216
WALKER, S. 141, 195, 234
WALL, L.A. 90, 212
WALLACE, A.L.C. 121
WALLACE, R. 165, 205
WALLACE, T.J. 71, 74
WALLENBERGER, F.T. 32
WALLENFELS, K. 10, 29
WALLING, C. 100
WALLIS, L.P. 102
WALSH, J.T. 25
WALTER, C. 6
WALTER, J. 206
WALSH, A.D. 168, 169
WALSH, Jr., W.M. 190
WALTON, H.F. 234
WANG, C.H. 22
WANG, M.T. 150
WANLESS, G.G. 85
WANNAGAT, U. 14
WANNINEN, E. 139
WANZLICK, H.W. 32
WARBURG, O. 41, 96
WARD, A.G. 126
WARD, B. 77
WARD, R. 150
WARING, W.S. 102
WARING, Jr., R.K. 190
WARSOP, P.A. 168, 169
WARTIOVAARA, V. 163
WASER, P. 200, 202
WASIELEWSKI, C.Z. (see E. Taschner et al), 192
WASSERMAN, E. 77
WASSERMAN, J. 25
WATANABE, T. 86
WATERS, W.A. 209, 234

WATSON, J.D. 41, 132
WAYLAND, H. 93
WEAVER, H.E. 134
WEBB, E.C. 223
WEBB, J.L. 234
WEBB, J.R. 144
WEBER, R. (see Th. Kauffmann et al), 32
WEEDON, B.C. 108
WEHRLI, H. 116
WEIBULL, B. 136
WEIGANG, Jr., O.E. 179
WEIGEL, F. 69
WEIGEL, H. 2
WEIL, J.A. 183
WEIMANN, N. 94
WEIN, J. 142
WEINER, M.A. 42
WEINHOUSE, S. 2
WEINSTOCK, J. 205
WEIRAUCH, K. 84
WEIS, K.H. 30
WEISS, J.J. 103
WEISS, P. 88
WEISSBERGER, A. 161
WEISSBLUTH, M. 165
WEISSMANN, G. 122
WEISSMANN, S. 23
WEISZ, H. 139
WEISZ, P.B. 4
WEITKAMP, H. 141, 144
WELCH, D.E. (see D. Seyferth et al), 42
WELLONS, J.D. 87
WELLS, A.F. 209
WELLS, P.B. 4
WELLS, P.R. 47
WELLS, W.W. 156
WELVART, Z. 43
WENDLANDT, W.W. 234
WENTWORTH, W.E. 73
WEREZAK, G.N. 82
WERNER, H. 27
WERTHEIM, G.K. 133
WEST, G.B. 102
WEST, R. 16
WEST, T.F. 52
WESTHEIMER, F.H. 101

WESTRUM, Jr., E.F. 112
WETTERHOLM, A. 138
WEYGAND, F. 28, 146, 192, 205
WHALLEY, E. 19
WHALLEY, W.B. 111, 155
WHARRY, D.M. 53
WHEELER, K.W. 95
WHEELER, O.H. 234
WHISTON, J. 53
WHITE, D. 38
WHITE, E.F.T. 90
WHITE, L.W. 94
WHITE, R.F.M. 195
WHITMORE, F.C. 234
WHITMORE, G.F. 37
WIBERG, E. 29
WIBERG, K.B. 135 234
WIBERLEY, S.E. 47, 222
WICHTERLE, O. 91
WICK, A.N. 198
WIDDER, R. 24
WIDOM, B. 5
WIEBLHAUS, V.D. 205
WIEDLING, S. 102
WIELAND, T. 28, 108, 114, 191
WIESNER, K. 111
WIJGA, P.W.O. 145
WILBRANDT, W. 199
WILCOX, W.R. 49
WILD, F. 234
WILD, W. 213
WILENSENS, L.C. 234
WILEY, P.F. 161, 235
WILEY, R.H. 161, 235
WILHELM, M. 205
WILKE, G. 27, 44
WILKINSON, 17, 18
WILKINSON, K. 86
WILLEMART, A. 34
WILLEMS, J.F. 69
WILLIAMS, A. 117
WILLIAMS, A.E. 185, 221
WILLIAMS, A.F. 139
WILLIAMS, C.M. 156
WILLIAMS, D.A. 235
WILLIAMS, D.H. 221, 222
WILLIAMS, F. 117

WILLIAMS, M.W. 192
WILLIAMS, N.R. 55
WILLIAMS, R.J.P. 6, 167
WILLIAMS, R.T. 155, 200
WILLIAMS, T.I. 65
WILLIAMS-ASHMAN, H.G. 122
WILLIAMS-DORLET, C. 165, 183
WILLIAMSON, A.G. 38
WILLIAMSON, D.G. 19, 169
WILLETT, J.E. 191
WILSKI, H. 94
WILSON, A. 72
WILSON, C.L. 139
WILSON, Jr., E.B. 110
WILSON, E.O. 122
WILSON, T.P. 81
WILZBACH, K.E. 21
WINBERG, K.B. 235
WINBURY, M.M. 235
WINCHESTER, J.W. 113
WINDEY, J.P. 64
WINDHOLZ, M. 32
WINDHOLZ, T.B. 32
WINSLOW, F.H. 178
WINTERINGHAM, F.P.W. 54
WINTOUR, M. (see J.R. Blair-West et al), 121
WIPKE, W.T. 116
WISE, E.N. 71
WITKOP, B. 29, 66
WITT, P.N. 200
WITTIG, G.W. 42, 69, 110, 115
WITTMANN, H.G. 96
WITZEL, H. 103, 146
WOBER, W. 146
WOHNSIEDLER, H.P. 83
WOLDBYE, F. 123
WOLF, A.P. 19
WOLF, D. (see Th. Kauffmann et al), 32
WOLF, L. 211
WOLF, W. 226
WOLFF, M.E. 47
WOLFGANG, R.L. 22
WOLFROM, M.L. 2
WOLFSBERG, 112, 113
WOLLENBERGER, A. 200
WOLLENWEBER, P. 214

WOLLMAN, S.H. 121
WOOD, H.C.S. 205
WOOD, J.L. 117, 132
WOODBINE, M. 170
WOODFORD, F.P. 170
WOODS, C.W. 151
WOODS, L.A. 67
WOODWARD, R.B. 109, 114, 140
WOOL, I.G. 142
WOOLF, L.I. 8
WORMAN, J.C. 25
WORTHINGTON, R. 55
WOTIZ, H.H. 24
WRIGHT, R.D. (see J.R. Blair-West et al), 121
WRIGHT, Jr., D.P. 151
WULFF, G. 188
WUNDERLICH, B. 81, 94
WUNSCH, E. 192
WYLUDA, B.J. 190
WYMAN, Jr. J. 20
WYNBERG, H. 16, 232
WYNNE-JONES, W.F.K. 164

YAGER, W.A. 77
YAMADA, R. 93
YAMAMOTO, A. 211
YANG, J.Y. 113
YANG, N.C. 116
YANKWICH, P.E. 112
YARSLEY, V.E. 235
YASHIN, Ya. I. 25
YASUDA, H. 87, 178
YATSIMIRSKII, K.B. 158
YEN, V.Q. 34
YON, J. 60, 166
YOSHIDA, N. 85
YOUNG, G.J. 153
YOUNG, G.T. 191, 192, 193
YOUNG, L.J. 177, 178
YURINA, M.J. 40
YURKEVICH, A.M. 130
YUKHNEVICH, G.V. 129

ZACHARIAS, B. 137
ZACHMANN, H.G. 70
ZAFFARONI, A. 180
ZAHN, H. 28
ZAHRADNIK, R. 138
ZAITSEV, V.M. 129
ZAKA'NYCZ, S. (see D. Dunbar et al), 65
ZAMBELLI, A. 86
ZAMENHOF, S. 155
ZAMYATINA, V.A. 131
ZARETSKAYA, I.I. 46
ZAUBERIS, D.D. 150
ZAUGG, H.E. 100
ZAVGORODNII, S.V. 233
ZAVITSANOS, P.D. 112
ZBINDEN, G. 152
ZBINDEN, R. 235
ZECHMEISTER, L. 68
ZEEGERS-HUYSKENS, T. 75
ZEISS, H.H. 44
ZELEZNICK, L.D. 134
ZEMLICKA, J. 218
ZENGEL, H. (see Th. Kauffmann et al), 32
ZERVAS, L. 191
ZETTNER, A. 8
ZIEGLER, I. 206, 207
ZIEGLER, K. 44, 60
ZIELINSKI, A.Z. 60
ZILVERSMIT, D.B. 22
ZIMMERMAN, H.E. 17, 116
ZIMMERMAN, H.K. 57
ZIMMERMAN, J. 87
ZIMMERMANN, G. 154
ZIMMERMANN, H.G. 145
ZINOV'EV, A.A. 128
ZIOMEK, J.S. 63
ZIRKLE, C.L. 152
ZIRKLE, R.E. 203
ZLATKIS, A. 24
ZOLLINGER, H. 19
ZOLOTOV, Yu.A. 127
ZRELOV, V.N. 162
ZUBOV, V.P. 89
ZUBROD, C.G. 7
ZUCCARELLO, R.K. 93

ZUCKERMAN, J.J. 14
ZUMAN, P. 1, 235
ZURCHER, R.F. 56, 58
ZUTTY, N.L. 178
ZWEIFEL, G. 100
ZWEIG, G. 62
ZWIETERING, P. 4
ZWOLENIK, J.J. 169

SUBJECT INDEX

AAAS Journal, 132
Aberrations, genetic, 157
Abnormal peptides, 192
Absolute configuration, 111
Absolute intensities
    of infrared absorption bands, 117
Absorption, 55
    by a rigid crystal, 119
    dipolar Debye, and phase change, 183
    infrared, (see Infrared absorption)
    optical, 119 209
    ultraviolet (see Ultraviolet absorption)
Absorption spectra
    electronic, of DCO radical, 168
    electronic, of HCO radical, 168
    of group 5 deuterides, halides, hydrides, 169
    of steroid hormones, 197
Abundance tables
    for mass spectrometry, 221
Acceptor properties
    quadripositive, of metals, 117
Accumulation
    of triglycerides by the liver, 199
Acetaldehyde
    polymerization, 85
Acetic acid
    hydrazino-, peptide derivatives, 192
    molecular association, 183
    triple ion equilibria, 164
Acetoacetic acid
    analysis in blood, 146
Acetogenins
    biosynthesis, 230
Acetone
    analysis, with salicylaldehyde, 146
    separation from blood, 146
Acetophenone
    derivatives dipole moments, 141
Acetylation
    sulfanilamide, 201
Acetylcholine
    receptor, protein nature 202
Acetylene, 52 60
Acetylene derivatives, 12, 34, 42, 50
    coupling, 16
    hydration, 100
    natural, 69
Acid base titrations
    nomogram for, 139
Acidic metabolites
    from deoxycorticosterone, 180
Acidity 32
Acid(s)
    fatty 170
    hydration, 127
    hydroxy, metal complexes of, 127
    α-keto, 35
    perchloric, 128
    ribonucleic, 232
    solvation, 127
    synthesis, 175
    unsaturated, 87
Acrolein, 232
    polymers 32
Acrylates 88
    copolymerization, 178
Acrylic acid 60
Acrylic polymers, 212
Acrylonitrile, 88
    copolymerization, 177
ACTH secretion 198
    pharmacological control 198
Actinides, 37
Actinoidin, 40
Actinomycin
    nucleic acid function, 104
Activated molecules, 101
Activation
    chemical, 118
    nuclear, 14
Activation analysis, 37
    nuclear, 132
Activation entropy, 19
Activators
    of drug enzymes, 201
Active agents 226
Active centers
    of enzymes, 101
Active esters
    in peptide synthesis, 191
Activity
    and structure in peptides, 202
    biological (see Biological activity)
    enzyme, and corticosteroids, 142
Actomyosin
    and cardiac glycosides, 200
Acylation 117 174-175, 230
    aliphatic, 175
    enzymatic and peptide bond formation, 201
    mechanism, 175
    with dicarboxylic acid derivatives, 175
    with polycarboxylic acid derivatives, 175
Adamantane, 49
Addition polymerization
    anionic mechanism, 160
    cationic mechanism, 160
    radical mechanism, 160
Addition reactions
    1,3-dipolar cycloadditions, 29
    free radical, 100
Adducts, 190
Adenylic acid, 41
Adipose tissue
    metabolism, effect of hormones, 122
    release of fatty acids from, 199
Adrenocortical factors, 122
Adrenocortical steroids
    and antiinflammatory action, 121
    biosynthesis, non-hormonal inhibitors, 181
    metabolic effects, 121

Adrenergic neurone blocking, 9
Adsorbents
    impregnated, for chromatography, 188
    with energy-heterogeneous surfaces, 158
Adsorption
    of dielectrics, 129
    of polymers, 70
    of semiconductors, 129
    potential theory, 158
Adsorption chromatography
    isotope fractionation, 113
Adsorption-flocculation reactions
    of macromolecules, 125
Adsorption reactions
    of macromolecules, 125
Advances
    in polarography, 126
Aeronomy
    chemical, 169
Aerosols, 224
Agar-agar layers
    high-voltage electrophoresis, 146
Agricultural chemistry
    Feigl's reactions, 139
Agriculture 210
Alcohol
    ethyl, rotational isomerism, 183
Alcohol dehydrogenases, 203
Alcohols
    addition to triple bonds, 130
    dehydrogenation, 158
    dielectric relaxation, 183
    ethylenic, 35
    for alkylation, 174
    hydrogen bonds, 34
    intermolecular association, 183
    O-nitration, 189
    oxidation, 141
    viscosity, 183
Aldehydes
    for alkylation, 174
    5-hydroxymethylfurfural, 125
    polymerization, 85, 224
    synthesis, 175
Aldosterone secretion
    control, 121
Algae
    cell-free preparations, 124
    photosynthesis, 124
Alicyclic compounds, 227
    analytical methods, 144
    from thiophene derivatives, 141
    gas chromatography, 144
    synthesis, 141
Aliphatic acids
    kinetics of dissociation, 144
Aliphatic acylation, 175
Aliphatic alcohols
    oxidation, 141
Aliphatic compounds
    reaction mechanisms, 209
Aliphatic hydrocarbons
    halogen derivatives, 209

Aliphatic mono-olefins, 160
Aliphatic substitution
    electrophilic, 123
Alka-1,3-dienes
    1-substituted, 127
Alkali
    instability, in 4-oxalysine, 192
Alkaloids, 35, 45, 52, 114, 222
    analysis, 188
    apocynacea, 45
    aporphine, 48
    biogenesis, 28, 29, 101
    bisbenzylisoquinoline, 104
    chromatography, 188, 214, 215
    diterpenes, 66
    diterpenoid, 123
    elaboration by Schmidt reaction, 111
    ergot, 78
    gas chromatography, 156
    identification, 188
    indole, 225
    indole, dimeric, 114
    lactonic, 49
    7-membered ring, 111
    phenolic, biogenesis, 101
    rauwolfia, 102
    separation 188
    steroid, mass spectrometry, 141
    tropane, gas chromatography, 146
    veratrum, 78
    vinca, 7
Alkanes, 209
    for alkylation, 173
    high molecular, nitration, 189
Alkenes
    alkylation, 174
    for alkylation, 173
    phosphene, 115
    substituted for alkylation, 173
Alkoxylvinyl esters
    as coupling reagents in peptide synthesis, 191
Alkyl, arenesulfonates, 127
    for alkylation, 174
C-alkylation
    of cyclic -ketoesters, 141
Alkylation, 173-174
    of alkenes, 174
    of aromatic compounds, 173, 174
    of saturated hydrocarbons, 174
    saturated ketones, 123
    unsaturated ketones, 123
    with alcohols, 174
    with aldehydes, 174
    with alkanes, 173
    with alkenes, 173
    with alkyl arenesulfonates, 174
    with alkynes, 174
    with carbonyl compounds, 174
    with dienes, 173
    with ethers, 174
    with haloalkanes, 174
    with inorganic esters, 174
    with ketones, 174

with olefins, 233
with substituted alkenes, 173
Alkylbenzenes
  mass spectra, 185
  ultraviolet spectra, 141
Alkyl cations
  rearrangements, 110
Alkylhydrazines, 130
Alkynes
  for alkylation, 174
Allene hydrocarbons, 35, 129
Alloxazines, 128
Allyl
  anion wave function 168
  cation, wave function, 168
  radical, wave function, 168
Allylics, 99
Alpha-chymotrypsin
  enzyme catalysis, 133
Alternant sytems
  conjugated, 172
Aluminum, 30, 44
  alkyl chlorides, in polymerization, 84
  in polymerization, 86, 89
Aluminum compounds
  aryl 125
  electrochemical properties, 130
  physicochemical properties, 130
Aluminum halides
  coordination compounds, 173
Amanita phalloides, 108
  peptides, 114
  toxicology, 108
Ambident ions, 32
Amides
  poly-, synthetic hetero-chain, 227
Amination, 175
Amine oxidases
  biological inactivation by, 201
Amines, 150
  acid-base equilibria, 57
  aromatic (ss Aromatic amines)
  biosynthesis, 155
  centrally active, 29
  gas chromatography, 25, 156
  metal, in solution, 167
  tertiary, analysis, 139
α-Amino acid anhydrides
  N-carboxy-N-trityl-, in peptide synthesis, 192
Amino acids, 35, 36
  analysis, by electrical conductivity, 146
  analysis, of peptides, 204
  analysis, of proteins, 204
  and peptides, 146
  basic, protection with tosyl group, 191
  biosynthesis, 155
  chromatography, 45
  dinitrophenyl, 45
  electrophoresis, 45
  fractionation, 188
  gas chromatography, 156, 188
  gel filtration, 188
  glycosyl esters, 193
  in enzymes, 101
  magnetochemistry, 77
  metabolism disorders, 157
  N-methylated, synthesis, 192
  microbial synthesis, 137
  protective groups, 191
  thin-layer chromatography, 188
  uncommon, in peptide synthesis, 192
  "unnatural", in peptide synthesis, 192
p-Aminobenzoic acid, 217
Aminoboranes, 30
γ-Aminobutyric acid, 142
Aminodeoxy sugars, 131
Aminoethyl-2-benzo-1,4-dioxane, 198
S-(β-Aminoethyl-)-L-cysteine
  in peptide synthesis, 192
α-Amino groups
  protection, in peptide synthesis, 191
Amino-plastics, 234
Aminopteridines, 205
Amino sugars, 10
Ammonia
  electronically excited states, 169
  liquid, 35
  reactions of metal halides with, 102
Ammonium perchlorate, 139
Ampholytes, 53
Amphoterism
  of HNO_3, 189
  of nitronium ion, 189
Analgesics, 65, 67
  narcotic antagonists, 152
Analog computer
  in enzyme synthesis, 166
Analysis, 222, 231, 234
  automatic structural, 144
  by flame photometry 225
  calorimetric, of bromine, 144
  chromatographic (see Chromatography)
  clinical, of natural substances, 145
  colorimetric, 222
  commercial methods, 232
  electron probe, 222
  functional group, 123, 223
    semi-micro methods, 222
  galvanic, 1
  gas chromatographic (see Gas chromatography)
  gravimetric, use of EDTA, 167
  group reactions in, 144
  infrared, of polyethylene, 145
  in organic chemistry, 143
  inorganic, complexing reagents 229
    liquid ion exchangers, 139
  in polymer research, 145
  luminescence, development, 143
  mass spectrometric (see Mass spectrometry)
  mercury drop electrodes in, 1
  modern, in food science, 143
  nuclear activation, 132
  of acetoacetic acid, in blood, 146
  of acetone, with salicylaldehyde, 146
  of active agents, 226
  of alkaloids, 188
  of amines, 139
  of amino acids, 204
    by electrical conductivity, 146

of anesthetics in blood, 156
of bromine, 144
of carbon, 144
   in mineral oils, 144
   rapid combustion apparatus, 144
of carbon compounds, 209
of chlorine content, 143
of coenzyme A esters, 188
of corticosteroids, 215
of creatinine, 146
of drug-receptor interactions, 202
of drugs, 224
of elements, 143
of fatty acids, 156
of functional groups, 143, 223
   by NMR, 144
of hormones, 121
of hydrogen, 144
   in mineral oils, 144
   rapid combustion apparatus, 144
of nitrogen content, 143
of nitro-group position, 189
of organic compounds, 234
of pesticide residues, 143
of pharmaceuticals, by tropaeolin method, 146
of polyethylene, 145
of polyphenol mixtures, 145
of pteridines, 207
of radioactive compounds, 188
of steroidal hormones, 121
of steroids, 188, 229
of sugars, 188
of synthetics, 145
of synthetic fibers, 143
of synthetic sulfur compounds, 162
of tertiary amines, 139
of thiourea, 143
of trace impurities, 111, 128
of vitamin B_6, 142
of xylene oxidation products, 176
optical activity measurement, 133
pharmacological, of central nervous action, 202
photometric titrations, 1
polarographic, (see Polarography)
powder method, X-ray, 126
precision, of multicomponent mixtures, 144
qualitative, 1
quantitative, 111
   of drugs, 224
radioactivation, 222
radiocarbon data, 132
rotational, of SO_2 bands, 168
semi-micro methods, 222
spectrophotometric, chelating agents in, 123
structural, at high pressure, 179
structural, by X-rays, 179
thermal methods, 234
thin layer chromatographic (see Thin layer chromatography)
titrimetric, 221
X-ray, 115, 144
   of natural products, 101, 115

Analytical chemistry, 139
   complexation in, 230
   enzymes in, 137
   ion exchange separations, 231
   of thiocarbamides, 143
Analytical methods
   catalytic reactions with hydrogen peroxides, 158
   for alicyclic compounds, 144
   gas, development, 143
Analytische Chemie
   Zeitschrift, 143-146
Androgenicity, 121
Androgenic steroids, 153
Androgens
   biosynthesis, 121
   inhibitors, 122
   interconversion, 121
   secretion, 121
   specific activities, 122
Anesthetics, 199
   in blood, 156
   local, 102
Anhydrase
   carbonic, 121
Anhydrides
   in peptide synthesis, 192
Anhydrocyanodine
   absolute configuration, 111
   stereochemistry, 111
   structure, 111
Anilines
   spectra, 63
Animal tissues
   free nucleotides in, 104
Animals, 122
   chemical communication, 122
Anion
   allyl, wave function, 168
Anion exchangers
   separation of hydroxy-acids, 138
Anionic copolymerization, 177
Anionic polymerization
   kinetics of propagation, 106
Anomolous relaxation
   in ESR spectra of radical ions, 183
Antagonists
   of folic acid, 207
Anthracyclinones, 68
Antialdosterone compounds
   action, 181
Antiandrogenic compounds, 180, 194
Antianginal drugs, 235
Antibiotics, 40, 68
   amino sugars from, 3
   biosynthesis, 111, 126
   iron-containing, 108
   macrolide, 117, synthesis, 126
   peptide-type, 108
   polyene macrolide, 138
   structure-activity relation, 108
   synthetic approach, 108

Antibodies, 66
Anticancer,
  chemotherapy, 7
Anticancer agents
  resistance to, 2
Anticonvulsant drugs, 102
Antidepressants, 152
Antidiabetic drugs
  release control, 198
Antiestrogenic compounds, 180
Antifertility agents, 67
Antifungicides, 68
Antigonadotrophic activity
  of progestinics, 199
Antihypertensives, 152
Antiinflammatory action
  adrenocortical steroids, 121
Antineoplastic, 7
Antioxidant effects
  in biochemistry, 107
  in physiology, 107
Antioxidant mechanisms
  in elastomers, 213
Antiovulatory compounds, 180
Antiprogestins, 194
Antisera
  to ACTH, 122
  to growth hormone, 122
  to TSH, 122
Antitussive drugs, 102
Antitussives, 9
Aphids
  coloring matters, 110
Aporphine alkaloids, 48
Appearance potential, 185
Applied chemistry
  reviews, 125
Approximation procedures
  in isotope effects, 112
Aquametry, 26
Archaeology
  and archaeometry, 126
  and physical sciences, 126
Archaeometry, 126
Arenes
  metal, 149
Arenesulfonates, 127
  alkyl, for alkylation, 174
Aromatic acids
  urinary, 156
Aromatic amines
  structure, 200
  toxicity, 200
  velocity of biological n-hydroxylation, 200
Aromatic carbonium ions
  intramolecular proton exchange, 113
  isotope effect, 113
Aromatic compounds, 160
  alkylation, 173, 174
  bridged, 232
  cycloalkylation, 174
  dehydrogenation condensation, 174
  fluorescence, 47
  free radicals, 77
  from lignin, 111
  fungal metabolism, 111
  hydrogen exchange, 174
  5-membered ring, 126
  meta-bridged, 47
  sulfonyl chlorides, 127
Aromatic diazonium compounds, 127
Aromatic hydrocarbons, 58, 172
  chromatography, 50, 214
  crystal structure, 210
  copolymers, 87
  isomerization, 174
  monocyclic, 129
  nonbenzoid aromaticity, 187
  photochemistry, 17
  polycyclic, 129
  proton complexes, 33
  ultraviolet spectra, 73
Aromaticity
  molecular orbital calculations, 187
  of nonbenzoid aromatics, 187
Aromatic ketones
  synthesis, 174
Aromatic ligands, 149
Aromatic molecules
  electron spin resonance, 77
  radio resistance, 165
Aromatic nitration
  positive poles, 189
Aromatic radical-ions
  electron spin resonance spectra, 117
Aromatic radicals
  electron spin resonance spectra, 117
Aromatic side chain effects
  on polypeptide structure, 166
Aromatic substitution
  directive effects, 19
  electrophilic, 105
  hydrogen isotope effect in 19
  nucleophilic, 105
  quantitative, 210
Aromatic systems
  nonbenzenoid, 31
  planarity, 19
Arrhenius equation
  deviations, 118
Arrhenius parameter effects
  evidence for tunnelling, 113
  in base-promoted elimination reactions, 113
Arrows
  in biochemistry, 165
Arsenic complexes, 32
Arsines, 14
Arylaluminum compounds, 125
Aryl methyl sulfides
  dipole moments, 140
Arylsulfonamides
  hypoglycaemic, 198
L-Ascorbic acid, 217
Association
  intermolecular, of alcohols, 183
  molecular, of acetic acid, 183
  of proteins, 20

Astatine, 14
Asymmetric catalysis
  theory, 158
Atisine, 123
Atomic absorption spectra, 1
Atomic absorption spectroscopy, 8
Atomic distances
  additivity, 69
Atomic structure
  and chemical bonding, 231
ATP, 41
Attractions
  Van der Waals dispersion, in DNA, 166
Automatic analysis
  for carbon, 144
  for hydrogen, 144
  for nitrogen, 144
Automaticity
  of cardiac muscle cells, 200
Automatic structural analysis, 144
Automatic X-ray diffractometer, 144
Autoradiography, 22
Azapyrimidines, 103
Azeotropy, 233
Azides
  inorganic, 117
Azide synthesis
  of peptides, side reactions, 191
Azoic dyes
  infrared absorption, 141
Azo groups
  infrared absorption, 141
Azotobacter
  nitrogen fixation by, 137
Azuridines, 161

Bacillus thurungiensis, 151
Bacteria
  in RNA synthesis, 104
Bacterial viruses
  genetic coding in, 125
Bacteriogenic fractionation
  of sulfur isotopes, 113
Baekeland, L.H., 54, 73
Bailey's
  Industrial Oils and Fat products, 233
Bands
  flame, of carbon monoxide, 168
  infrared, absolute intensities, 117
Bark, 163
Bases
  pseudo, heterocyclic, 12
  pyrimidine, and hydroxyl free radicals, 166
Base strength
  and molecular orbitals, 187
Bent bonds, 132
Benzazepines, 34
Benzene
  triplet state, 77
  voltage-current charactertistics, 164

Benzene derivatives
  equilibration, 110
  gas-phase halogenation, 110
  o-nitro, substituent interactions, 118
Benzodiazepines, 79
2-Benzo-1,4-dioxane
  aminoethyl-, 198
Benzoic acid
  substituent effects, 140
Benzotriazin-4-one
  3-amino-, 27
Benzyloxycarbonyl groups
  influence, on racemization, 192
Best's Safety Maintenance Directory, 226
Betuline
  triterpenes, NMR, 141
Bidentate chelates, 159
Bifunctional compounds
  polyflurinated, 128
Biguanide
  β-phenethyl-, degradation products, 198
Biguanide compounds
  in hepatectomized dogs, 199
  in hepatopancreatectomized dogs, 199
Bile acids
  gas chromatography, 156, 188
  metabolism, 199
  thin-layer chromatography, 188
Bile pigments
  biogenesis, 155
Biocatalyst modelling, 130
Biochemical calorimetry, 112
Biochemical dimensions, 123
  in photosynthesis, 123
Biochemical hydroxylation
  electronic aspects, 166
Biochemically important compounds, 131
Biochemical macromolecules
  at very low temperature, 166
Biochemical mechanisms
  and reversible reactions, 166
Biochemical pathology, 157
Biochemical systems
  electrochemical aspects, 164
  glucose-glucose oxidase, 164
  theory, 166
Biochemistry, 224
  antioxidant effects, 107
  arrows, 165
  electronic aspects (International Symposium), 165-167
  electrons, 165
  Feigl's reactions, 139
  of biotin, 142
  of ferrichrome compounds, 167
  of β-1,3-glucans, 125
  of inositol lipids, 142
  of nucleic acids, 225
  of progesterone, 142
  orbitals, 165
  quantum, 5
  Rapid Mixing and Sampling Techniques (collective volume), 208
  separations, 228

Biodegradability, 39
  of hydrocarbons, 54
Bioflavonoids, 217
Biogenesis (see also Biosynthesis)
  of alkaloids, 101, 155
  of bile pigments, 155
  of carbohydrates, 155
  of chlorophyll, 155
  of conjugation products, 155
  of detoxication products, 155
  of fatty acids, 170
  of heme, 155
  of lignins, 155
  of lipids, 155
  of natural compounds, 155
  of phenolic alkaloids, 101
  of plant products, 114
  of proteins, 155
  of purine nucleotides, 155
  of pyrimidine nucleotides, 155
  of rubber, 155
  of terpenes in plants, 155
  of vitamins, 217
Biological activity
  and chemical constitution, 217, 231
  and chemical properties, 202
  and molecular shape, 201
  and molecular size, 201
  and physical properties, 202
  polypeptide hormones, 108
Biological complexity
  and radio-sensitivity, 134
Biological implications
  of gas chromatography, 133
Biologically active microbial metabolites
  structure, 110
Biological oxidation
  of fatty acids, 170
Biological specificity
  in protein molecule interactions, 202
Biological systems
  and excitation migration, 165
  copper in, 167
  intermolecular forces, 166
  metal chelates, 159
  metal ion catalysis in, 149
  phosphorylation in, 136
Biological transfer reactions, 135
Biology
  Isotope Mass Effects (Symposium), 112
  molecular, intermolecular interactions, 166
  paramagnetic resonance, 190
  Perspectives in (collection of papers), 194
Bioluminescence, 50
Biomedical applications
  of gas chromatography, 156
Biomedical Frontiers
  in Medicine (collective volume), 157
Biophysics
  Progress in, and Molecular Biology
    (collection of reviews), 203
Biopolymers
  electrical conductivity, 92
  optical properties, 165

Biopterin, 206
Biosynthesis
  of acetogenins, 230
  of adrenocortical steroids, 181
  of alkaloids, 101
  of amino acids, 155
  of androgen, 121
  of antibiotics, 111, 126
  of carotenoids, 155
  of fatty acids, 107
    unsaturated, 133
  of fungal metabolites, 155
  of lipids, 229
  of macrolide antibiotics, 126
  of nucleic acids, 155
  of phenolic plant products, 155
  of steroids, 155, 230
  of tannins, 155
  of terpenes, 230
  of tetrapyrrole, 227
  of vitamin A, 155
  of vitamins, 224
    water-soluble, 155
Biotin, 217
  biochemistry, 142
Biradicals
  trapping, in solution, 190
Birefringence
  streaming of polymers, 92, 93
Bisbenzylisoquinoline alkaloids, 104
Block, Richard J.
  memorial symposium, 152
Block copolymerizations, 177
Block copolymers
  containing ethylene, 177
Blood
  analysis of acetoacetic acid, 146
  anesthetics in, 156
  separation of acetone, 146
  transport of fatty acids, 199
Blue green algae
  cell-free preparations, 124
  photosynthesis, 124
Bohr, Niels
  memorial lecture, 101
Bombybol, 98
Bond energies, 39, 71, 117, 140
Bonding, 223
  and atomic structure, 231
  and electronic structure, 231
  bent bonds, 132
  geometry of molecules, 230
  hydrogen (see Hydrogen bonding)
  in organophosphorous compounds, 115
  sulfur-sulfur, 135
Bond length
  of carbon-carbon, 137
Bond orders, 172
Bonds, 23, 50
  bent, 132
  carbon-metal, 42
  electronic charges, 222
  hyperconjugation, 141
  in polymers, 87

magneto-optical study, 34
metal-ligand, 159
OH, infrared characterization, 129
organometallic compounds, 126
unsaturated, 145
Bones, 142
Borane, deca-, 14
Boranes, 157
amino-, 30
Borazines, 14
Boron
carbon-hydrogen system, 139
high temperature reactions, 31
organo-, 233
Boron compounds
polymeric, 131
toxicology, 157
unsaturated, 131
Boron halides
coordination compounds, 173
Boron hydrides, 227
ionic, 157
Boron-nitrogen chemistry, 151
Box model
in conjugated systems, 171
Bradykinin, 66
and vaso-dilating polypeptides, 202
Bridged structure
of $C_2H_6^+$, 168
Brittleness
in high polymers, 178
Bromine
calorimetric sub-microanalysis, 144
Bromonium ions, 72
Butadiene, 27, 86, 90
polymerization, 85
Butenolides, 49

^{13}C hyperfine interactions
in semiquinones, 183
^{14}C, 96
$C_2H_6^+$
bridged structure, 168
C_3 molecule
spectrum, 168
^{14}C glucose
hormonal regulation of carbohydrate
metabolism, 121
Cadmium, 44
Cahn electrobalance, 216
Calciferol, 218
Calciphyloxis
and parathyroid glands, 122
Calcitonin
and control of plasma calcium, 122
Calcium
and cardiac glycosides, 200
plasma, control, 122
variation, 113
Calorimetric analysis
of bromine, 144

Calorimetry, 112, 154
biochemical, 112
inorganic reactions, 112
mixing processes, 112
non-reacting systems, 112
organic compounds, 112
of ordering transitions, 112
of phase transitions, 112
of polymers, 94
solution, 112
Cambridge
chemistry at, 1901-1910, 101
Cancer, 37
biochemistry, 157
with purine antimetabolites, 157
Capillary chromatography, 154, 226
Caprolactam, 60
Carbamide, 90
Carbamyl phosphate, 132
Carbanions
structure, 110
Carbenes, 13, 47, 104, 226
methylene, 106
Carbinyl
methylcyclopropyl-, derivatives, 114
Carbohydrated metabolism
disorders, 157
Carbohydrates, 3, 36, 52, 53, 82, 190, 224
amino sugars, 3, 10
biognesis, 155
crystal structure, 3
electrophoresis, 3
gas chromatography, 3, 156
infrared spectra, 3
metabolism, 121
NMR spectra, 3
photochemistry, 3
reactions, 29
sugar nucleotides, 10
thiosugars, 3
Carbolines, 13
Carbon^{14}, 72
early history, 132
lowlevel counting, 26
Carbon
analysis, in mineral oils, 144
automatic, 144
rapid combustion apparatus, 144
boron-hydrogen system, 139
carbon bonds, hyperconjugation, 141
length, 137
divalent, 225
hydrogen bonds, hyperconjugation, 141
isotope effects, pyrolysis, 112
metal bond, 42
oxides, 33
polyatomic molecules, energy calculations, 187
saturated, nucleophilic substitution at, 222
subgroup, vibrational spectra, 129
Carbon compounds, 209
analysis, 209
crystallography, 209

isotopically labelled, 209
optical absorption, 209
shape, 225
stereochemistry, 209
wave mechanics, 209
Carbonium ions, 105, 160
intramolecular proton exchange, 113
isotope effect, 113
Carbonic anhydrase
in steroid-responsive tissues, 121
Carbon monoxide
flame bands, 168
Carbon tetrachloride, 29
Carbonyl compounds, 117
catalysis, 105
charge distributions, 141
condensations, with phosphite esters, 115
α-β-epoxyketones, 116
for alkylation, 174
mechanism, 105
olefination stereospecific, 115
Paterno-Buchi reaction, 116
photochemistry, 116
Carboranes, 74
Carboxyl groups
protection, in peptide synthesis, 191
Carboxylic acids
catalytic synthesis, 131
derivatives, acylation with, 175
infrared spectra, of complexes, 127
of salts, 127
Koch synthesis, 30
spectra, 76
N-Carboxy-N-trityl- -amino acid anhydrides
in peptide synthesis, 192
Carcinogenesis
mechanism, 165
Cardiac glycosides, 199, 200
and actomyosin, 200
and calcium, 200
and ion transport, 200
in automaticity, of cardiac muscle cells, 200
in contractile activity, of cardiac muscle cells, 200
inotropic action, 200
metabolic action, 200
metabolism, 200
Cardiac metabolism
effect of digitalis, 200
Cardiac muscle cells
automaticity, 200
contractile activity, 200
Carotenoids, 16
biosynthesis, 155
Carrin system, 32
Cartilage, 142
Cata-condensed hydrocarbons, 171
Catalog
of NMR spectra, 221
Catalysis, 4, 31, 73, 221, 230
acid, 12
acid-base, 30
and catalysts, 230

and ESR spectra, 91
and inhibition, 221
and kinetics, 158
asymmetric, theory, 158
biocatalyst modelling, 130
by iron, of mercaptoacetate autoxidation, 150
by metals on pyridines, 12
carbonyl group reactions, 105
dinitriles, reduction, 130
electrical conductivity, 158
enzymatic, 60, 133
theory, 158
enzyme-metal ion activation, 159
Fischer-Tropsch synthesis, 158
heterogeneous, 38
hydrogen waves, 129
infrared spectroscopy, 134
mercaptoacetate autoxidation, 150
metal ion, in biological systems, 149
in nucleophilic reactions, 149
in perester synthesis, 136
in peroxide reactions, 136
multiplet theory, 158
of polymerization, 45, 98
olefins in presence of metal oxides, 128
organometallic, 30
pyridoxal, 232
semiconducting oxide catalysts, 158
Catalysts, 54
and catalysis, 230
carbonic anhydrase, 121
for Friedel-Crafts reaction, 173
in polymerization, 90
Lewis acid, 173
metal oxide, for olefin copolymerization, 177
metallic, 99
stereospecific, 84
Ziegler-Natta, 84
Catalyst surfaces, 216
Catalytic combustion
ionization detectors, 176
Catalytic hydrogenation
by pentacyanocobaltate, 150
of furan compounds, 128
Catalytic phenomena
enzyme-metal ion activation, 159
metal complexes, 159
Catalytic properties
of dielectrics, 129
of semiconductors, 129
Catalytic reactions
complex formation, 158
with hydrogen peroxide, 158
Catalytic redistribution
of hydrogen, in unsaturated cyclic hydrocarbons, 127
Catalytic reduction
of dinitriles, 130
Catalytic synthesis
from alcohols, 131
from alkenes, 131
from carbon monoxide, 131
of carboxylic acid esters, 131
of carboxylic acids, 131

Catecholamines
  gas chromatography, 25
  transport of fatty acids, 199
Catena compounds, 32
Cationic copolymerization, 177
Cationic graft polymerization, 160
Cationic polymerization, 160
  high energy radiation, 160
  mechanism, vinyl monomers, 129
Cationic reactions
  of polymers, 160
Cations
  allyl, wave function, 168
  alkyl, rearrangements, 110
  hydrolysis equilibria, 137
Cats
  neurogenic influence, 199
Cell-nuclear compounds
  under ionizing radiation, 165
Cells, 108
  macromolecular synthesis, 114
  mitosis, 114
Cellular level
  pharmacological effects, 200
Cellular sites
  of RNA synthesis, 103
Cellulose, 3, 53, 82, 90, 99, 163, 233
  copolymers, 87
  degradation, 39
  enzymic hydrolysis, 153
  grafting on, 87
  ion exchangers, 128
  pyrolysis, 94
  radiation effects, 56
  structure, 82
Cellulose acetate, 235
Cellulose ethers, 235
Cellulose nitrate, 235
Cellulose plastics, 235
Celluloses
  regenerated, 235
Cell wall
  gram-negative, 134
Centrifugal chromatography, 62, 214
Centrifugation
  density gradient, 87
Cephalins, 44
Cerium compounds
  oxidation by, 150
Characterization
  by starch gel electrophoresis, 121
  of organic compounds, 234
Charge distribution
  in carbonyl compounds, 141
  in thiocarbonyl compounds, 141
Charges
  electronic, 222
Charge transfer, 73, 74
  complexes, 211
Chelates, 47, 149
  bidentate, 159
  flame photometry, 65
  metal, precipitation, 139
  multidentate, 159
  of copper (II), 167
  of nickel (II), 167
  polymers, 129
Chelating agents, 31
  and metal chelates, 159
  in spectrophotometry, 123
Chemical aeronomy, 169
Chemical analysis (see Analysis)
Chemical bonding (see Bonding)
Chemical bonds (see Bonds)
Chemical communication
  among animals, 122
Chemical constitution
  and biological activity, 217, 231
Chemical data
  for industry, 137
  for science, 137
Chemical Education
  Journal (cumulative indices), 235
Chemical engineering, 221
Chemical equilibria (see Equilibria)
Chemical kinetics (see Kinetics)
Chemical mutagenesis
  specificity, 103
Chemical nomenclature, 222
Chemical progress, 123
Chemical properties
  and biological activity, 202
Chemical reactions (see Reactions)
Chemical Reviews
  Russian, 127
Chemical Society
  Proceedings, 101
Chemical terminology
  German-English, 224
Chemiluminescence
  in solution, 115
Chemisorption, 225
Chemistry
  analytical (see Analytical chemistry)
  clinical, 8, 225
  inorganic, 225
    progress, 102
  Journal of the Royal Institute, 126
  medicinal, 222
    progress, 102
  modern, 226
  nuclear, 228
  organic (see Organic chemistry)
  physical organic, progress, 105
  progress, 135
  pure and applied, 108
    reviews, 125
  quantum (see Quantum chemistry)
  solid state, 101
Chemosterilants, 151
Chemotherapeutic agents
  structure-activity relationship, 40
Chemotherapy, 7, 9, 40
  anticancer, 7

cancer, 157
clinical, 7
prophylaxis, 7
viruses, 7
with purine antimetabolites, 157
Chlorine analysis
of non-volatile compounds, 143
Chlorine dioxide
ultraviolet spectrum, 168
Chlorohydrins, 60
Chlorophyll
biogenesis, 155
Chloroplasts, 4
photophosphorylation, 124
Chloroprene, 90
Choline, 218
Cholinergic receptor, 202
Chromatography, 61, 62, 224
adsorption, 113
capillary, 154, 226
centrifugal, 62, 214
circular, 214
column partition, 79
gas, (see Gas chromatography)
ion-exchange (see Ion exchange chromatography)
linear elution adsorption, 1
liquid, permanent columns, 145
mobilities, of steroid hormones, 197
of alkaloids, 188
of amino acids, 45
of aromatic compounds, 50
of fatty acids, 170
of gastric hormones, 8
of lipids, 15, 107
of oligonucleotides, 103
of polar lipids, 189
of polynucleotides, 103
of proteins, 128
of steroids, 228
of sulfur compounds, 162
on $AgNO_3$-SiO_2 gel thin layers, 145
on impregnated adsorbents, 188
paper, (see Paper chromatography)
thin-layer (see Thin-layer chromatography)
Chromium complexes
in redox reactions, 150
Chromyl chloride
compounds with hydrocarbons, 211
Chugaev, L.A., 72
Cinchona, 35
Circular chromatography, 214
Circular dichroism, 23
in stereochemistry, 101
of dissymmetric α,β-unsaturated ketones, 168
of transition metal complexes, 123
Cis-trans isomerization
kinetics, 106
Clathrates, 151, 190
and other condensed phases, 118
molecular interactions, 118
Cleavage
non-enzymatic, of peptide chains, 193
Clinical analysis
of natural substances, 145

Clinical chemistry, 8, 225
Clinical medicine
diagnostic methods, 157
Clinical methodology
of radioisotopes, 22
CNS drugs, 95
CO_2
in oxygen determination, 112
Coatings, 52, 125
Codewords
in RNA, 134
Coding
genetic, 36, 125
Coefficients
partition, of steroid hormones, 197
Coenzymes, 36, 234
A group esters, analysis, 188
interaction with steroid hormones, 180
pyridine nucleotides, 10
Q group, 10
ubiquinones, 107
vitamin B_{12}, 111
Collagen, 126
Collision theory
reactions in liquids, 228
Colloids, 83
Colorimetric analysis, 126, 222
Column chromatography
of oligonucleotides, 103
of polynucleotides, 103
Column technology
for gas chromatography, 176
Combustions
calorimetry, 112
for radioisotopes, 22
inorganic reactions, 112
organic compounds, 112
Comonomer distribution
in copolymers, 213
Commercial methods
of analysis, 232
of synthesis, of α,β-unsaturated aldehydes, 130
of α,β-unsaturated ketones, 130
Communication
chemical, among animals, 122
Competitive deuteration
of toluene and toluene- , , -d_3, 113
Competitive scavenger method
in radiation chemistry, 119
Complexation
in analytical chemistry, 230
Complex compounds
cyanides, 223
dibenzofurans, 136
structure, 226
theory, 226
Theory and Structure of, (international symposium), 211
Complexes
5-coordinate, 167
dipole moments, 76
donor-acceptor, 108
intermediate, 173
macrocyclic ligand, 123

metal, 221
   of hydroxy acids, 127
     substitution reactions, 135
     with olefins, 27
molecular, 221
multidentate ligand, 123
of arsenic, 32
of carboxylic acids, 127
of peptides, 108
of phosphorus, 32
of proteins, 108
of transition metals, 5
   magnetic properties, 102
proton, of hydrocarbons, 33
selenium, 129

Complex formation
  of ethylene diamine, with mercury (II) ions, 167
  pi-complex, 113
  secondary isotope effects, 113
  stepwise, 112
  thermochemistry, 112

Complexing reagents
  structure behavior, 229

Complex ions
  of chromium, 149

Composition tables, 233

Computers
  analog, in enzyme synthesis, 166
  for chemical problems, 138

Condensation
  dehydrogenation, of aromatic compounds, 174
  polymers, 212
  reactions, of carbonyl compounds, 115

Condensed phases
  clathrates, molecular interactions, 118
  equilibria in, at high pressures, 179

Conductivity, 72
  electrical, in solid proteins, 165
    of amino acids, 146
  of biopolymers, 92
  thermal, of polymers, 94, 95

Conductors
  organic, 6

Configuration, 73

Configurational statistics
  of polymer chains, 177

Conformation
  of peptides, 193
  of steroids, 58

Conformational analysis, 42, 46

Conformational equilibria
  in substituted ethanes, 110

Congress lectures, 110

Conjugated molecules, 131
  electronic spectra, 229
  free-electron theory, 229
    (collection of papers), 171-172
  quantum theory, 166
  radioresistance, 165

Conjugated systems
  electron transport, 165
  large ring, 111

Conjugation
  and non-bonded interactions, 141
  of π-electrons, in polyphenyls, 187
  of glucuronides, 200
  of sulfates, 200
  of mercapturic acid, 200

Conjugation energies
  in cyclic compounds, 140

Conjugation products
  biogenesis, 155

Connective tissue
  effects of corticosteroids, 122

Consecutive processes
  kinetics, 106

Constitution
  and biological activity, 231
  of organic compounds, 144
  of rifamycins, 111

Contact angles
  in wettability, 151

Contractile activity
  of cardiac muscle cells, 200

Contraction
  of muscles, electronic aspects, 166

Control
  metabolic, of timing, 132
  of insects, 151
  of plasma calcium, 122
  pharmacological, of ACTH secretion, 198

Conversion
  enzymatic, of cyanide to thiocyanate, 201
  inter-, of androgens, 121
  of oxygen, into CO_2, 112

Convex molecules, 5

Coordinate complexes
  5-, reactivity, 167
    stabilization, 167
    stereochemistry, 167

Coordination
  model, in non-aqueous solvents, 102
  of chemical processes in living cells, 108

Coordination chemistry, 31
  Essays in (collection of reviews), 167
  IUPAC conference, 182
  of metal dioximes, 136
  of tetrapyrrole pigments, 159
  Plenary lectures, 108

Coordination compounds, 227
  infrared spectra, 228
  of aluminum halides, 173
  of boron halides, 173
  of gallium halides, 173
  polarography, 102
  role of oxygen in, 211

Coordination polymerization
  in metal alkoxides, 211

Copolymerization, 160, 177
  anionic, 177
    kinetics, 106
  block, 177
  cationic, 177
  containing acrylates, 178
  containing acrylonitrile, 177
  containing methacrylates, 178
  containing styrene, 177
  containing styrene derivatives, 177
  containing vinyl chloride, 177
  containing vinylidene chloride, 177
  graft, 177
  high pressure, containing ethylene, 178
  of olefins, 177
  reactivity ratios, 178
  theory, 177
Copolymers, 86, 87, 213
  block, containing ethylene, 177
  configuration, 86
  containing ethylene, 177
  containing propylene, 177
  of α-olefins, 177
Copper
  in biological systems, 167
  magnetic moments, 49
Copper (II) chelates
  with Schiff bases, 167
Copper compounds
  of ketoimines, 150
Copper phthalocyanine
  electron spin resonance spectra, 190
Correlations
  kinetic isotope effects, 112
  mass-spectral, 227
  spectra-structure, 229
  structure-function, in enzyme action, 133
Cortinoids, 218
Corrins
  synthesis, 111
Corrosion
  inhibitors, 138
Corticoids
  pharmacology, 146
Corticosteroids
  analysis, 215
  and enzyme activity, 142
  chromatography, 215
  effect on connective tissue, 122
  effect on lysosomes, 122
  excretion, in Guinea pigs, 199
  in rat plasma, 199
Corticotropins
  structure-function relations, 108
Cost engineering
  in chemical industry, 221
Coumarins, 79
Counter-current distribution, 62
  of fatty acids, 170
  of lipids, 189
Coupling
  of vibrational and electronic motions, 102
  spin-spin, in NMR, 125

Coupling reactions
  influence of benzyloxycarbonyl groups, 192
  in peptide synthesis, 191
  with alkoxylvinyl esters, 191
  with dicyclohexylcarbodiimide, 192
Cracking
  stress-, in high polymers, 178
Craig distribution
  tables, 137
Creatinine
  analysis, 146
Cross-linking, 81
  of polymers, by radiation, 133
Cryogenics
  thermodynamics, 38
Crystallography, 78, 209, 225
Crystals, 23, 38, 54, 71, 119
  growth, inhibitors, 138
  liquid, 128
  melting mechanisms, 5
  of polymers, 81
  rigid, optical absorption, 119
  structure, of aromatic hydrocarbons, 210
    of carbohydrates, 3
  Van der Waals, 119
  violet, 99
Cyanides
  complex, 223
  enzymatic conversion, to thiocyanates, 201
Cyanine dyes, 161
Cyanuryl chloride
  2,4,6-trichloro-1,3,5-triazine, 130
Cycliacylation, 175
Cyclic compounds
  conjugation energies, 140

  desipeptides, synthesis, 193
  β-diketenes, nitration, 189
  inorganic, silicon containing, 128
  β-ketoesters, C-alkylation, 141
  macro-, synthesis, 131
  5-membered, structure, 140
  6-membered, structure, 140
  molecular orbital theory, 141
  mono-, aromatic hydrocarbons, 129
  19-norsteroids, 130
  oxygen containing, 160
  perhydroindene, 129
  poly-, aromatic hydrocarbons, 129
  purines, 125
  pyrazolidine-3,5-diones, 129
  silicon containing, 128
  stereoselectivity of reactions, 101
  strain energies, 140
  terpenes in plants, 155
  unsaturated, 127
Cyclizations, 31
  by radical reactions, 123
  of olefinic acids to ketones and lactones, 118
Cycloalkylation
  of aromatic compounds, 174
Cyclobutanes, 72, 129
  synthesis, 129

Cycloheptane, 34
Cyclol tripeptide
  synthetic, 193
Cyclooctatetraene, 30
Cyclopentadiene, 91
Cyclopentadienyl
  metal, 149
Cyclopentene
  polymerization, 33
Cyclopeptides, 108
  hydroxyacyl-, tautomerism, 193
  of amanita phalloides, 108
  toxicology, 108
Cyclopolymers
  diolefin, 213
Cyclopropanes
  infrared spectroscopy, 141
  NMR spectroscopy, 141
  structure, 141
Cyclopropene, 49
Cysteamine
  radiation protection, 119
Cystein
  S-(β-aminoethyl-)-L-, in peptide
    synthesis, 192
Cysteine
  peptides, 191
  residues, cleavage of peptide chains at, 193
  thiol group protection, 191
Cystine peptides, 191

Data
  thermodynamic, at National Bureau of
    Standards, 138
Dating
  radiocarbon, 126
DCO radical
  electronic spectra, 168
Deamino-oxytocin
  crystalline, 133
Debye absorption
  dipolar, and phase change, 183
Decaborane, 14
Decarboxylation
  enzymic, of acetoacetic acid, 101
β-Decay
  chemical changes during, 129
Decomposition
  isotope effects, 113
  ^{15}N isotope effects, 113
  of organic ions, 185
  tritium isotope effects, 113
Defects
  of crystals, 119
Definitions
  food, 224
Degradation
  of high polymers, 178
  of polyethylene, 213
  of quaternary amines, 113
  thermal, of polymers, 227
    of polyolefins, 145

Degradation products
  of β-phenethyl biguanide, 198
Dehydrogenase
  liver alcohol, inhibitors for, 202
Dehydrogenation
  condensation, of aromatic compounds, 174
  of alcohols, 158
Demethylation
  of drugs, 201
De Novo protein synthesis
  in vitro, 104
Deoxycorticosterone
  in vitro conversion to acidic metabolites, 180
Deoxyribonucleic acid, 23, 41, 65
  conductivity, 92
  electronic properties, 6
  label, HT$_3$DR, 114
  quantum chemistry, 5, 166
  replication, incorporation errors, 166
  proton transfer, 166
  structure by X-ray scattering, 103
  triplet states, 165
  Van der Waals dispersion attractions, 166
Depsopeptide analogues
  of glutathione, 193
  of ophthalmic acid, 193
Desaminooxytocin, 194
Desipeptides
  cyclic, synthesis, 193
Desulfuration, 201
Detection
  of diastereoisomers, in dipeptides, 192
  of racemization, by "two spots" method, 192
  of radioactive compounds, 188
Detection devices
  for gas chromatography, 176
Detectors
  ionization, 63
    catalytic combustion, 176
Detergents, 53, 54
Detoxication
  mechanism, in vivo, 200
  products, biogenesis, 155
Deuteration
  of toluene and toluene-α,α,α-d$_3$, 113
Deuterides
  of group 5 elements, 169
Deuterium Arrhenius parameter effects
  base-promoted elimination reactions, 113
Deuterium isotope effects
  in enzymic formate oxidation, 114
  in organophosphorous compounds, 114
  kinetics, 114
Deuterium solvent isotope effects
  in acid-base equilibria, 112
Diagnostic methods
  in clinical medicine, 157
Dialdehydes
  cyclization, 31
Dialysis
  differential, 133
Diastereoisomers
  detection, in dipeptides, 192
1,4-Diazine
  absorption spectrum, polarization, 169

Diazirines
   photolysis, 116
Diazoethane, 85
Diazomethane, 12
Diazonium compounds
   aromatic, 127
Diazonium salts
   reaction with nucleic acids, 146
   reaction with nucleotides, 146
Dibenzofurans
   complex, 136
Dicarbonyl compounds
   photochemistry, 116
Dicarboxylic acid derivatives
   acylation with, 175
1,3-Dichlorobut-2-ene
   preparations with, 129
Dichroism
   and polymers, 181
   circular (see Circular dichroism)
Dictionary
   German-English, for chemists, 228
Dicyanamide
   spectra, 63
Dicyclohexylcarbodiimide
   for coupling reactions, 192
Dielectric behavior
   of polyethylene, 183
Dielectric constants, 76
Dielectric increments
   of peptides, 193
Dielectric relaxation
   and viscosity, in alcohols, 183
Dielectric saturation
   in solids, containing H bonds, 183
Dielectrics
   radiation effect, on adsorption properties, 129
      on catalytic properties, 129
Dienes
   alka-1,3-, 1-substituted, 127
   for alkylation, 173
   iron carbonyls, 16
   photochemistry, 116
   synthesis, 228
Dienophiles
   containing hetero-atoms, 131
Diethylmalonate-d,t
   tritium exchange in, 113
Diethyl methyl-d_3-malonate-t
   tritium exchange in, 113
Differential dialysis, 133
Differential polarography
   of disulfides, 162
   of elemental sulfur, 162
Diffraction
   methods, 196
   neutron, 221
   X-ray, of heterocyclic compounds, 195
Diffractometer
   X-ray, 144

Diffusion, 40
   thermal, 154
Digitalis
   in cardiac metabolism, 200
Dihydroxydiarylmethane, 28
β-Diketenes
   cyclic, nitration, 189
Dimers
   excited, 110
   of indole alkaloids, 114
   vibronic states, 120
Dinitriles
   catalytic reduction, 130
Dinitrogen tetroxide
   reaction with n-hexane, 189
2,4-Dinitrophenylhydrazones
   chromatography, 214
Diolefin cyclopolymers, 213
Dioles
   1,2-ene, 48
1,4-Dioxane
   aminoethyl-2-benzo-, 198
Dioximes
   metal, coordination chemistry, 136
   vicinal, 123
Dipeptides
   diastereoisomers in, by paper chromatography, 192
Diphenylpicrylhydrazl
   electron spin resonance, 190
Dipolar Debye absorption
   and phase change, 183
Dipole moments, 76, 196, 227
   and infrared intensity studies, 181
   microwave procedure, 141
   of acetophenone derivatives, 141
   of aryl methyl sulfides, 140
   of electronically excited molecules, 168
   of heterocyclic systems, 195
Dipole orientation
   and viscous flow, 183
Discharges
   electric, mass spectrometry of ions from, 185
Discussions
   of the Faraday Society, 168-169
Disinfection, 53
Disorders
   of carbohydrated metabolism, 157
Dispersion attractions
   Van der Waals, in DNA, 166
Displacement
   nucleophilic, at sulfur-sulfur bond, 135
Dissociation, 184
   energies, of fluorides of first row elements, 169
      of hydrides of first row elements, 169
   equilibria in solution, 105
   extraction, of meta- and para-cresols, 182
   in electronically excited molecules, 119
   kinetics, of aliphatic acids, 144
   of proteins, 20
Distillation, 25, 154
Distribution
   counter-current, of lipids, 189

Craig, tables for, 137
   tissue, and duration of drug action, 201
Disulfide groups
   in enzymes, 137
Disulfides
   differential polarography, 162
Disulfur difluoride, 56
Disymmetric ketones
   α, β-unsaturated, circular dichroism, 168
Diterpenoid alkaloids
   atisine, 123
Dithiocarbamic acids
   nomenclature, 143
Divalent carbon, 225
Divinylacetylene, 130
D_2O
   in tissue culture cells, 114
   macromolecular synthesis in cell exposed to, 114
   mitosis in cells exposed to, 114
   X-ray irradiation, 114
Documentation, 59, 65
   atomic, 71
   of spectra, 63
Dogs
   liver, inhibitors, 201
   neurogenic influence, 199
Domagk
   Gerhard, 50
Donor-acceptor complexes
   physical properties, 108
   structure, 108
Donor atom
   oxidation-reduction potential, 159
Dose-determined effects
   of HT_3DR, as DNA label, 114
      on liver cell replication time, 114
Dreiding models, 72
Drug
   binding protein, 202
Drug action
   duration, and tissue distribution, 201
      metabolic factors, 200
   time courses, biological inactivation by, 201
Drug enzymes
   activators, 201
Drug metabolism, 37, 40, 67
   comparative biochemistry, 201
   individual variations, 201
   inhibitors, 201
   physiological impairment, 201
   species differences, 201
Drug receptors, 80
   interactions, 108, 202
Drugs, 9, 66, 67, 80, 152 (see also Pharmaceuticals)
   absorption, 78
   analysis, 78
   and membranes, 200
   antianginal, 235
   anticonvulsant, 102
   antitussive, 102
   as activators, of drug enzymes, 201
   as inhibitors, of drug metabolism, 201
   demethylation, 201
   hazards, 132
   hydrolysis, by enzymes, from microsomes, 201
   in mobilization, of lipids, 199
   in nervous system regulation, 199
   in synthesis, 199
   methylation, 201
   neurotropic, action on ACTH secretion, 198
   release of free fatty acids, 199
   steroid, 221
   transport, 9
   urinary excretion, 201
Drying
   intensive, 101
Dual-labelled compounds
   isotope fractionation, 113
Dyes, 27, 32, 57, 125
   for foods, 78
   interaction, with high molecular weight substances, 129
   polycondensation, 27

Eel
   electric, 202
Elastic skeleton
   in free-electron theory, 172
Elasticity
   of macromolecular materials, 136
Elastomers, 87
   antioxidant mechanisms, 213
   polyurethane, 94
Electric resonance
   magnetic and, and relaxation (collection of papers), 183
Electric tissue
   of electric eel, 202
Electrical conductivity
   in solid proteins, 165
   of amino-acids, 146
Electrical discharges
   ions from, mass spectrometry, 185
   reactions, 14
Electrically conducting solutions, 130
Electrobalance, 216
Electrochemical groups
   reversible, in macromolecular systems, 125
Electrochemistry, 164
   of aluminum compounds, 130
   of biochemical systems, 164
   of heterocyclic compounds, 195
   of polycyclic acids, 50
   of semi-conductors, 127
Electrodeposition
   isotopic exchange reactions, 138
Electrodes, 1
Electrolysis
   free radical reactions, 127
   Kolbe synthesis, 136
   of organic compounds, 127
Electrolytes, 39
   π-, 5
   hydrated, 134

in biochemistry, 165
one - theory, 5
Electron acceptors, 75
   electronic spectra, 110
   electron-transfer interaction, 110
Electron affinity, 32
π-Electron conjugation
   in polyphenyls, 187
Electron diffraction
   low-energy, 132
Electron donors, 75
   electronic spectra, 110
   electron-transfer interactions, 110
Electron impact spectroscopy, 168
Electron microscopy, 83, 133
Electron probe microanalysis, 222
Electron spin relaxation
   of aromatic free radicals, 183
Electron spin resonance, 6, 38, 46, 49, 56, 77, 79, 119, 232
   anomalous relaxation, of coupling constants, 183
   anomalous sign, of coupling constants, 183
   at low fields, 190
   at low temperatures, 190
   in DPPH, 190
   in polymers, 91
   of biological systems, 65
   of copper phthalocyanine, 190
   of free radicals, 19
   of plants, 6
   of polymers, 70
   of radical ions, 183
   of -radiolysis, 65
   radiation protection, 119
   second-order splitting, 183
π-electron states
   resonance formulation, 187
Electron transfer reactions, 81
   donors and acceptors, 110
   theory, 38
Electron transport
   through conjugated systems, 165
Electronic aspects
   of biochemical hydroxylation, 166
   of biochemistry (international symposium), 165-167
   of muscle contraction, 166
   of pharmacology, 167
Electronic charges
   in bonds, 222
Electronic excitation, 119
Electronic levels, 115
Electronic orbitals, 128
Electronic spectra, 136
   and hydrogen bonding, 140
   and quantum chemistry, 231
   electron transfer interaction, 110
   of conjugated molecules, 229
   of crystals, 38
   of DCO radical, 168
   of HCO radical, 168
   of HCO_2 radical, 168
   of heterocyclic compounds, 195
   theory, 228
Electronic structure
   and bonding, 231
   and periodic law, 232
   of first row fluorides, 169
   of first row hydrides, 169
   of HCO_2 radical, 168
   of hemoglobin, 5
   of hemoproteins, 5
   quantum theory, 229
Electronically excited states
   and Kerr effect, 168
   dissociation processes, 119
   energy transfer phenomena, 119
   in gas phase, 168
   of ammonia, 169
Electrophilic substituent effects
   in gas phase, 140
Electrophilic substitution
   aliphatic, 123
   and MO calculations, 187
   aromatic, 105
Electrophoresis, 45
   high-voltage, in agar-agar layers, 146
   of carbohydrates, 3
   of gastric hormones, 8
   of metal complexes, 62
   of pituitary hormones, 121
   paper, 154
   starch-gel, 121
   thin-layer, 214
   zone, 53, 61
Elemental analysis, 143
Elements
   colorimetric analysis, 222
Elimination reactions
   base-promoted, 113
Emetine, 34
Enamines, 16
Endocrine system
   and the stomach, 142
Energetics
   of propellants, 231
Energy
   calculations, for polyatomic carbon molecules, 187
   conversion reactions, of photosynthesis, 124
   exchange reactions, 119
   low - electron diffractions, 132
   transformation, 99
Energy transfer
   dissociation processes, 119
   in photosensitization, 17
   in solution, 18
   molecular exciton model, 119
Engineering
   chemical, 221
   cost, 221
Entropy, 19

Enzymatic
  acylation, and peptide bond formation, 201
  alteration, of nucleic acid structure, 124
  catalysis, theory, 158
  conversion, of cyanide to thiocyanate, 201
  decarboxylation, of acetoacetic acid, 101
  hydrolysis, 30
Enzyme synthesis
  induction mechanisms, 166
  of nucleic acids, 203
  repression mechanisms, 166
Enzymes, 1, 10, 60, 66, 223
  action, 133
  active center, 41
  activity, and corticosteroids, 142
  amino-acid sequence, in active centers, 101
  carbonic anhydrase, 121
  catalysis, 133
  α-chymotrypsin, 133
  deuterium isotope effects, 114
  disulfide groups, 137
  drug, activators, 201
  drug-hydrolyzing, isolation and
    solubilization, 201
  formate oxidation, 114
  for polypeptide inactivation, 201
  from microsomes, 201
  glucose analysis, 8
  inactivation, by ionizing radiation, 119
  in analytical chemistry, 137
  inhibitors, 234
  interaction, with steroid hormones, 180
  isotope effects, 75, 114
  kinetics, 6
    of inactivation, 119
  lipid, 15
  mechanism of action, 6
  microsomal, oxidation by, 200
    reduction by, 200
  nucleic acid structure alteration, 134
  phototransferases, 36
  radiation inactivation mechanisms, 165
  resins, 96
  structure, 10
  thiol groups, 137
Enzyme synthesis
  induction mechanisms, 166
  of nucleic acids, 203
  repression mechanisms, 166
Enzymorphology
  estradiol-17β, 181
Epithelium
  germinal, in man, 122
Epoxides, 160
  cycloaliphatic, 54
α,β-Epoxyketones
  photochemistry, 116
  related carbonyl systems, 116
Epoxy resins, 212
Equilibria, 227
  acid-base, in dioxane-water, 112
  hormonal - cycle, in rats, 198

hydrolysis, of cations, 137
  in acid solution, 135
  in condensed systems, at high pressures, 179
  in gases, at high pressures, 179
  in liquid sulfur dioxide, 105
  molecular orbital studies, 104
  quasi-, theory of mass spectra, 185
  thermodynamic, 154
  triple-ion, 164
Ergot alkaloids, 28
Escherichia coli
  nucleases, 103
Essays in Coordination Chemistry (collection
  of reviews), 167
Essential fatty acids, 218
Esterase action, 201
Esterification
  of steroids, 199
Esters
  active, in peptide synthesis, 191
  alkoxylvinyl, as coupling reagents, in
    peptide synthesis, 191
  glycosyl, of amino-acids, 193
  long-chain, mass spectrometry, 185
  of carboxylic acids, synthesis, 131
  of coenzyme A, analysis, 188
  of inorganic acids, for alkylation, 174
α-Esters
  of N-substituted glutamic acid, synthesis, 191
Estradiol-17
  action of stroid hormones, 181
Estrogens
  chromatography, 62
  gas chromatography, 24
  inhibitors, 122
  specific activities, 122
Estrones, 46
Etard reaction
  mechanism, 140
Ethane derivatives
  NMR of conformational equilibria, 110
Ethanol
  rotational isomerism, 183
Esterification
  of steroids, 199
Ethers
  cellulose, 235
  derivatives, of steroidal hormones, 180
  for alkylation, 174
Ethylamine derivatives
  containing acetothiolo groups, 136
  containing acetoxy groups, 136
  containing halogy mercapto groups, 136
Ethyl-2-benzo-1,4-dioxane
  amino-, 198
Ethylene
  high pressure copolymerization, 178
  in block copolymers, 177
  oxides, 136, 161
  polymerization, 27, 84
Ethylenediamine
  complex formation, with mercury (II) ions, 167

Ethylene-propylene
  copolymers, 177
Ethylenic alcohols, 35
Exchange
  hydrogen, 19
    in aromatic compounds, 174
  isotopic, 231
  of oxygen, 4
Exchange reactions
  iodide, 138
  multidentate ligand complexes, 123
Excitation
  migration, and biological systems, 165
Excited states
  dissociation of dimers, 110
  formation of dimers, 110
  in gas phase, 168
  of ammonia, 169
  of aromatic ketones, 116
  wave functions, 168
Exciton
  contribution to optical rotation of polymers, 120
  model, symposium, 119
  molecular - model for molecular aggregates, 119
  theory, by second quantization, 120
  Wannier, in simple Van der Waals crystals, 119
Experimental methods, 160
  in gas reactions, 227
Exploration
  of space, scientific instruments in, 133
Explosives
  chemistry and technology, 233
Extraction
  dissociation, of meta- and para-cresols, 182
  isothermal - gas chromatography, 145
  liquid-liquid, 33
    of acids, 137
    of water, 137
  of acids, 127
  of salts, 127
  tables, 137

Factor
  intrinsic, 142
Fallout, 37
Familial goiter
  metabolic errors, 121
Faraday effect, 23
Faraday Society
  Discussions, 168-169
Far infrared spectroscopy, 117
Fast reactions
  in solution, 135
Fats, 15, 107 (see also Lipids)
  industrial, 233
Fatty acids, 170, 227
  biogenesis, 170
  biological oxidation, 170
  biosynthesis, 107
  branched-chain, 107
  essential, 218
  nitrogen derivatives, 170
  oxidation, 107
  ozonolysis, 107
  physical properties, 227
  release from adipose tissue, 199
  separation, 23
  sulfur derivatives, 170
  synthesis, 170
  transport in blood, 199
  unsaturated, 31
    biosynthesis, 133
Feigl's reactions, 139
Ferrichrome compounds
  biochemistry, 167
Ferrocene, 92
Fertility
  control, 153
  effects of steroids on, 122
Fibers
  heat resistant, 32
  synthetic, 221
Ficin, 73
Filipin
  structure, 138
Filtration
  gel, 188
Fischer-Tropsch synthesis
  catalysis, 158
Flame bands
  of carbon monoxide, 168
Flame photometry, 65
  analysis by, 225
Flames
  ions from, mass spectrometry, 185
  kinetics in, 117
Flash
  desorption, 4
  of genius, 224
  photolysis, 126
Flavin
  free radicals, 166
Flavonoids, 52
Flavor, 53
Flindersia species, 125
Flocculation
  reactions of macromolecules, 125
Fluorescence, 18, 75
  for protein structure, 165
  of aromatic compounds, 47
  of polymers, 70
  of steroid hormones, 197
Fluorides
  of first row elements, 169
Fluorine, 11
  by exchange, 11
  compounds, 128
  displacement, 11
  hypofluorites, 48
  nitrogen derivatives, 11
  nuclear magnetic resonance, 46
  organic derivatives, 11
  oxygen, 48

Fluorocarbons, 65
    derivatives of metal, 16
Fluoropolymers, 212
Folic acid, 36, 68
    antagonists, 207
    derivatives, 206
Food, 56
    colors, chromatography, 215
    definitions, 224
    odors, 131
    research, 11
    standards, 224
    vitamin B_6 analysis, 142
Force constants
    in small molecules, 110
    secondary isotope effects, 113
Forces
    intermolecular, in biological systems, 166
Foreign literature
    of chemistry, 132
Formaldehyde, 234
Formic acid, 4
Formulas
    molecular, computation for mass spectrometry, 227
Fraction collection
    in gas chromatography, for infrared analysis, 176
Fractionation, 108
    bacteriogenic, 113
    by gel filtration, 188
    of amino-acids, 188
    of macromolecules, 132
    of particles, 132
    of peptide hormones, 108
    of polysaccharides, 188
    of protein hormones, 108
    of proteins, 188
    of sulfur isotopes, 113
Free-electron network model
    for conjugated systems, 171
Free-electron theory
    of conjugated molecules, 229
Free radical reactions, 232
    in preparative organic chemistry, 232
    of carbon compounds, 209
    steric effects, 123
Free radicals
    absorption spectra, 77
    alkyl, rearrangements, 110
    and unstable molecules, 133
    atomic, 49
    biological, 77
    by -radiation, 183
    flavin, 166
    from desoxyribonucleic acid, 183
    hydroxyl, and pyrimidine bases, 166
    in electrolysis, 127
    in oxidative phenol coupling, 99
    intramolecular, 32
    mass spectrometry, 184
    nitration mechanism, 140
    sulfur, 32

Freeze-drying methods, 11
Friedel-Crafts and Related Reactions (collective volume), 173-175
Fries reaction, 174
Fruits, 11
Fuel system cells, 152
Fuels
    liquid, 235
Fulvenes, 27
Functional group analysis, 222, 223
    by photometric titration, 123
    by proton magnetic resonance, 144
Fungal metabolites
    aromatic compounds from lignin, 111
    biosynthesis, 111, 155
    paper chromatography, 137
    sclerotiorin group, 111
    structure, 111
Fungichromin
    structure, 138
Fungicides, 39, 40
Furan compounds
    catalytic hydrogenation, 128
    hydrogenolysis, 128

Galactans, 82
Gallium, 14
    halides, coordination compounds, 173
Galvanic analysis, 1
Gamma radiation
    in desoxyribonucleic acid, 183
Gas analysis, 143
Gas reactions
    kinetics, 226
Gas chromatography, 1, 24, 25, 36, 50, 53, 56, 62, 64, 65, 78, 121, 126, 128, 145, 154, 156, 226, 227
    absorption agents, 50
    accuracy, 139
    argon detector, 24
    biological implications, 133
    biomedical applications, 156
    column patterns, 25
    coupling of thin-layer chromatography to, 145
    cross ionization detector, 24
    detection devices, 176
    flame ionization detector, 24
    fraction collection for infrared analysis, 176
    hydrogen flame detector, 26
    in petroleum industry, 176
    isothermal-extraction, 145
    literature guide, 231
    mass flow rate detectors, 24
    mass spectra, 25
    of alicyclic compounds, 144
    of alkaloids, 146, 156, 188
    of amines, 156
    of amino-acids, 156, 188
    of bile acids, 156, 188
    of carbohydrates, 3, 156
    of coenzyme A esters, 188
    of fatty acids, 156

of glycerides, 52
of lipids, 107, 131
of peptides, 28
of racemization, in peptide synthesis, 192
of radioactive compounds, 188
of steroidal hormones, 121
of steroids, 188, 229
of sugars, 24, 188
of trace impurities, 128
of tropane alkaloids, 146
of urinary aromatic acids, 156
of xylene oxidation products, 176
preparative, 64
Proceedings of, (international symposium), 176
pyrolysis-, 145
radioactivity detection, 29
specific retentions, of solutes, 24
tables, 226
theoretical plates, 24
ultrasonic detectors, 24
Gas counting, 22
Gas phase
electronically excited species, 168
oxidation, of hydrocarbons, 231
radical rearrangements, 118
photochemistry, 17
Gas reactions, 227
kinetics, 226
methylene, 105
Gases
equilibria in, at high pressures, 179
noble, 138
potential theory of adsorption, 158
Gastric hormones, 8
Gelatin
macromolecular chemistry, 234
Gel filtration
of amino-acids, 188
of peptides, 188
of polysaccharides, 188
of proteins, 188
Genetic aberrations
biochemical basis, 157
Genetic code, 23, 66, 96, 132
bacterial viruses, 125
plant viruses, 125
Genetic control
of hemoglobin synthesis, 132
Genius
the flash of, 224
Geometry
molecular, bonding, 230
German-English
chemical terminology, 224
dictionary, for chemists, 228
Germanium, 48
quadripositive, acceptor properties, 117
Germinal epithelium
in man, 122
Gibberellic acid, 78
Gibberellins, 56
Glands
parathyroid, 122
β-1,3-Glucans, 125

Glucose
^{14}C-labelled, in carbohydrate metabolism, 121
enzymatic determination, 8
glucose oxidase, electrochemistry, 164
metabolism, 225
mutarotation, 149
β-Glucosidases, 153
Glucuronide conjugation, 200
Glutamic acid, 48
α-esters of N-substituted-, 191
Glutathione
depsopeptide analogs, 193
Glycerides, 52
vegetable oil, structure, 107
Glycolipids
plant, separation, 189
Glycopeptides
model compounds, 193
Glycoproteins, 58
Glycosides, 3
cardiac (see Cardiac glycosides)
pteridine-, 205
stereochemistry, 199
structure modification, 200
Glycosyl esters
of amino-acids, 193
Goiter
familial, metabolic errors in, 121
Gold catalyst, 85
Gonodotropins
pituitary, effect of steroids on, 122
release, 198
Graft copolymerization, 177
Graft copolymers, 82
Grafting, 87, 88, 90
Gram-negative cell wall
lipopolysaccharide, 134
Graphite, 71
Gravimetric analysis
use of EDTA, 167
Grignard reagents, 16, 43
mechanism, 135
structure, 135
Grignard solutions, 43
Griseofulvin, 117
Group displacement law, 101
Group 5
deuterides, 169
halides, 169
hydrides, 169
Group reactions
in analysis, 144
Growth factors
microbic, 108
Growth hormone, 66
studies with antisera, 122
Growth regulators, 39
Guanidine
amino-, 29
Guniea pigs
excretion of corticosteroids, 199
Gums, 11

HCO radical
  electronic spectra, 168
HCO$_2$ radical
  electronic spectrum, 168
  electronic structure, 168
Halides
  boron, 173
  of group 5 elements, 169
Haloalkanes
  for alkylations, 174
Haloalkylations, 174
Halocarbenes, 100
Halocyclopropanes, 100
Halogenated compounds
  photochemistry, 17
Halogenation, 175
  of benzene derivatives, 110
Halogens, 73
  detection, by flame ionization, 24
  exchange for fluorine, 11
  oxidation with, 55
Halothane, 54
Hammett equation, 13
  theoretical interpretations, 105
Handbook
  infrared bands, 233
  of paper chromatography, 225
Hazards, 55, 73
  in vapor-phase nitration, 189
  of new drugs, 132
Health Physics, 22
Heats of formation, 112
Heme
  biogenesis, 155
Heme compounds
  reaction rates, 106
Heme protein
  Mossbauer resonance, 165
Hemicelluloses, 3, 163
Hemoglobins, 6, 8, 20, 36
  electronic structure, 5
  synthesis, 132
    genetic control, 132
Hemoproteins, 5, 6
Hemoveratrylamine, 34
1,3,4,6,7,9b-Heptazaphenalene, tri-1,-
  3,5-triazine sym-heptazine, 131
Herbicides, 30
Heteroaromatic compounds
  free radical reactions, 12
  reacivity, 195
Hetero-chain polyamides, 227
Heterocycles
  nitrogen, 73
Heterocyclic chemistry, 12, 13
  Physical Methods in (collection of reviews),
    195
Heterocyclic compounds, 31, 42, 161
  aziridines, 161
  cyanine dyes, 161
  electrochemical properties, 195
  electronic spectra, 195
  ethylene oxides, 161
  ethylene sulfides, 161
  furan compounds, 128
  5-hydroxymethylfurfural, 125
  imidazole derivatives, 125
  infrared spectra, 195
  β-lactones, 161
  3-membered rings, 161
  4-membered rings, 161
    2 hetero-atoms, 161
  nuclear magnetic resonance spectra, 195
  nuclear quadrupole resonance, 195
  oxaziranes, 161
  oxetanes, 161
  oxygen containing, 125
  purine synthesis, 125
  pyrazolidine-3,5-diones, 129
  pyrazolidones, 161, 235
  pyrazolines, 161
  pyrazolones, 235
  solubility, 195
  thietane, 161
  trimethyleneimines, 161
  X-ray diffraction, 195
Heterocyclic systems
  dipole moments, 195
Heterogels, 28
Heteropolymers, 28
Hexachlorobutadiene, 130
n-Hexane
  reaction with dinitrogen tetroxide, 189
Hexoses
  deuterated, 114
  isotope effects, 114
H'-H' spin-spin coupling
  long range, in NMR, 125
High polymers, 224, 230
  (series of reviews), 177-178
  brittleness, 178
  chemical modification, 178
  crystalline olefin polymers, 230
  degradation, 178
  infrared spectroscopy, 235
  irradiation, 178
  molecular weights, 229
  permeability, 178
  polydispersity, 229
  polyurethanes, 231
  stabilization, 178
  statistics, 178
  tests, 178
  vibrational spectra, 228
High pressures, 39, 52, 57
  copolymerization, containing ethylene, 178
  equilibria, in gases, 179
    incondensed systems, 179
  kinetics, 179
  measurement, 179
  Physics and Chemistry (collection of articles),
    179
  polymerization by, 90
  production, 179
  radiofrequency spectroscopy, 179
  research, 179

shock waves, 179
spectroscopy, of liquids, 179
   of solids, 179
High temperatures
  reactions at, 135
  thermodynamics, 131
Higher plants
  constituents, 230
Hirschberg, Y., 71
HNO_3
  amphoterism, 189
Hofmann degradation
  ^{14}Carbon isotope effects, 113
  of quanternary amines, 113
  tritium isotope effects, 113
Hoffman-Loffler reaction, 47
Holarrhena alkaloids
  chromatography, 214
Homocyclic systems
  synthesis, 111
Homopolymerization, 81
Honey comb, 96
Hormonal equilibria cycle
  in rats, 198
Hormonal steroids (Proceedings of International Congress), 180-181
  metabolism, 180
  synthesis, 180
Hormones
  ACTH, 122
  adrenocortical factors, 122
  analogs, C-6 substituted, 180
  and plasma calcium, 122
  and vitamin B_6, 142
  antagonists, 122
  calcitonin, 122
  corticosteroids, 122
  fractionation, 108
  gastric, antisera, 122
  growth, antisera, 122
  hypophysical lutenizing, secretion, 122
  hypothalmic factors releasing, 122
  in adipose tissue, 122
  in regulation of carbohydrate metabolism, 121
  insect, 28
  non-steroid, 36
  parathyroid, 122
  peptide, 108
  pituitary, 121, 122
  polypeptide, 108
  protein, 108
  regulation, of human sebaceous gland, 121
  release control, 198
  research, 121
  secretion, neuroendocrine regulation, 122
  steroid (see Steroid hormones)
  testicular, 122
  thyroid, biosynthesis, 198
    secretion, 198
  TSH, 122
  Vitamins and - (annual collection of reviews), 142
Hot atom reactions, 33
  of energetic tritium, 19

Hot synthesis
  with radioactive tritium, 128
Houben-Hoesch synthesis, 174
HT_3DR
  as DNA label, 114
  dose-determined effects, 114
Human sebaceous gland
  regulation by steroidal hormones, 121
Humic acid, 72
Hyaluronic acid, 41
Hydrated crystals
  water molecules in, 125
Hydrated electron, 134
Hydration
  of acids, 127
  of isobutene, 113
  of salts, 127
Hydrazide
  sodium, 32
Hydrazines
  monoalkyl derivatives, 130
Hydrazinoacetic acids
  synthesis, of peptide derivatives, 192
Hydrazyl-type radicals
  paramagnetic resonance, 183
Hydrides
  boron, 227
  of first-row elements, 169
  of group 5 elements, 169
Hydroboron ions, 157
Hydrocarbon ions
  stereochemistry, 168
Hydrocarbons
  allene, 129
  aromatic, (see Aromatic hydrocarbons)
  biodegradability, 54
  bond-energy term values, 117
  cyclic, molecular orbital theory, 141
    unsaturated, 127
  gas chromatography, 25
  gas phase oxidation, 231
  hydrogenation, 4
  oxidation, 4
  polycyclic, 222
  radiation chemistry, 117
  radiolysis, 233
  saturated, alkylation, 174
    isomerization, 174
    vapor-phase nitration, 189
Hydrochloric acid, 139
Hydrocortisone, 67
Hydrogen
  analysis, automatic, 144
    in mineral oils, 144
    rapid combustion apparatus, 144
  boron-carbon system, 139
  exchange, in aromatic compounds, 174
  halides, addition to olefins, 31
    solubility, 75
  isotope effects, in isobutene hydration, 113
  isotopic exchange, 231
  mobile, 34
  - oxygen system, photochemistry, 17
  replacement, 231

separation, 49
solution, in palladium, 150
sulfide, reactions with ketones, 28
Hydrogenation
  catalytic, 128
    by pentacyanocobaltate, 150
  hydrolytic, of polysaccharides, 158
  in solutions, 232
Hydrogen bonding, 61, 77, 211
  and electronic spectra, 140
  and solvent effects, 181
Hydrogen bonds, 34, 76, 96
  in solids, dielectric saturation, 183
Hydrogenolysis
  of furan compounds, 128
Hydrolysis
  equilibria, of cations, 137
  of alkyl arenesulfonates, 127
  of aromatic sulfonyl chlorides, 127
  of aryl arenesulfonates, 127
  of drugs, by enzymes from microsomes, 201
  trypic, metals in, 166
Hydrophobic bonds, 76
Hydrotropy, 60
Hydroxy-acids
  metal complexes, 127
  separation, with anion exchangers, 138
Hydroxyacylcyclopeptides
  tautomerism, 193
Hydroxyapatite
  as adsorbent, 188
  for thin-layer chromatography, 188
N-Hydroxyethylenediamine
  cleavage, 149
4-Hydroxy-3-keto-Δ^4-steroids
  biological properties, 180
Hydroxyl free radicals
  and pyrimidine bases, 166
Hydroxylamines, 57
Hydroxylation
  biochemical, electronic aspects, 166
n-Hydroxylation
  biological, of aromatic amines, 200
Hydroxyl group
  determination, 26
5-Hydroxymethylfurfural, 125
5-Hydroxytryptamine
  in pregnancy, 199
Hyperconjugation, 61
  carbon-carbon bonds, 141
  carbon-hydrogen bonds, 141
Hypertension, 152
Hypnotics, 95
Hypochromism
  theories, in polynucleotides, 120
Hypofluorites, 48
Hypoglycaemic arylsulfonamides, 198
Hypohalites
  photochemistry, 18
Hypothalmic factors
  releasing pituitary hormone, 122

Identification, 222
  of active agents, 226
  of alkaloids, 188
  of steroids, 188
  of sugars, 188
  spectrometric, 232
  tables, 224
Imidazoles
  synthesis of purines from, 125
Imidogens, 49
Imines, 48
Impact
  electron, spectroscopy, 168
Impurity effects
  in organic semi-conductors, 165
Inactivation
  biological, by amine oxidases, 201
    by time courses, of drug action, 201
  of polypeptides, by enzymes, 201
  radiation-mechanisms, in enzymes, 165
Inclusion compounds, 35, 190
Incorporation errors
  in DNA replication, 166
Increments
  dielectric, of peptides, 193
Indanedione, 35
Index
  cumulative, J. Chem. Ed., 235
  to Reviews, Symposia, Volumes and Monographs, 226
  to Synthetic Methods of Organic Chemistry, 233
Individual variations
  in drug metabolism, 201
Indoles, 12, 67
  alkaloids, 225
  dimeric, alkaloids, 114
  Fisher synthesis, 48
Induction mechanisms
  in enzyme synthesis, 166
Industrial
  applications, of radioisotopes, 229
  fats, 233
  oils, 233
Inert gases
  compounds, 123
Information retrieval, 132
  systems, 133
Information transfer
  and nucleic acids, 103
Infrared analysis
  fraction collection, in gas chromatography, 176
  of polyethylene, 145
Infrared spectra, 63
  in analysis, 78
  interpreted, 233
  of alcohols, 35
  of carbon subgroup compounds, 129
  of carboxylic acid complexes, 127
  of carboxylic acid salts, 127
  of coordination compounds, 228
  of heterocyclic compounds, 195
  of inorganic compounds, 228
  of polyatomic molecules, 224
  of simple molecules, 181
  of solids, 181

Infrared spectroscopy, 6, 141, 196, 222, 223, 225, 230, 233
   and Molecular Structure (edited volume), 181, 223
   band handbook, 233
   band intensities, 117
   chemical applications, 230
   far, 117
   in catalysis research, 134
   in CsI region, 64
   of active agents, 226
   of azo groups, 141
   of azoic dyes, 141
   of cyclopropanes, 141
   of high polymers, 235
   of metal-diphenylcarbazone complexes, 211
   of OH bonds, 129
   spectral perturbations, 110
   theory, 233
Inhibitors, 221
   enzyme, 234
   for liver alcohol dehydrogenase, 202
   hormone antagonists, 122
   in chemotherapy, 201
   in dog liver, 201
   in enzymes, 6
   metabolic, 234
   nonhormonal, of adrenocortical steroid biosynthesis, 181
   of drug metabolism, 201
   of specific activity, of androgen, 122
      of estrogen, 122
   of spermatogenesis, 122
   selective, of photosynthesis, 124
Inorganic analysis
   complexing reagents, 229
   liquid ion-exchangers, 139
Inorganic chemistry, 225
   calorimetry, 112
   preparative, 226
   progress, 102
Inorganic compounds
   azides, 117
   cyclic, containing silicon, 128
   electronic states, 102
   esters, for alkylation, 174
   infrared spectra, 228
   nitrates, 118
   nitrato-compounds, 118
   vibrational states, 102
Inositol lipids
   biochemistry, 142
Inotropic action
   of cardiac glycosides, 200
Insects
   chemical attractants, 132
   chemical resistance, 54
   sex attractants, 151
Insertion reactions
   in peptide synthesis, 192
Instrumental analysis
   photometric titrations, 123
Instrumentation, 71
   for process control, 221
   scientific, in space exploration, 133
   ultrahigh vacuum, 133
Intensity
   quantitative, and dipole moments, 181
Intensity theory
   for infrared spectra, of polyatomic molecules, 224
Intensive drying, 101
Interactions
   ^{13}C-hyperfine, in semiquinones, 183
   drug-receptor, 108, 202
   intermolecular, in molecular theory, 166
   molecular, in clathrates, 118
   non-bonded, and conjugation, 141
   of dyes, with high molecular weight substances, 129
Interaction theory
   of localized electronic excitations, 119
Interconversion
   of androgens, 121
Interdiffusion
   in polymer systems, 130
Interference spectroscopy
   for infrared, 110
   near infrared, 110
Intermediary metabolism, 181
Intermediate complexes, 173
Intermetallic compounds, 133
   non-stoichiometry in, 150
Intermolecular
   association, of alcohols, solvent effects, 183
   forces, 75
      in biological systems, 166
   interactions, in molecular biology, 166
International Union
   of Pure and Applied Chemistry, (international conferences), 182
Interstellar effects
Intrinsic factor, 142
Inverted steroids
   biological activities, 180
   preparation, 180
Iodide exchange reactions, 138
Ion exchange
   kinetics, 130
Ion-exchange chromatography, 45, 65, 145, 231
   cellulose ion-exchangers, 128
   liquid ion-exchangers, in inorganic analysis, 139
   of pteridines, 206
   of triterpenoids, 188
   on cellulose, 128
   sulfonated cation exchangers, 130
Ionic
   boron hydrides, 157
   movements, in nervous system, 41
   solvation, 5
Ionization, 184
   and dissociation equilibria, 105
   constants, 195
   detectors, catalytic combustion, 176
   energies, of fluorides of first-row elements, 169
      of hydrides of first-row elements, 169
   potentials, 105
Ionizing radiation
   dinitrogen tetroxide, 189

effects on nucleic acids, 103
  enzyme inactivation kinetics, 119
  n-hexane, 189
  in polymerization, 28
  luminescence under, 165
  permissable exposure to, 132
Ions
  ambident, 32
  bromonium, 72
  carbonium, 105
  decompositions, 185
  hydroboron, 157
  hydrocarbon, stereochemistry, 168
  in gases, 72
  in non-aqueous solvents, 164
  in solid hydrocarbons, 17
  magnetic, in nucleic acid pseudocrystals, 190
  mass spectrometry, 185
  - molecule reactions, 184, 185
  negative, mass spectra, 185
  nitronium, amphoterism, 189
  optical absorption by, 119
  radical, ESR spectra, 183
  rearrangements, 185
  sources, 1
  transport, and cardiac glycosides, 200
Ioxazoles, 13
Iproniazid
  in pregnancy, 199
Iron compounds
  antibiotics, 108
  carbonyls, 16
  complexes with isonitriles, 149
  porphyrin, Mossbauer resonance, 165
  tricarbonyl complexes, 71
Irradiation
  of high polymers, 178
Irradiation effects
  on adsorption properties, of dielectrics, 129
    of semiconductors, 129
  on catalytic properties, of dielectrics, 129
    of semiconductors, 129
  on tissue culture cells, 114
Irreversible processes
  thermodynamics, 234
Isoalloxazines, 128
Isobiopterin, 206
Isobutene, 90, 160
  hydration, 113
  hydrogen isotope effects, 113
Isocoumarins, 49
Isocyanates, 89
Isocyanides
  in synthesis, 136
Iso-hetero polyanions, 108
Isomerism, 136
  rotational, in ethyl alcohol, 183
    proton magnetic resonance, 183
Isomerization, 27
  cis-trans, kinetics, 106
  of aromatic hydrocarbons, 174
  of saturated hydrocarbons, 174
  photochemical, of polyenic systems, 115

Isoprene, 59
  polymerization, 85, 89
Isoprenoid structure
  perfumes with, 127
Isoquinolines, 34
Isothermal-extraction gas chromatography, 145
Isothiazoles, 30
Isothiocyanates
  natural derivation, 110
Isotope effects, 18, 19, 113
  abnormal tritium, 113
  acid-base equilibria, 112
  and bonding, 112
  approximation procedures, in theoretical
    calculations, 112
  aromatic carbonium ions, 113
  carbon, 113
  chlorine, 113
  criteria of mechanism, 114
  deuterated hexoses, 114
  deuterated nucleic acids, 114
  deuterated proteins, 114
  deuterium, 112, 114
  deuterium solvent, 112
  dioxane-water mixtures, 112
  enzymic formate oxidations, 114
  hydrogen atom addition, 113
  in decomposition, 113
  in hexamethylbenzene, 113
  in H_2O-D_2O mixtures, 112
  in isobutene hydration, 113
  in p-nitrophenethyltrimethyl ammonium iodide, 113
  in phase equilibria, 75
  in pyrolytic decomposition, 112
  in 3-center reactions, 112
  kinetic, 112
  kinetic deuterium, 114
  kinetic hydrogen, 113
  nitrogen, 113
  on reaction rates, 74
  secondary, 105, 113
  systems containing water, 112
  theoretical calculations, 112
  tritium, 113
  two-phase liquid systems, 112
Isotope exchange, 26, 231
  and hydrogen replacement, 231
  by electrodeposition, 138
Isotope fractionation
  computation, 113
  detection, 113
  in adsorption chromatography, 113
  of dual-labelled compounds, 113
Isotope labelling
  of natural products, 114
Isotope mass effects
  in biology, 112
  in chemistry, 112
  IUPAC conference, 182
  symposium, 112
Isotopes
  biological effects, 75

^{48}Ca/(total Ca), 113
^{14}C, 132
  in polymerization, 48
  oxygen determination, 112
  radiocarbon data, 132
  radiocarbon dating, 126
  separation, 75
  sulfur, 113
  tracers, 173
IUPAC
  international conferences, 182

Journal
  of Chemical Education (cumulative indexes), 235
  Royal Institute of Chemistry, 126
  Science, 132

Kallidin, 66
Kerr effect, 168
α-Keto-acids, 35
β-Ketoesters
  cyclic, C-alkylation, 141
Ketones, 29
  alkylation, 123
  aromatic, 116
    excited states, 116
    synthesis, 174
  cyclic, photochemistry, 17
  dissymmetric α,β-unsaturated, circular dichroism, 168
  for alkylation, 174
  from cyclization of olefinic acids, 118
  non-conjugated, 116
  photochemistry, 116
  pyrazolidine-3,5-diones, 129
  reactions, with hydrogen sulfide, 28
  stereospecificity, 116
  Willgerodt-Kindler reaction, 30
3-Keto-Δ^4-steroids
  4-hydroxy-, biological properties, 180
Kinetics, 71, 73, 106, 154, 223, 229
  at high pressures, 179
  at liquid-gas interface, 158
  at liquid-liquid interface, 158
  cis-trans isomerization, 106
  deuterium isotope effects, 114
  enzymic formate oxidation, 114
  equilibria, 227
  hydrogen isotope effects, in isobutene hydration, 113
  in flames, 117
  in shock tubes, 133
  isotope effects, and bonding, 112
    in 3-center reactions, 112
  modern, 223
  consecutive processes, 106
  of dissociation, of aliphatic acids, 144
  of electrode reactions, by potentiostatic methods, 164
  of enzyme inactivation, by ionizing radiation, 119
  of gas reactions, 226
  of germinal epithelium, in man, 122
  of ion exchange, on sulfonated cation exchangers, 130
  of propagation, copolymerization, 106
    anionic polymerization, 106
  of redox systems, 158
  rates, 227
  surfaces, 4
Koch reaction, 30
Kolbe synthesis, 136
Kynurenine, 8

Labelled compounds
  radioactive tritium, 128
  vitamin B_6 group, 142
Laboratory techniques
  in organic chemistry, 235
Lactation
  pharmacological control, 198
Lactones
  β, 161
  from cyclization of olefinic acids, 118
  in alkaloids, 49
  in polymers, 89
Laminar silicates
  organic derivatives, 128
Langerhans
  of the pancreas, 198
Large ring systems
  conjugated, 111
Lasers, 23, 126
Latex, 83
Law
  group displacement, 101
  periodic, and electronic structure, 232
LCAO calculations
  with ω-technique, 141
Lead
  organo- compounds, 234
  quadripositive, acceptor properties, 117
  tetraacetate, 141
Lecithins, 44
Less Common Means of Separation (symposium), 182
Leucopterins, 205
Lewis acid catalysts
  in non-aqueous solutions, 173
Lewis acids, 173
Lexan polycarbonate, 94
Life processes
  and π-molecular orbitals, 187
Ligand(s)
  complexes, macrocyclic, 123
    multidentate, 123
  coordinated, 149
  - field, 65
  in metal complexes, 149
  oxidation-reduction potentials, 159
  reactions, 123
Light
  absorption, by rigid crystals, 119
  reactions, in photosynthesis, 123

Ligninines, 44
Lignins
  aromatic compounds related to, 111
  biogenesis, 155
  wood, 163
Linear bifunctional compounds
  polyfluorinated, 128
Line spectra
  structure, 229
Lipids, 15, 36, 44, 45, 54, 107, 131
  bacteria, 15
  biogenesis, 155
  biosynthesis, 229
  chromatography, 214
  composition, of tissue culture cells, 114
  counter-current distribution, 189
  crystallography, 54
  fatty acids, 53
  functional, 124
  gas chromatography, 107, 131
  glyco-, 41
  inositol, biochemistry, 142
  in photosynthesis, of higher plants, 124
  lecithins, 80
  mobilization, effects of drugs, 199
  of higher plants, 124
  of tubercle bacillus, 32
  oxidation, 60
  phosphoinositides, 15
  phosphorus derivatives, 15
  polar, chromatography, 189
  separations, 188
  soaps, 47
  structure, 15
  sulfolipids, 15
  triglycerides, 15
Lipopolysaccharides
  of gram-negative cell wall, 134
Lipoproteins
  egg, 107
  intracellular, 107
  serum, 107
Liquid chromatography
  permanent columns, 145
Liquid crystal, 128
Liquids, 72, 74, 76
  cohesion, 75
  deuterium isotope effects, 112
  exchange, 62
  fuels, 235
  interaction in, 75
  ion exchangers, in inorganic analysis, 139
  partition methods, 154
  reactions in, collision theory, 228
  scintillation counting, 21
  separations, with polymer films, 182
  two-phase systems, containing water, 112
  viscosity, 75
Lithium, 42, 43, 81, 86
  in polymerization, 85
Liver
  alcohol dehydrogenase, 202
  cell replication time, in growing rat, 114
  dogs, inhibitors, 201
  triglycerides, 199

Living
  cell, coordination of chemical processes, 108
  molecules, 210
  polymers, and applications, 137
Local anesthetics, 102
Long-chain esters
  mass spectrometry, 185
Low energy
  electron diffraction, 132
Low-temperature methods, 11
Luminescence, 65, 91
  analysis, 143
  of cell-nuclear compounds, 165
Lung
  pharmacology, 202
Lysine
  4-oxa-, alkali instability, 192
Lysosomes
  effects of corticosteroids on, 122

Macrocyclic compounds
  synthesis, 131
Macrocyclic ligand complexes
  synthesis, by ligand reactions, 123
Macrolide antibiotics, 117
  biosynthesis, 126
  polyene, 138
Macromolecules, 28, 39, 56, 59, 134
  biochemical, at very low temperatures, 166
  elasticity, 136
  fractionation, 132
  gelatin, 234
  natural organic, 226
  polymers, cross-linking, 133
  radiation effects, 37
  reversible electrochemical groups, 125
  ribonucleic acids, 103, 232
  separation, 132
  swelling, 136
  synthesis, cells exposed to D_2O, 114
Magnesium, 43
  organo- compounds, 34
Magnetic
  and Electric Resonance and Relaxation,
    (collection of papers), 183
  circular dichroism, 23
  fields, superconducting, NMR spectra in, 134
  ions, in nucleic acid pseudocrystals, 190
  properties, 5
    of transition metal complexes, 102
  relaxation, nuclear, 6
  resonance, 232
    proton, of rotational isomerism, 183
    spectroscopy, 196
  rotatory dispersion, 23
  susceptibility, of hemoproteins, 6
Magnetochemistry, 76
Magneto-optical study
  of bonds, 34
Maintenance directory
  and safety techniques, 226
Mammalian metabolites
  paper chromatography, 137

Man
  germinal epithelium in, kinetics, 122
Manganous oxalate
  carbon isotope effects, 112
  pyrolytic decomposition, 112
Marine environments
  adrenocortical factors in adaptation to, 122
Mass effects
  isotope, IUPAC conference, 182
Mass spectra, 1, 222
  correlations, 151, 227
  interpretation, 222
  negative-ion, 185
  of alkylbenzenes, 185
  of terpenes, 185
  quasi-equilibrium theory, 185
Mass spectrometry, 36, 46, 56, 144, 196, 214, 221, 227
  abundance tables, 221
  collection of reviews, 184
  computation, of molecular formulas, 227
  high resolution, 114, 185
  mass tables, 221
  of free radicals, 184
  of ions, from electric discharges, 185
  of ions, from flames, 185
  of long chain esters, 185
  of natural products, 114, 185
  of organic ions, (collection of reviews), 185
  of organic radicals, 185
  of steroid alkaloids, 141
  structure determination, 16, 222
Materials
  X-ray studies, 224
Mathematics
  for chemists, 226
Matrix experiments
  infrared spectral perturbations, 110
Measurement
  standards and units, 126
Mechanics
  of aerosols, 224
Mechanisms, 19
  acylation, 117, 175
  addition of polymerization, 160
  aliphatic compounds, 209
  alkylation, of ketones, 123
  and silicon, stereochemistry, 232
  antioxidant, 213
  azapyrimidines, 103
  biochemical, and reversible reactions, 166
  carbonyl group reactions, 105
  carciogenesis, 165
  catalytic redistribution, of hydrogen, 127
  cationic polymerization, of vinyl monomers, 129
  criteria, deuterium isotope effects, 114
  cyclization, by radical reactions, 123
  detoxication, in vivo, 200
  electrolysis, 127
  energy conversion reactions, of photosynthesis, 124
  energy transfer, 119
  enzymic decarboxylation, of acetoacetic acid, 101
  Etard reaction, 140
  fast elementary steps, 108
  free radical, 127
  Grignard reagents, 135
  hydrolysis, of alkyl arenesulfonates, 127
    of aromatic sulfonyl chlorides, 127
    of aryl arenesulfonates, 127
  induction, in enzyme synthesis, 166
  molecular, of radiation effects, 103
  neighboring group participation, 117
  nucleophilic displacement, at sulfur-sulfur bond, 135
  organophosphorous compounds, 114
  oxidation, 233, 234
  oxidation-reduction, 123, 135
  oxidative thermal degradation, of polymers, 129
  phenol-dienone rearrangements, 127
  photochemical reactions, 135
  photophosphorylation, in chloroplasts, 124
  photosensitized oxygen transfer reactions, 116
  radiation inactivation, in enzymes, 165
  radiation protection, by cysteamine, 119
  radiobiological, 203
  redox systems, 158
  repression, in enzyme synthesis, 166
  selective inhibitors, of photosynthesis, 124
  solvolysis, 233, 234
  stabilization, of polymers, 129
  steric effects, free radical reactions, 123
  steroid action, 180
  substitution reactions, of metal complexes, 135
  transfer, of radiation-induced unpaired spins, 119
Mechanochemistry
  of polymers, 221
Mechano-thermal degradation
  of polyethylene, 213
Medicinal chemistry, 102, 222
Medicine
  Biomedical Frontiers in, (collective volume), 157
Melting
  of crystals, 5
  of polymers, 70, 94
  point, determination, 154
    tables, 234
Membranes
  and drugs, 200
Mental diseases, 52
Mercaptans
  addition to triple bonds, 130
  auto-oxidation, 150
Mercaptoacetate
  auto-oxidation, 150
Mercapturic acid
  conjugation, 200
Mercury, 43
  complex formation, with ethylene diamine, 167
  deoxymercuriation, 45
  drop electrode, 1
  photosensitized reactions, 106
Mesityl oxide, 47
Mesomerism, 136
Messenger ribonucleic acid, 103
Metabolism
  adrenocortical steroids, 121
  and duration of drug action, 200

carbohydrated, disorders, 157
cardiac, effect of digitalis, 200
disorders, of amino-acids, 157
   of proteins, 157
   phenylalanine, 8
   tryptophan, 8
   tyrosine, 8
errors, in familial goiter, 121
inhibitors, 234
intermediary, 181
of adipose tissue, effects of hormones on, 122
of amino-acids, 157
of bile acids, 199
of carbohydrates, 121
of cardiac glycosides, 200
of drugs, (see Drug metabolism)
of glucose, 225
of hormonal steroids, 180
of proteins, disorders, 157
of steroid hormones, thyroid interaction with, 180
of steroids, 180
pancreatic islet tissue, 121
species differences, 201
timing control, 132

Metabolites
  acidic, from deoxycorticosterone, 180
  biosynthesis, 111
  fungal, 111
    biosynthesis, 155
    chromatography, 137
  mammalian, chromatography, 137
  structure, 111

Meta-cresol
  separation, from para-cresol, 182

Metals
  alkali, in polymerization, 85
  alkaline organometallic compounds, 42
  binding, 78
  catalysis, 30, 99
    in biological systems, 149
    in perester synthesis, 136
    in peroxide reactions, 136
    on pyridines, 12
  in tryptic hydrolysis, 166
  ligand bonds, 159
  nitrates, 14
  organic stabilizers, for polyolefin plastics, 212
  organometallic compounds, 42
  organo-, silyl derivatives, 29
  oxides, olefin reactions, 128
  oxide catalysts, for olefin copolymerization, 177
  polymerization by, 45, 85
  reactions, 71
  transition, 48, 71, 73
    peroxy derivatives, 14

Metal chelates, 149
  biological systems, 159
  EDTA, 159
  flame photometry, 65
  optical phenomena, 159
  precipitation, 139
  quasi aromatic, 149

Metal complexes, 57, 62, 135, 221
  ligands in, 149
  of hydroxy-acids, 127
  of olefinic ligands, 27
  peptides, 108
  proteins, 108
  substitution reactions, 135
  transition, 123

Metal compounds, 52
  alkoxy-metals, 60
  amines, in solution, 167
  arenes, 149
  arylaluminum compounds, 125
  boron containing, 157
  carbonyls, 33, 59, 117
  cyclopentadienyls, 149
  dioximes, coordination chemistry, 136
  halides, reaction with aliphatic amines, 102
    reaction with ammonia, 102
  hydrides, non-stoichiometry in, 150
  hydroxy-acid complexes, 127
  metallocenes, 135
  metallophosphines, p-substituted, 115

Metallic contacts
  magnetic properties, 158
  structure, 158

Metastasis, (see Cancer), 7

Methacrylates, 87
  copolymerization, 178
  polymerization, 85

Methides
  quinone, 118

Methoden der Organischen Chemie, 186

Methyl
  structure, 187

Methyl cyanide
  rotational structure, 187

Methylation
  of drugs, 201

(Methylcyclopropyl)-carbinyl derivatives
  solvolysis rates, 114

Methylene, 17, 105
  and carbenes, 106
  spectra, 63
  structure, 187

Methylthioethyl group
  for carboxyl group protection, 191
  in peptide synthesis, 191

Michaelis-Arbusow reaction, 115

Microanalysis
  by X-rays, 143
  electron probe, 222
  qualitative, 231

Microbalance
  torsion, 216
  vacuum, 216

Microbial metabolites
  structures of biologically active, 110

Microbial synthesis, 55
  of amino-acids, 137

Microbic growth factors
  iron-containing antibiotics, 108

Microorganisms
  biosynthesis, of fatty acids, 133

Microscope
  electron, 133
Microscopy
  field emission, 4
Microsomal enzymes
  oxidation by, 200
  reduction by, 200
Microsomes
  drug-hydrolyzing enzymes from, 201
Microwave, 38
  procedure, for dipole moments, 141
    for relaxation times, 141
  spectroscopy, 65, 110, 232
Migration
  excitation, and biological systems, 165
Mineral acid solutions
  equilibria in, 135
Mitochondrin basis, 227
Mitosis
  cells exposed to D_2O, 114
Mixing processes
  calorimetry, of non-reacting systems, 112
Mixtures
  H_2O-D_2O, solvent isotope effect, 112
  multicomponent, precision analysis, 144
Mobilities
  chromatographic, of steroid hormones, 197
Modern
  analysis, in food science, 143
  chemistry, 226
  kinetics, 223
  Methods, in the Analysis of Organic Compounds, (international symposium), 143
  polarographic methods, 231
Molecular
  aggregates, exciton model, 119
  aspects, of mutations, 165
  association, 75
    of acetic acid, 183
  basis, of structure and function, 227
  biology, intermolecular interactions, 166
  Biology, Progress in Biophysics and, (collection of reviews), 203
  complexes, 221
  electronic structure, quantum theory, 229
  formulas, for mass spectrometry, 227
  interactions, in clathrates, 118
  mechanisms, of radiation effects, 103
  reactions, reversible, and biochemical mechanisms, 166
  rearrangements, terpenes, 101
  shape, and biological activity, 201
  sieves, 65, 74
  size, and biological activity, 201
  spectroscopy, 144
  theories, of memory, 133
  vibrations, physicochemical problems, 110
  weights, determination, 196
    of high polymers, 229
Molecular orbitals, 72, 73
  and base strength, 187
  and life processes, 187
  calculations, and aromaticity, 187
    and electrophilic substitution, 187
  equilibria, 104
  in Chemistry, Physics and Biology, (collective volume), 187
  reaction rates, 104
  theory, 172
    in strained cyclic hydrocarbons, 141
    of bent bonds, 141
Molecular structure, 72, 226
  and infrared spectroscopy, 223
    (edited volume), 181
  and spectroscopy, 226
    (IUPAC conference), 182
  models, 72
  of petroleum, 185
  (special lectures), 110
Molecules
  activated, 101
  convex, 5
  electronic orbitals, 128
  geometry, and bonding, 230
  unstable, 133
Monoalkylhydrzines, 130
Monographs, 221-235
Monomers
  organosilicon, synthesis, 229
  potential, polyfluorinated linear bifunctional compounds, 128
Morphine
  chromatography, 214
Morpholine, 35
Morphology
  of polymers, 230
Mossbauer effects, 14, 28, 133
Mossbauer resonance
  in iron porphyrin, 165
  in heme protein, 165
Mucopolysaccharides, 8
Multicomponent mixtures
  precision analysis, 144
Multidentate
  chelating agents, stereochemistry, 159
  ligand complexes, exchange reactions, 123
Multiplet theory
  of catalysis, 158
Muscle contraction
  electronic aspects, 166
Muscular distrophy, 8
Mutagenesis
  specificity, 103
Mutations
  molecular aspects, 165
Myoglobin, 20
  protein structure, 132

Name reactions, 227
Naphthaldehydes, 77
Naphthoquinone, 47
Narcotic antagonists, 152
National Bureau of Standards
  thermodynamic data, 138
Natural pigments, 128
Natural products
  biogenesis, 155
  isotope labelling, 114
  mass spectrometry, 114, 185

nuclear magnetic resonance spectra, 141
oxygen-ring, 223
polycyclic, synthesis, 114
structure, 222
(symposia), 114
ultraviolet spectra, 231
X-ray analysis, 101, 115
Neber rearrangement, 48
Negative ion
  mass spectra, 185
Neighboring group participation, 117
Neoplasm, 7
Nervous action
  central, pharmacological analysis, 202
Nervous system
  ionic movements, 41
  regulation, 199
  sympathetic, fatty acid transport, 199
Network model
  of sigma electron densities, 171
Neuroendocrine regulation
  of hypophysical luteinizing hormone secretion, 122
Neurogenic influence, 199
Neuromuscular block, 102
Neuronal function, 142
Neurones
  adrenergic blocking, 9
Neurophysin, 194
Neurotoxin
  tarichatoxin-tetrodotoxin, 133
Neurotropic drugs
  action of ACTH secretion, 198
Neutron diffraction, 221
Neutrons
  effects of liquids, 76
New Biochemical Separations
  (comprehensive series of articles), 188-189
Nickel
  in oil, 26
Nickel chelates
  with Schiff bases, 167
Niels Bohr
  memorial lecture, 101
Niobium halides
  reaction with pyridine, 150
Nitrates
  inorganic, 118
Nitration
  aromatic, positive poles, 189
  free radical mechanism, 140
  of cyclic -diketenes, 189
  of high molecular n-alkanes, 189
  of paraffins, 189
  vapor-phase, hazards, 189
    of saturated hydrocarbons, 189
O-Nitration
  of alcohols, 189
Nitrenes, 29
Nitric acid, 139
Nitriles
  di-, catalytic reduction, 130
  polymerization, 89

Nitrites
  photochemistry, 18
Nitrobenzenes
  o-substituted, 118
  substituent interactions, 118
Nitrocompounds, 53
  (international symposium), 189
  photochemistry, 116
  poly-, 48
  unsaturated, 116, 229
Nitro groups
  as proton acceptors, 141
  in hydrogen bonding, 141
  position, analysis, 189
Nitrogen
  active, 47
  analysis, automatic, 144
    of non-volatile compounds, 143
  derivatives, of fatty acids, 170
  fixation, biological, 137
    by azotobacter, 137
  fluorides, 11
  silicon derivatives, 14
  sulfur compounds, 29
Nitrogen compounds, 160
  aminodeoxy-sugars, 131
  amino sugars, 3
  azides, inorganic, 117
  dinitriles, 130
  1,3,4,6,7,9b-heptazaphenalene, tri-1,3,5-triazines, 131
  monoalkylhydrazines, 130
  osotriazoles, 3
  1,3,4-oxadiazole derivatives, 131
  pyrazolidine-3,5-diones, 129
  2,4,6-trichloro-1,3,5-triazines, 130
Nitroheptane
  secondary, nitro group position, 189
Nitromethane, 31
Nitrones, 49
Nitronium ion
  amphoterism, 189
p-Nitrophenethyltrimethyl ammonium iodide, 113
Nobel prizes
  chemistry, 136, 137, 138
  medicine, 136
  physiology, 136
Noble gases, 138
  compounds, 96, 123, 134
Nomenclature, 222
  of dithiocarbamic acids, 143
  of peptides, 193
Nomogram
  for acid-base titrations, 139
Non-auqeous
  solutions, aluminum compounds in, 130
  solvents, coordination model, 102
    ions in, 164
  systems, electrically conducting, 130
Nonbenzenoid aromatic compounds
  aromaticity, 187
Non-bonded interaction
  and conjugation, 141

Noncovalent bonds
  in proteins, 204
Non-degenerate electronic states
  electronic and vibrational motions, 102
Nonelectrolytes, 38
Non-enzymatic cleavage
  of peptide chains, 193
Non-glycolytic pathways
  of glucose metabolism, 225
Non-hormonal activities
  of steroids, 181
Non-hormonal inhibitors
  of adrenocortical steroid biosynthesis, 181
Non-reacting systems
  calorimetry, 112
Nonsteroids, 181
Non-stoichiometric compounds
  (collective volume), 190
  (symposium), 150-151
19-Norsteroids, 130
Novabiocin, 66
Nuclear
  activation analysis, 132
  chemistry, 228
  phenomena, 37
  quadrupole coupling, 5
  quadrupole resonance, of heterocyclic compounds, 195
  reactions, 223
  recoil, in solids, 130
  resonance, 50
  ribonucleic acid, 103
  spin relaxation, of aromatic free radicals, 183
  transformations, in solids, 130
Nuclear magnetic resonance spectra
  catalog, 221
  of carbohydrates, 3
  of cyclopropanes, 141
  of ethane derivatives, 110
  of fats, 64
  of fluorine, 46
  of heterocyclic compounds, 195
  of hydrocarbons, 63
  of natural products, 141
  of organometallic compounds, 46
  of polymers, 70
  of steroids, 58, 221
Nuclear magnetic resonance spectroscopy, 27, 37, 38, 39, 46, 50, 52, 56, 221
  double, 46, 47
  in superconducting magnetic fields, 134
  long range H^1-H^1 spin-spin coupling, 125
  relaxation times, 76
  spin-spin coupling, 125
Nucleases
  of escherichia coli, 103
Nucleic acids, 39, 103, 225
  aza purines, 12
  aza pyrimidines, 12
  biochemistry, 225
  biosynthesis, 155
  deuterated, 114
  enzymatic synthesis, 203
  fractionation, 103
  fully deuterated, isotope effects, 114
  function, and actinomycin, 104
  information transfer, 103
  infrared spectra, 6
  photochemistry, 227
  plant virus, 103
  preparation, 103, 225
  pseudocrystals, magnetic ions in, 190
  radiation chemistry, 203
  radiation effects, 103
  reaction with diazonium salts, 146
  structure, enzymatic alteration, 134
  sugar nucleotides, 3
  thermal effects on, 6
Nucleophilic
  compounds, 32
  displacement, sulfur-sulfur bond, 135
  reactions, 149
  substitution, aromatic, 105
    at saturated carbon, 222
Nucleophilicity, 38
Nucleoproteins, 2
  radiation chemistry, 203
Nucleosides, 228
  amine-, 67
Nucleotides, 41, 228
  chromatography, 62
  free, in animal tissues, 104
  purine, biogenesis, 155
  pyridine, 10
  pyrimidine, biogenesis, 155
  reaction with diazonium salts, 146
  sugar, 10
Nucleus, 23
Nutrition, 53
Nylon, 87

Odors
  of foodstuffs, 131
Odorants
  chromatography, 214
OH bonds
  infrared spectroscopy, 129
Oils
  analysis, 64
  industrial, 233
Olefins, 31, 81
  α-, copolymers, 177
  acids, cyclization, to ketones and lactones, 118
  additions, 17
  alkylation with, 233
  complexes, with metals, 27
  copolymerization, 177
  crystalline polymers, 178, 230
  formation, by phosphor carbonyls, 115
  hydration, 100
  poly-, 35
  polymers, 230
    crystalline, 178, 230
  reactions, with metal oxides, 128

with titanium compounds, 30
   Staudinger, 115
   thiylation, 128
Oligonucleotides
   column chromatography, 103
Ophthalmic acid
   depsopeptide analogs, 193
Optical
   absorption, by pair of ions, 119
   activity, 133
      in polymerization, 84, 85
      of polymers, 51, 93
   crystallography, 225
   phenomena, in metal chelates, 159
   properties, of biopolymers, 165
      of interstellar dust, 171
      of molecular crystals, 119
   rotation, 39, 196
      of polymers, 120
   rotatory power, 117
Optical rotatory dispersion, 54, 56
      and circular dichroism, 101
      of transition metal complexes, 123
Optically active compounds
   tertiary phosphenes, 115
Orbitals, 73
   in biochemistry, 165
   Molecular, in Chemistry, Physics and
      Biology (collective volume), 187
   molecular, and base strength, 187
   π-molecular, and life processes, 187
Organic chemistry, 234
   and quantum chemistry, 137
   current topics, 223
   index to reviews, symposia, volumes and
      monographs, 226
   laboratory techniques, 235
   modern structural theory, 223
   Physical Methods in, (collective volume), 196
   physical progress, 105
   progress, 104
   selected works, 228
   synthetic methods, 233
   theoretical viewpoints, 230
Organic compounds
   characterization, 234
   constitution, 144
   identification, 222
   spectrometric identification, 232
Organic preparations, 223, 229
   free radical reactions, 232
Organischen Chemie
   Methoden, 186
Organoboron compounds, 233
   unsaturated, 131
Organolead chemistry, 234
Organometallic compounds, 16, 43
   arylaluminum compounds, 125
   bond theory, 126
   fluorocarbon derivatives, 16
   metal carbonyls, 117
   organic phosphor carbonyls, 115
   organophosphorous compounds, 115
   phosphene alkenes, 115
Organophosphorous compounds
   bonding, 115
   (symposium), 115
   synthesis, from p-substituted metallo-
      phosphines, 115
Organosilicon monomers
   synthesis, 229
Organotin compounds, 27, 137
Orientation
   dipole, and viscous flow, 183
Origin
   of life, 99
   steric, of secondary isotope effects, 113
Oscillographic polarography, 131
Ovarian steroids
   secretion, in vivo, 122
   synthesis, in vivo, 122
Oven
   ring, 139
1,3,4-Oxadiazole
   derivatives, 131
4-Oxalysine
   alkali instability, 192
Oxazines
   1,3-, 12
Oxaziranes, 161
Oxetanes, 161
Oxidases
   amine, biological inactivation by, 201
Oxidation
   by microsomal enzymes, 200
   enzymic formate, deuterium isotope effects, 114
   fatty acid, 107
   gas-phase, 118
      of hydrocarbons, 231
      radical rearrangement, 118
   mechanisms, 233, 234
   of aliphatic alcohols, 141
   of hydrocarbons, 231
   wet, of organic compositions, 139
Oxidation-reduction
   mechanism, 123, 135
   potentials, donor atom function, 159
      ligand function, 159
Oxido-reductases
   stereospecificity, by diamond lattice
      sections, 114
Oxides
   polymerization, 224
Oximes
   chromatography, 215
Oxygen
   compounds, 160
   conversion, into CO_2, 112
   heterocycles, 125
   isotopic exchange, 4
   ring compounds, 223
   transfer reactions, 116
      stereoselectivity, 116
Oxytocin, 194
   deamino-, crystalline, 133
Ozone
   effect on plants, 39

Ozonolysis
  of fatty acids, 107

Paints, 54
Palladium
  solutions of hydrogen in, 150
Pancreas
  langerhans, 198
Pancreatic islet tissue
    functional characterization, 121
    metabolic pathways, 121
Papaverine
  chromatography, 214
Paper chromatography, 61, 154, 225
  of dipeptides, 192
  of fungal metabolites, 137
  of mammalian metabolites, 137
  of phenolic steroids, 146
  of polar lipids, 189
  peak broadening, 145
  silicic acid impregnation, 189
Paper electrophoresis, 154
Papermaking
  synthetic fibers, 221
Para-cresol
  separation, from meta-cresol, 182
Paraffins
  nitration, 189
Paramagnetic resonance, 229
  in biological materials, 190
  (international conference), 190
  of hydrazyl-type radicals, 183
  of magnetic ions, 190
  of nucleic acid pseudo-crystals, 190
  phosphorescent molecules, 123
Paramagnetism, 6, 77
Parathyroid gland
  calciphyloxis, 122
Particles
  fractionation, 132
  separation, 132
  size, 23
Partition coefficients
  of steroid hormones, 197
Partition methods
  liquid-phase, 154
Patents, 152
Paterno-Buchi reaction, 116
  carbonyl compounds, 116
Pathology
  biochemical, 157
Peak-broadening
  in paper chromatography, 145
Pedler lecture
  amino-acid sequence, in active centers, 101
Pelletierine, 45
Penicillins, 9, 40, 78, 80, 152
Pentacyanocobaltate
  hydrogenation with, 150
Pentazoles, 13

Peptide(s), 37, 57, 204
  aminoacyl incorporation into, 114
  and amino-acids, 146
  bond formation, and enzymatic acylation, 201
  chemical properties, 193
  chromatography, 45
  cleavage, at cysteine residue, 193
    at serine residue, 193
  conformations, 193
  cyclic desi-, synthesis, 193
  cyclo-, 108
  cysteine, 191
  cystine, 191
  derivatives, of hydrazino acetic acids, 192
  dielectric increments, 193
  (fifth European symposium), 191-193
  fractionation, 188
  gel filtration, 188
  glyco-, 193
  hormones, biological activity and structure, 108
    fractionation, 108
  hydroxyacylcyclo-, tautomerism, 193
  hydroxyacyl incorporation into, 114
  intramolecular rearrangements, 114
  metal complexes, 108
  nomenclature, 193
  non-enzymatic cleavage, 193
  of amanita phalloides, 114
  physical properties, 193
  physiologically active, 202
  poly-, (see Polypeptides)
  proline, synthesis, 192
  protected, reaction with sodium in liquid ammonia, 191
  protection methods, 191
  structure, 23
  tri-, synthetic cyclol, 193
Peptide synthesis, 28, 111, 191, 204
  abnormal, 192
  alkoxylvinyl esters, 191
  amino blocking groups, 16
  α-amino group protection, 191
  carboxyl group protection, 191
  N-carboxy-N-trityl-α-amino acid anhydrides, 192
  coupling methods, 191
  gas chromatography of racemization, 192
  influence of benzyloxycarbonyl groups, 192
  insertion reactions, 192
  racemizations, with dicyclohexylcarbodi-imide, 192
  "two spots" method, for racemization detection, 192
  uncommon amino-acids, 192
  with active esters, 191
  with S-(β-aminoethyl-)-L-cystein, 192
  with azides, side reactions, 191
  with "unnatural" amino-acids, 192
Peptide-type antibiotics, 108
Peptolides, 66
  synthesis, 193
Peracids, 60

Perchloric acid, 128
Perchlorylation, 175
Peresters
  synthesis, metal ion catalyzed, 136
Perfluoralkyl groups, 11
Perfumes, 54
  synthesis, 127, 131
Perhydroindene systems
  stereochemistry, 129
Periodic law
  and electronic structure, 232
Perkow reaction, 115
Permanent columns
  for liquid-chromatography, 145
Permiability
  of high polymers, 178
Peroxidase, 231
Peroxides
  metal-ion catalyzed reactions, 136
  reaction, with carbohydrates, 3
Peroxy derivatives
  of transition metals, 14
Personnel
  selections, 126
Perspectives in Biology
  (collection of papers), 194
Perturbations
  infrared spectral, 110
  in matrix experiments, 110
Pesticide residue
  analysis, 143
Pesticides, 23, 62
Petroleum
  chemicals, 210
  molecular structure, 185
  solvents, physical chemistry, 230
pH
  measurement, 1
  variation of half-wave potential, 111
Pharmaceutical
  chemistry, plenary lectures, 108
  preparations, 131
Pharmaceuticals (see also Drugs)
  analysis, by tropaeolin method, 146
Pharmacokinetics, 40
Pharmacological
  activity, and structure, 201
  analysis, of central nervous action, 202
  control, of ACTH secretion, 198
    of antidiabetic drug release, 198
    of hormone release, 198
    of lactation, 198
    of prolactin secretion, 198
  effects, at cellular level, 200
    at subcellular level, 200
  importance, of pyrazolidine-3,5-diones, 129
  Meeting, Proceedings of the First International, 198
Pharmacology
  electronic aspects, 167
  of corticoids, 146
  of rauwolfia alkaloids, 102
  of the lung, 202

Phase change
  of frozen liquids, 183
Phellonic acid, 34
Phenalene, tri-1,3,5-triazine
  1,3,4,6,7,9b-heptaza-, sym-heptazine, 131
Phenethyl biguanide
  degradation products, 198
Phenolic
  alkaloids, biogenesis, 101
  plant products, biosynthesis, 155
  plastics, 95
  steroids, paper chromatography, 146
Phenol(s)
  coupling, 99
  -dienone rearrangements, 127
  oxidation, 30
  rearrangement, 127
Phenotiazines, 67, 79, 152
Phenoxazones, 104
Phenylalanine, 8
Phosgene, 29, 60
Phosphanes
  poly-, 29
Phosphate(s), 52
  carbamyl, 132
Phosphatides
  separation, by thin-layer chromatography, 146
Phosphatidic acids, 44
Phosphene alkenes, 115
  synthetic utility, 115
Phosphenes
  tertiary, optically active, 115
Phosphines, 14, 55, 102
  ligands, fluorine containing, 149
  metallic-, p-substituted, 115
Phosphite esters
  condensation of carbonyl compounds with, 115
Phospholipids
  physical chemistry, 203
  plant, separation, 189
  synthesis, 107
Phosphonitrilic compounds, 118
  derivatives, 118
Phosphonium salts, 33
Phosphorescence, 18, 123
  and paramagnetic resonance, 123
Phosphorous compounds
  preparation from elemental phosphorus, 115
  thermochemical properties, 117
Phosphorus, 35, 51
  complexes, 32
  compounds, 186
    with PN bond, 44
  diphosphorus, 71
  displacement reactions, 27
  flame ionization detection, 24
  fluorine derivatives, 149
  halides, 71
  lipid derivatives, 15
  organic derivatives, 14
  spectra, 63
S-Phosphorylated thiols
  biochemistry, 136

chemistry, 136
synthesis, 136
Phosphorylation, 16, 33
 in biological systems, 136
 in photosynthesis, 136
 rationale, 101
Phosphorylides
 stereospecific carbonyl olefination, 115
Photoalexins, 39
Photobiology, 6
Photochemistry, 10, 17, 74, 116, 133, 135, 169
 activation, in biochemistry, 208
 and spectroscopy, 230
 cell protection, 6
 chemiluminescence, 115
 electronic transitions, 115
 isomerization, in polyenes, 115
 mechanisms, 17
 of aromatic hydrocarbons, 17
 of atoms, 169
 of carbohydrates, 3
 of carbonyl compounds, 116
 of cycloheptanes, 34
 of diazirines, 116
 of dicarbonyl compounds, 116
 of dienes, 116
 of , -epoxyketones, 116
 of halogenated compounds, 17
 of hydrogen-oxygen system, 17
 of hypohalites, 18
 of iron porphyrins, 211
 of ketones, 116
 of manganese porphyrins, 211
 of metal carbonyls, 33
 of nitrites, 18
 of nitrogen compounds, 17
 of non-conjugated ketones, 116
 of nucleic acids, 227
 of proteins, 227
 of sulfur compounds, 17
 of unsaturated nitro compounds, 116
 of uranyl compounds, 230
 rearrangements, 17
 solid state, 116
 (symposium), 115
Photochromism, 17
Photodissociation, 18
Photoelectricity, 91
Photography, 60
Photoionization, 18
Photolysis
 flash, 126
Photometric titrations, 1, 79
 organic functional group analysis, 123
Photometry
 flame, 225
 selenium analysis, 139
Photooxidation, 18
Photophosphorylation
 in chloroplasts, 124
Photosensitization, 17
 mercury, 106

Photosynthesis, 10, 39, 41, 75, 98, 124, 203, 210
 biochemical dimensions, 123
 energy conversion reactions, 124
 functional lipids, of higher plants, 124
 of algae, 124
 of higher plants, 124
 phosphorylation in, 136
 photophosphorylation, in chloroplasts, 124
 respiration during, 124
 selective inhibitors, 124
 two light reactions, 123
Phototransferase, 36
Phototropy, 32
Phthalic anhydride, 53
Phthalocyanines, 73, 228
 copper, electron spin resonance spectra, 190
 semiconductivity, 117
Physical methods
 in Heterocyclic Chemistry, (collection of reviews), 195
 in organic chemistry, 154, 232
  (collective volume), 196
 of separation, 221
Physical organic chemistry
 progress, 105
Physical properties
 and biological activity, 202
Physical science
 statistics, 225
Phytochromes, 10
Pickling
 inhibitors, 138
Pigmentation
 role of pteridines in, 207
Pigments, 125
 natural, 128
Pimricin
 structure, 138
Pinene, 54
 α-, 72
Piperazine, 35
Piperidine, 35
Pituitary
 gonadotropin, effect of steroids on, 122
  release, 198
 hormone, characterization, 121
  hypothalmic factors releasing, 122
Plants
 chemistry, 39
 constituents, 230
 electron spin resonance spectra, 6
 glycolipids, separation, 189
 higher, constituents, 230
 phenolic products, biosynthesis, 155
 phospholipids, separation, 189
 phytochromes, 10
 products, biogenesis, 114
 viruses, genetic coding in, 125
 virus nucleic acids, 103
Plasma
 calcium control, 122
 rat, corticosteroids, 199
Plasmalogens, 107

Plasticizing
　of polymers, 130
Plastics
　amino-, 234
　cellulose, 235
Plastids, 57
Plenary lectures
　coordination chemistry, 108
　pharmaceutical chemistry, 108
　photochemistry, 115
　thermochemistry, 115
　thermodynamics, 112
Polarimetry, 73
Polarizabilities
　of electronically excited molecules, 168
Polarization, 76
　of 1,4-diazine absorption spectrum, 169
Polar lipids
　chromatography, 189
Polarography, 1, 44, 58, 69, 126, 129
　analysis, 235
　behavior, of coordination compounds, 102
　differential, of elemental sulfur, 162
　methods, 231
　of pteridines, 207
　oscillographic, 131
Poles
　positive, in aromatic nitration, 189
Polyacetals, 59
Polyacetylenes
　and related compounds in nature, 104
Polyallomers, 87
Polyamides
　synthetic hetero-chain, 227
Polyamino-acids
　electron spin resonance spectra, 119
　radiation effects, 119
Polyanions, 232
　and polycations, 232
　iso-hetero, 108
Polyatomic molecules
　energy calculations, 187
　intensity theory, 224
　structure, inexcited states, 168
Polyatomic radicals
　structure, in excited states, 168
Polyazeotropy, 233
Polyazine
　electric properties, 91
Polybutene, 94, 95
Polycarbonates, 212
Polycarboxylic acid derivatives
　acylation with, 175
Polycations, 232
Polycondensation, 90
Polycyclic hydrocarbons, 222
Polydicyaboacetylene, 91
Polydispersity
　of high polymers, 229
Polyenes, 160
　free-electron theory, 172
　macrolide antibiotics, 138
　photochemical isomerization, 115

Polyesters
　thermal analysis, 94
　urethanes, 88
　unsaturated, 222
Polyethylene, 81, 83
　degradation, 213
　dielectric behavior, 183
　grafting, 90
　thermal history, 94
Polyfluorinated compounds
　linear bifunctional, as potential monomers, 128
Polymerization, 76, 174
　anionic, 45, 85
　　kinetics of propagation, 106
　by cationic hydride shifts, 85
　by insertion, 57
　by isotopes, 48
　by metals, 45
　cationic, 88
　　of vinyl monomers, 129
　conversional, 84
　in mesomorphic phase, 91
　of aldehydes, 224
　of cyclopentene, 33
　of oxides, 224
　of vinyl acetate, 99
　of vinyl chloride, 59
　of vinyl monomers, 28
　phase transitions in, 88
　radical, 127
　solid phase, 92
　stereospecific, 50, 85, 86, 89, 92, 98
　vinyl, 92
　vinylic, 89
Polymers, 38, 54, 59, 66, 70, 81-95, 224, 228
　acrolein, 32
　analysis, 145
　and dichroism, 181
　at high temperatures, 212
　bio-, optical properties, 165
　birefringence, 92
　boron containing, 131, 213
　cationic reactions, 160
　chelate, 129
　configurational statistics, 177
　coordination, 62
　cross-linking, by radiation, 133
　crystalline olefin, 230
　crystallinity, 89
　crystallization, 94
　crystals, 81
　double chain, 81
　electron exchange, 92
　ethylene, 27
　exiton contribution, to optical rotation, 120
　fiber-forming, 28
　films, for liquid separations, 182
　fluorescence, 70
　from vinyl isocyanate, 212
　glass transition in, 70, 94, 95
　grafted, 86
　high, (see High polymers)

infrared spectroscopy, 235
in pharmaceutical preparations, 131
interdiffusion in, 130
living, and applications, 137
macromolecules, 125, 134
mechanics, 52
mechanochemistry, 221
molecular weights, 229
morphology, 83, 230
nuclear magnetic resonance spectra, 70
olefin, 230
  crystalline, 178
optical activity, 51
optical rotation, 120
oxidative thermal degradation, 129
particle size, 93
plasticizing, 130
polydispersity, 229
polyurethanes, 231
popcorn, 90
rheology, 231
self-diffusion in, 130
stabilization, 129
stereoregular, 51
syndiotactic, 86
thermal analysis, 26, 93
thermal degradation, 227
thermal stability, (international symposium), 212-213
thermodynamics, 86
thermostable, 57
vibrational spectra, 228
viscosity, 70
Polymethacrylates, 212
Polymethylacrylate, 212
Polymorphism, 88
Polynucleotides
  column chromatography, 103
  helical, 120
  hypochromism in, 120
  spectral properties, 120
  symmetry properties, 120
Polyolefins
  birefringence, 93
  electrical properties, 178
  stabilization, 213
  thermal degradation, 145
Polypeptides
  hormones, biological activity, 108
    chemical structure, 108
  inactivation, by enzymes, 201
  structure, 202
    aromatic side chain effects, 166
  vaso-dilating, and bradykinin, 202
Polyphenols
  analysis, 145
Polyphenyls
  π-electron conjugation, 187
Polyphosphoric acid, 45
Polypropylene, 94
  oxidation, 213

Polysaccharides, 36, 65, 82
  fractionation, by gel filtration, 188
  pneumococcal, 3
  solutions, 3
  yeast, 111
  wood, 28
Polystyrenes, 81
  thermal decomposition, 212
Polytetrafluoroethylene, 212
Polythene
  discovery, 210
Polyurethanes, 231
  elastomers, 94
Polyvinylanthraquinone, 92
Polyynes, 128
Porphyrins
  biosynthesis, 41
  iron, 211
    Mossbauer resonance, 165
  manganese, 211
  metal, 6
Positive poles
  in aromatic nitration, 189
Potassium, 31, 32
Potential
  appearance, of organic molecules, 185
  theory, of adsorption of gases and vapors, 158
Potentiometry, 45
Potentiostatic methods, 164
  kinetics of electrode reactions, 164
Powder method
  X-ray diffraction analysis, 126
Prebiological stage
  of earth, 131
Precision analysis
  of multicomponent mixtures, 144
Pregnancy
  5-hydroxytryptamine, 199
  in rats, 198
  iproniazid, 199
Preparations, 223, 229
  inorganic, 226
Proceedings
  Chemical Society, 101
  of First International Pharmacological Meeting, 198-202
Process control
  instrumentation, 221
Progestagens
  "retro", actions of, 180
Progesterone
  biochemistry, 142
Progestinics
  antigonadotrophic activity, 199
  given by mouth, 199
Progress
  in chemistry, 135
  in inorganic chemistry, 102
  in medicinal chemistry, 102
  in nucleic acid research, 103
  in physical organic chemistry, 105
  in reaction kinetics, 106

Prolactin secretion
  pharmacological control, 198
Proline peptides
  side reactions, in synthesis, 192
  synthesis, 192
Propargyl, 35
Propellants, 59
  energetics, 231
Propenal
  ultraviolet spectrum, 169
Propionaldehyde, 57
Propylene, 84
  oxide, 54, 86
Propynal
  absorption spectrum, 169
Prostate gland
  protein synthesis, 122
  ribonucleic acid synthesis, 122
Protected peptides
  reaction with sodium in liquid ammonia, 191
Protection
  in peptide synthesis, 191
  of α-amino groups, 191
  of basic amino-acids, with tosyl group, 191
  of carboxyl groups, 191
  of thiol groups, in cysteine, 191
Protective groups, 16
  for amino-acids, 191
Proteins, 6, 20, 36, 37, 204
  biogenesis, 155
  chromatography, 128, 215
  (collective volume), 204
  deuterated, 114
  drug binding, 202
  fractionation, 188
  gel filtration, 188
  heme, Mossbauer resonance, 165
  hemoglobin synthesis, 132
  hormones, fractionation, 108
  hydrolysates, thin-layer chromatography, 188
  interactions, by fluorescence, 165
    with small molecules, 202
  isotope effects, 114
  metabolism disorders, 157
  metal complexes, 108
  myoglobin, 132
  non-covalent bonds in, 204
  nuclear, 2
  of gastric secretion, 8
  photochemistry, 227
  radiation, 6
  radiation-induced unpaired spins, 119
  resins, 96
  separations, 188
  solid, electrical conduction, 165
  structure, 20, 23, 132, 202, 204
    by fluorescence, 165
  sulfur in, 204
Protein synthesis, 41, 96, 122, 132, 204
  De Novo, in vitro, 104
  in prostate gland, 122
  in yeast, 111

  ribonucleic acid, codewords, 134
    involvement, 132
Protolytic reactions
  rate constant, 106
Proton
  acceptors, nitro groups, 141
  acids, 173
  magnetic resonance, functional group
    analysis, 144
    of rotational isomerism, 183
  resonance spectra, of pteridines, 205
  transfer, 30
    in desoxyribonucleic acid replication, 166
Provitamin, 217
Pseudocrystals
  nucleic acid, magnetic ions in, 190
Pseudopregnancy
  in rats, 198
Psychopharmaca, 199
Psycho-pharmaceutics, 57
Psychotherapeutic agents, 29
Pteridines, 31, 205-207
  biologically active, 205
  glycosides, 205
  (international symposium), 205-207
Pterin, 205
Pure and applied chemistry, 108
  reviews, 125
Pure substances
  trace element determination, 111
Purines
  aza, 12
  nucleotides, biogenesis, 155
  synthesis, from imidazoles, 125
Purification
  zone melting, 154
Purity
  determination, 154
Pyrazolidine-3,5-diones
  pharmacological importance, 129
  synthesis, 129
Pyrazolidones, 235
  derivatives, 235
Pyrazolones, 235
  derivatives, 235
Pyridines, 4, 27, 35
  metals and, 12
  nucleotides, 10
  polymerization, 89
  reactions with metal halides, 150
Pyridinium salts, 28
Pyridoxal catalysis, 232
Pyrimidines
  and hydroxyl free radicals, 166
  aza, 12
  base function, in ribonuclease reaction, 103
  nucleotides, biogenesis, 155
Pyrolysis
  -gas chromatography, 145
  of polymers, 95
Pyrolytic decomposition
  in manganous oxalate, 112

Pyrophosphoric acid, 42
Pyrroles, 12, 48, 76

Q-e scheme
  of high polymers, 177
Q-e values
  tabulation, 178
Quadrupole resonance
  nuclear, 195
Qualitative analysis, 231
  of thiourea, 143
Quantitative analysis
  masking of reactions, 111
  of drugs, 224
  of polyphenol mixtures, 145
  promotion of reactions, 111
  weak bases, 105
Quantization
  exiton theory, 120
Quantum
  biochemistry, 5
  chemistry, and electronic spectra, 231
    and organic chemistry, 137
  mechanics, 5
  theory, 38
    of conjugated molecules, 166
    of molecular electronic structure, 229
    of reactivity, 187
  yields, in chloroplasts, 4
Quarterly Reviews, 117
Quasi-equilibrium
  theory, of mass spectra, 185
Quinazol-4-one
  3-amino-, 27
Quinine
  chromatography, 214
Quinolines, 4, 27
Quinolinium compounds
  preparation, 127
Quinones, 32, 47
  methides, 118
Quinoxaline, 12
Quinazolines, 12

Racemization
  detection, by "two spots" method, 192
  gas chromatography, in peptide synthesis, 192
  influence of benzyloxycarbonyl groups, 192
  of peptides, 28
  with dicyclohexylcarbodiimide, 192
Radiation
  chemicals, 223
  cross-linking of polymers, 133
  decomposition, 21
  effects, 37
    electron spin resonance spectra, 119
    in man, 97
    molecular mechanisms, 103
    on cellulose, 56, 82
    on nucleic acids, 103
    on polyamino-acids, 119
  gamma, in desoxyribonucleic acid, 183
  inactivation mechanisms, in enzymes, 165
  ionizing, 119, 132, 189
    effect on proteins, 6
  luminescence of compounds, 165
  measurement, 216
  of high polymers, 178
  permissible exposure, 132
  protection, by cysteamine, 117
  polymerization by, 90
  radiomimetric chemicals, 223
  research, 119
  unpaired spins, in proteins, 119
Radiation chemistry, 38
  alpha radiolysis, 130
  biological effects, 96
  competitive scavenger method, 119
  enzyme inactivation, by ionizing radiation, 119
  fundamental processes, 169
  of carbonic acid, 99
  of hydrocarbons, 117
  of nucleic acids, 203
  of nucleoproteins, 203
Radicals, 18
  allyl, wave function, 168
  aromatic, electron spin resonance spectra, 117
  bi-, trapping in solution, 190
  cyclization reactions, 123
  DCO, electronic spectra, 168
  HCO, electronic spectra, 168
  HCO_2, electronic spectrum, 168
  HCO_2, electronic structure, 168
  hydrazyl-type, paramagnetic resonance, 183
  identification, by electron spin resonance
    spectra, 19
  ions, aromatic, 117
    electron spin resonance spectra, 183
  mass spectrometry, 185
  polyatomic, structure in excited states, 168
  polymerization, 127
    secondary reactions, 127
  rearrangement, in gas-phase oxidation, 118
  tropyl, spectrum, 169
Radioactivation analysis, 222
Radioactive compounds, 188
  tritium-labelled, 128
Radioactivity
  detection, by gas chromatography, 29
  radiocarbon dating, 126
Radiobiological mechanisms, 203
Radiobiology
  cellular, 37
Radiocarbon
  data, 132
  dating, 126
Radiochemistry, 14
Radiofrequency spectroscopy
  at high pressures, 179
Radioisotopes, 53, 67, 78, 229
  ^{14}C, 72, 96, 132
  human use, 97

industrial applications, 229
instrumentation, 22
^{99m}Tc, 97
Radiolysis
  alpha, 130
  electron spin resonance spectra, 65
  of hexane, 58
  of hydrocarbons, 233
  polymer formation, 88
Radiometric titrations, 139
Radiomimetric chemicals, 223
Radiopolymerization, 88, 89
Radioresistance
  of aromatic molecules, 165
  of solid conjugated molecules, 165
Radiosensitivity
  biological complexity, 134
Raman spectroscopy, 63, 196, 222
Rapid combustion
  apparatus, for carbon analysis, 144
    for hydrogen analysis, 144
Rashig's reaction, 69
Rate
  plasma, corticosteroids, 199
  pregnancy, 198
Rauwolfia
  alkaloids, 102
  chemistry and pharmacology, 102
X-ray emission spectroscopy, 26
Reaction rates, 5, 227
  halogen atom reactions, 106
  heme compounds, 106
  methods, 1
  molecular orbital studies, 104
  of tritium exchange, in buffered solutions, 113
    in diethyl malonate-d,t, 113
    in diethylmethyl-d_3-malonate-t, 113
  protolytic reactions, 106
  solvolysis, 114
    of (methylcyclopropyl)-carbinyl derivatives, 114
Reactions, 73, 112
  acylation, 117, 230
  acyloin condensation, 49
  additions, of hydrogen halides, 31
  alcohol addition to triple bonds, 120
  alcoholation, 35
  alcoylation, 42
  aldolic, 42
  alkylation, of ketones, 123
  α-amidoalkylations, 100
  anionotropic, 74
  autoxidations, 53
  biological transfer, 135
  calculations, 5
  calorimetry, of combustions, 112
  catalysis, 221
  cationotropic, 74
  3-center, 112
  chlorination, 48
  collision theory, in liquids, 228

coupling, of acetylenic compounds, 16
  with dicyclohexylcarbodiimide, 192
cyclization, 47
dehydrogenations, with N-haloimides, 47
deoxymercuriation, 45
deuterium isotope effects, as criteria of mechanism, 114
diene synthesis, 46
displacement, 27
E_2, 57
electrochemical reduction, 50
electrophilic, 35
electrophilic additions, 52
energy conversion, of photosynthesis, 124
enzymatic, 1
equilibria, 227
Etard, mecnaism, 140
exchange, ligand complex, 123
fast, 38, 135
Feigl's, 139
free radical, 32, 127, 209, 232
  in preparations, 232
  of heteroaromatic compounds, 12
  steric effects, 123
Friedel-Crafts, 173-175
Fries, 174
Gas, 227
  kinetics, 226
Grignard, 16
group, in analysis, 144
halogenation, 45
high temperature, 135
homolytic, 74
Houben-Hoesch, 174
hydration, 100
  of acetylene, 34
hydrogenations, 4
  hydrocarbons, 4
hydrogenolysis, 47
in electrical discharges, 14
inhibition, 221
in liquids, collision theory, 228
inorganic, 112
in presence of metal oxides, 128
in quantitative analysis, 111
in shock waves, 224
iodide exchange, 138
ion-molecule, 184, 185
insertion, in peptide synthesis, 192
isotope effects, correlation, 112
isotope exchange, by electrodeposition, 138
Knoevenagel-Claisen, 59
Koch, 30
light, in photosynthesis, 123
mercaptan addition to triple bonds, 130
mercury photosensitized, 106
methylation, biological, 10
Michaelis-Arbusow, 115
multidentate ligand complexes, 123
name, 227
nuclear, 223

nuclear recoil, 14
nucleophilic, 32, 35
nucleophilic substitution, 13, 34, 49
   of aliphatic compounds, 209
   of atoms, 169
   of carbonyl groups, 11
   of cellulose, 153
   of cyclic compounds, stereoselectivity, 101
   of cyclobutanes, 129
   of diazomethane, 12
   of dyes, with high molecular weight substances, 129
   of fatty acids, 170
   of fluorine compounds, 11
   of Grignard compounds, 43
   of N-haloimides, 47
   of heterocyclic compounds, 12
   of hydrazide, 32
   of macromolecules, 125
   of methylene, 106
   of molecules, 169
   of muscle, 99
   of olefins, 128
   of organometallic compounds, 42, 43
   of organophosphorous compounds, 114
   of phenols, 127
   of ohotosynthesis, 124
   of protected peptides, with sodium in liquid ammonia, 191
   of steroids, 223
   oxidation, 53, 55, 71
     of hydrocarbons, 4
     of n-paraffins, 60
     of phenols, 30
     with N-haloimides, 47
   oxidation-reduction, 123
   oxidative condensations, 31
   oxidative coupling, 54
   oxidative phenol coupling, 99
   oxygen transfer, 116
   Perkow, 115
   permethylation, 29
   peroxide, metal ion catalyzed, 136
   peroxyester, 31
   phenol-dienone rearrangements, 127
   phosphorylation, 16, 33
   photochemical, 116
   polymerization, 56, 66, 81-95
     cationic, 70
     of ethylene, 57
   preparative inorganic, 226
   radical cyclizations, 123
   Rashig's, 69
   recoil, 33
   reversible, and biochemical mechanism, 166
   ring substitution, 149
   Scholl, 174
   secondary, radical polymerization, 127
   substitution, electrophilic, 123
     in metal complexes, 135
   thermodynamic equilibria, 154
   thiylation, of olefins, 128
   Willgerodt-Kindler, 30
   with lead tetraacetate, 141
   Wittig, 100, 115, 117
     for synthesis, 140
     stereochemistry, 140
   Ziegler polymerization, 81
Reactivity
   and selectivity, 173
   aromatic ketones, excited states, 116
   heteroaromatic, 195
   quantum theory, 187
   ratios, in copolymerization, 178
Reagents
   colorimetric, 126
   complexing, and structure behavior, 229
Rearrangements, 27
   Chapman, 100
   molecular, of terpenes, 101
   of alkyl cations, 110
   of free alkyl radicals, 110
   of ions, 185
   of ketoxime-O-sulfonates, 48
   of phenols, 127
   phenol-dienone, 127
   radical, in gas-phase oxidation, 118
Receptors
   acetylcholine, protein nature, 202
   cholinergic, 202
   drug, interactions, 202
Redox reactions, 150, 158
   kinetics, 158
   mechanism, 158
Reduction
   by microsomal enzymes, 200
   Reformatzky, 44
Regenerated celluloses, 235
Regulation
   of carbohydrate metabolism, 121
Relaxation
   dielectric, and vicosity in alcohols, 183
   electron spin, of aromatic free radicals, 183
   nuclear spin, of aromatic free radicals, 183
   times, microwave procedure, 141
Release
   control, of steroid hormones, 180
   of antidiabetic drugs, 198
   of fatty acids, 199
   of hormones, 198
   of triglycerides, by the liver, 199
Replication
   of DNA, incorporation errors, 166
     proton transfer, 166
Repression mechanisms
   in enzyme synthesis, 166
Residues
   pesticide, analysis, 143
Resins
   enzyme, 96
Resistance
   radio, of aromatic molecules, 165
     of solid conjugated molecules, 165
Resolution, 47
Resonance, 136
   electron spin, 232
   magnetic, 232

magnetic and electric, and relaxation
(collection of papers), 183
magnetic-spectroscopy, 196
Mossbauer, in heme protein, 165
in iron porphyrin, 165
nuclear quadrupole, 195
of π-electron states, 187
paramagnetic, 229
(international conference), 190
hydrzyl-type radicals, 183
proton magnetic, of rotational isomerism, 183
theory, 172
Respiration
during photosynthesis, 124
Reticulocites, 96
"Retro" progestagens
actions, 180
Reversible
electrochemical groups, macromolecular systems, 125
reactions, and biochemical mechanisms, 166
Reviews
of pure and applied chemistry, 125
Russian Chemical, 127
symposia, volumes and monographs, index, 226
Rheology, 87
of polymers, 231
Rheo-optical properties
of polymers, 93
Riboflavin, 207
Ribonucleases, 41
applications, 103
chemical nature, 103
in takadiastase, 103
properties, 103
reactions, pyrimidine base function, 103
Ribonucleic acids, 36, 41, 72, 96, 232
codewords, 134
in protein synthesis, 132, 134
macromolecular structure, 103, 232
messenger, 103
nuclear, 103
synthesis, cellular sites, 103
in prostate gland, 122
regulation, in bacteria, 104
Rifamycins, 40
constitution, 111
Rigid crystal
absorption of light, 119
Ring compounds, 223
aromatic, 5-membered, 126
5-membered, 42
strained, 27
Ring oven, 139
Rockets, 73
Rotation, 73
optical
Rotational
analysis, of SO_2 bands, 168
isomerism, in ethyl alcohol, 183
proton magnetic resonance, 183
structure, of methyl cyanide, 187
Rotenoids, 68

Royal Institute
of Chemistry, (collection of reviews), 210
Rubber, 59, 68
biogenesis, 155
ethylene-propylene copolymers, 177
Russia, 71

Safety
hazards, of new drugs, 132
in chemical laboratories, 235
maintenance, 226
measures, 235
permissible exposure to ionizing radiation, 132
techniques, 226
Salicylaldehyde
for acetone analysis, 146
Saline water, 150
Salts
hydration, 127
of carboxylic acids, 127
solvation, 127
Saturated compounds
hydrocarbons, alkylations, 174
isomerization, 174
vapor-phase nitration, 189
ketones, alkylation, 123
Saturation
dielectric, in solids containing H bonds, 183
Schiff
bases, 167
polybases, photoelectric properties, 91
Schmidt reaction
elaboration of alkaloids, 111
7-membered ring alkaloids, 111
Scholl reaction, 174
Scientific
instruments, in space exploration, 133
Sclerotiorin group
fungal metabolites, 111
Sebaceous gland
regulation, by steroidal hormones, 121
Second quantization
exiton theory by, 120
Secondary
isotope effects, on -complex formation, 113
force constant changes, 113
steric origin, 113
nitroheptane, nitro group position analysis, 189
phosphines, preparation and properties, 102
reactions, in radical polymerization, 127
Secretion
of ACTH, 198
pharmacological control, 198
of androgens, 121
of normal ovary, 122
of polycystic ovary, 122
of prolactin, pharmacological control, 198
ovarian steroid, 122
Selective inhibitors
of photosynthesis, 124

Selectivity
    and reactivity, 173
Selenazoles, 13
Selenium
    complexes, 129
    compounds, 30
    organic derivatives, 13
    photometric determination, 139
Self-diffusion
    in polymer systems, 130
Semiconductors, 72, 228
    electrochemistry, 127
    impurity effects, 165
    oxide catalysts, electrical conductivity, 158
    phthalocyanenes, 117
    polymers as, 91
    radiation effect, on adsorption properties, 129
        on catalytic properties, 129
Semimetals, 29
Semimicro analysis, 222
Semiquinones
    ^{13}C hyperfine interactions, 183
Separations, 221
    by chromatography, 188
    by dissociation extraction, 182
    chemical methods, 221
    in biochemistry, 228
    ion-exchange, 145, 231
    Less Common Means of, (symposium), 182
    liquid, with polymer films, 182
    of alkaloids, 188
    of fatty acids, 170
    of hydroxy-acids, 138
    of lipids, 188
    of macromolecules, 132
    of meta- and para-cresols, 182
    of particles, 132
    of phosphatides, 146
    of plant glycolipids, 189
    of plant phospholipids, 189
    of proteins, 188
    of radioactive compounds, 188
    of steroids, 188
    of sugars, 188
Serine residues
    cleavage, of peptide chains at, 193
Serum
    creatinine analysis, 146
    lipoproteins, 107
Sesquiterpenes
    synthesis, 118
Sex attractants
    insect, 151
Shape
    of carbon compounds, 225
    of molecules, and biological activity, 201
Shock tube, 210
    kinetics in, 133
Shock waves
    at high pressures, 179
    reactions in, 224
Side-chain effects
    aromatic, on polypeptide structure, 166

Silanes
    polymers, 89
Silica gel
    for thin-layer chromatography, 145
Silicates
    laminar, organic derivatives, 128
Silicic acid
    impregnated paper, for chromatography, 189
Silicon
    compounds, cyclosilanes, 16
        laminar silicates, 128
    cyclic compounds, inorganic, 128
    mechanism, 232
    nitrogen compounds, 14, 42
    organic derivatives, 14
    organometallic, 29
    organo-monomers, synthesis, 229
    quadripositive, acceptor, properties, 229
    stereochemistry, 232
Silicones, 69
Silver nitrate
    for thin-layer chromatography, 145
Size
    of molecules, and biological activity, 201
Small molecules
    force constants, 110
Sodium, 31, 32
    in liquid ammonia, reaction with protected peptides, 191
Solid state
    in chemistry, 101
    physics, magnetic resonance, 232
        Mossbauer effect in, 133
Solubility
    of heterocyclic compounds, 195
Solubilization
    of drug-hydrolyzing enzymes, 201
Solutions
    fast reactions, 135
    hydrogenation, 232
    of mineral acids, equilibria, 135
    of non-conjugated ketones, photochemistry, 116
    of non-reacting systems, calorimetry, 112
    of polysaccharides, 3
    oxidation-reduction reactions, 123
    structure, 76
Solvation, 57
    ionic, 5
    of acids, 127
    of salts, 127
Solvents, 46
    effects, and hydrogen bonding, 181
        in deoxyribonucleic acid, 166
        on intermolecular association, of alcohols, 183
        on Van der Waals dispersion attractions, 166
    extraction, 47
    for Friedel-Crafts reaction, 173
    isotope effects, in H_2O-D_2O mixtures, 112
    non-aqueous, 164
        coordination model, 102
    petroleum, 230
    polarity, 77
Solvolysis

mechanisms, 233, 234
   rates, deuterium substitution, 114
     of (methylcyclopropyl)-carbinyl derivatives, 114
Space exploration
   scientific instruments, 133
Special lectures
   molecular structure and spectroscopy, 110
Species differences
   in drug metabolism, 201
Specification
   of stereospecificity, by diamond lattice
     sections, 114
   of some oxido-reductases, 114
Specificity
   biological, in protein-small molecule interactions, 202
   of chemical mutagenesis, 103
Spectra, 63
   (see also Electronic, Infrared, Raman, Ultraviolet,
     Spectroscopy, etc.)
   data, 234
   structure correlation, 229
Spectrofluorometry, 67
Spectrometry
   identification, of compounds, 232
Spectrophotometry, 4, 53
   role of chelating agents, 123
Spectropolarimetry, 56
Spectroscopy, 234
   and molecular structure, (IUPAC conference), 182
   and molecular structure, 226
   and photochemistry, 230
   and structure, 226
   applications, in far infrared, 110
     in near infrared, 110
   atomic emission, 40
   diffuse reflectance, 28
   electronic, 4, 210
   forbidden transitions, 5
   interference, 110
   of liquids, at high pressures, 179
   of solids, at high pressures, 179
   of uranyl compounds, 230
   special lectures, 110
   vacuum, 143
Spermatogenesis
   inhibition, 122
   recovery, 122
Spherulites, 83
Spinal cord
   depressants, 95
Spin relaxation
   electron, of aromatic free radicals, 183
   nuclear, of aromatic free radicals, 183
Squalane, 212
Stabilization
   of 5-coordinate complexes, 167
   of high polymers, 178
   of polymers, 129
Standards
   food, 224
   of measurements, 126
Starch gel
   electrophoresis, of pituitary hormone, 121

Statistics
   configuration, of polymeric chains, 177
   in physical science, 225
   of high polymers, 178
Staudinger
   olefin formation, 115
   organic phosphor carbonyls, 115
Steacie
   memorial lecture, 101
Stereochemistry, 27, 78, 173
   and mechanism, 232
   and silicon, 232
   anhydroryanodine, 111
   circular dichroism in, 101
   of carbon compounds, 209
   of 5-coordinate complexes, 167
   of glycosides, 199
   of hydrocarbon ions, 168
   of multidentate chelating agents, 159
   of perhydroindene systems, 129
   optical rotatory dispersion in, 101
Stereoselectivity
   of photosensitized oxygen transfer reactions, 116
   reactions of cyclic compounds, 101
Stereospecific reactions, 33
   carbonyl olefination, with phosphorylids, 115
   diamond lattice sections, 114
   of ketones, photochemistry, 116
   of oxido reductases, 114
Steroid hormones, 121
   absorption spectra, 197
   action, 181
     on estradiol-17, 181
   chromatographic mobilities, 197
   effects on fertility, 122
     on pituitary gonadotropin, 122
   ether derivatives, 180
   fluorescence spectra, 197
   human, 121
     gas chromatography, 121
   in human sebaceous gland, regulation, 121
   interaction, with coenzymes, 180
     with enzymes, 180
   metabolism, thyroid interaction with, 180
   Physical Properties, (series of articles), 197
   release control, 180
   synthesis, 181
     control, 180
   ultraviolet absorption, 197
Steroids, 34, 52, 181, 222, 231
   action, 180
   adrenocortical, 121
     biosynthesis, 181
   aldosterone, 121
   alkaloids, mass spectrometry, 141
   analysis, 156, 188, 229
   androgenic, 152
   bases, chromatography, 214
   biologically active, 221
   biosynthesis, 155, 230
   chromatography, 214, 228
   conformation, 46
   dynamics, 180

effects, on fertility, 122
   on pituitary gonadotropin, 122
esterification, 199
etherification, 199
far ultraviolet spectra, 64
gas chromatography, 229
identification, 156, 188
inverted, biological activities, 180
   preparations, 180
17-keto-, 46
metabolism, 180
non-hormonal activities, 181
19-nor-, 32, 130
normal ovary, secretion in, 122
nuclear magnetic resonance spectra, 58, 221
ovarian, secretion in vivo, 122
   synthesis, 122
phenolic, paper chromatography, 146
polycystic ovary, secretion in, 122
reactions, 223
responsive tissues, carbonic anhydrase, 121
separation, 156, 188
synthesis, 45, 155
thin-layer chromatography, 188
with hormone-like activities, 180
Sterols, 66
  acetates, separation, 145
Stibines, 14
Stilboestrol
  chromatography, 215
Stomach
  and endocrine system, 142
Strain energies
  in cyclic compounds, 140
Stress-cracking polymers, 178
Structure
  analysis, at high pressures, 179
    automatic, 144
  by X-rays, 179
  and activity, 60, 78, 80
    in peptides, 202
    of antibiotics, 108
  and function, in corticotropins, 108
  and pharmacological activity, 201
  and reactivity, 105
  anhydroryanodine, 111
  atomic, and bonding, 231
  behavior, of complexing reagents, 229
  biologically active microbial metabolites, 110
  bridged, of C_2H_6, 168
  deoxyribonucleic acid, by X-ray scattering, 103
  determination, by mass spectrometry, 16
  donor-acceptor complexes, 108
  electronic, and bonding, 231
    and periodic law, 232
    of first row fluorides, 169
    of first row hydrides, 169
    of HCO_2 radical, 168
    quantum theory, 229
  in enzyme action, 133
  isoprenoid, 127
  macromolecular, of ribonucleic acids, 103, 232

  mitochondrin basis, 227
  molecular, and spectroscopy, 226,
    (IUPAC conference), 182
    of petroleum, 185
  of aromatic amines, 200
  of carbanions, 110
  of complex compounds, 226
  of cyclopropanes, 141
  of electronically excited species, 168
  of glycosides, modification, 200
  of Grignard reagents, 135
  of line spectra, 229
  of liquid crystals, 128
  of 5-membered rings, 140
  of 6-membered rings, 140
  of methyl, 187
  of methylene, 187
  of natural products, 222
  of nucleic acids, enzymatic alteration, 134
  of polyatomic molecules, in excited states, 168
  of polyatomic radicals, in excited states, 168
  of polypeptide hormones, 108
  of polypeptides, 202
    aromatic side-chain effects, 166
  of polyphosphinates, 213
  of polyphosphine oxides, 213
  of proteins, 132, 202, 204
  of tetrodotoxin, 114
  of yeast polysaccharides, 111
  quantum theory, 229
  rotational, of methyl cyanide, 187
  spectra correlation, 229
  theory, modern, 223
  valence theory, 135
Strychnine
  synthesis, 140
Styrenes, 50, 88, 90, 160
  copolymerization, 177
  polymerization, 84, 85
    on wood, 82
Styrolene, 87
Subcellular level
  pharmacological effects, 200
Sub-microanalysis
  of bromine, 144
Substituent effects
  by electronic spectroscopy, 210
  electrophilic, in gas-phase, 140
  in aromatic nitration, 189
  in benzoic acid, 140
  in o-substituted nitrobenzenes, 118
  of positive poles, 189
Substitution reactions
  electrophilic, aliphatic, 123
    and molecular orbital calculations, 187
    aromatic, 105
  in metal complexes, 135
  nucleophilic, aromatic, 105
    at saturated carbon, 222
Successive labelling
  isotope effect studies, 113
  reaction mechanism, 113

Succinimide
  N-phenyl-, 35
Sugars
  aminodeoxy-, 131
  analysis, 188
  gas chromatography, 24
  identification, 188
  nucleotides, 3
  reactions, 11
  separation, 188
  structure, 55, 222
Sulfa drugs, 152
Sulfanilamide
  acetylation, 201
Sulfates
  conjugation, 200
Sulfides
  aryl methyl, dipole moments, 140
Sulfolipids, 15
Sulfonamides, 54
  metabolism, 40
Sulfonated cation exchangers
  kinetics of ion exchange on, 130
Sulfonates
  alkyl arene, 127
Sulfonation, 175
Sulfonylation, 175
Sulfonyl chlorides
  aromatic, 127
Sulfonylurea compounds, 152, 198, 199
Sulfur
  allotropes, 49
  differential polarography, 162
  free radical reactions, 32
  in proteins, 204
  isotopes, bacteriogenic fractionation, 113
  neighboring group effect in, 30
  nitrogen compounds, 29
  radicals, thermochemistry, 140
  reactions, 74
  valence shell expansion, 140
Sulfur compounds, 23, 110, 160
  alkyl arenesulfonates, 127
  and proteins, 119
  aromatic sulfonyl chlorides, 127
  aryl arenesulfonates, 127
  chromatography, 162
  ethylene sulfides, 161
  fatty acid derivatives, 170
  for selenium analysis, 139
  from petroleum distillates, 162
  in gas turbine engines, 162
  mercaptans, 59, 130
    oxidation, 71
  molecular spectra, 110
  photochemistry, 17
  quantum chemistry, 138
  sulfites, 48
  sulfur-sulfur bond, 99
    nucleophilic displacement, 135
  synthetic, analysis, 162
  thermochemistry, 140
  thiolactones, 131
  thio-sugars, 3
  thioxo compounds, 69
  thiylation, of olefins, 128
Sulfur dioxide
  bands, rotational analysis, 168
  liquid, dissociation equilibria, 105
    ionization equilibria, 105
Sulfurous acid, 78
Superconducting magnetic fields
  nuclear magnetic resonance spectroscopy
    in, 134
Surface
  agents, 60
  chemistry, 38
  coatings, 125
Svensk Kemisk Tidskrift
  review journal, 136-138
Swelling
  of macromolecular materials, 136
Sydnones, 45, 49
Sym-heptazine
  chemical properties, 131
  1,3,4,6,7,9b-heptazaphenalene, tri-1,3,5-
    triazine, 131
  molecular structure, 131
Symposia
  exiton model, 119
  isotope mass effects, 112
  natural products, 114
  organic photochemistry, 115
  organophosphorous compounds, 115
  reviews, volumes and monographs, index, 226
Synthesis (see also Reactions)
  by ligand complexes, 123
  catalytic, of carboxylic acid esters, 131
    of carboxylic acids, 131
  compounds labelled with radioactive tritium, 128
  control, of steroid hormones, 180
  cross-linking, of polymers, 133
  effects of drugs on, 199
  enzymatic, of nucleic acids, 203
  enzyme, induction mechanisms, 166
    repression mechanisms, 166
  Fischer-Tropsch, catalysis, 158
  from p-substitued metallophosphines, 115
  hot, 128
  Houben-Hoesch, 174
  in the prostate gland, 122
  isocyanides in, 136
  Kolbe electrolytic, 136
  macromolecular, in cells exposed to
    D_2O, 114
  microbial, of amino acids, 137
  of acids, 175
  of aldehydes, 175
  of alicyclic compounds, 141
  of aliphatic compounds, from thiophene, 162
  of aromatic ketones, 174
  of corrins, 111
  of cyclic desipeptides, 193
  of cyclobutanes, 129

of depsopeptide analogs, of glutathione, 193
   of ophthalmic acid, 193
of dienes, 131, 228
of α-esters, of N-substituted glutamic acid, 191
of fatty acids, 170
of heme, 228
of hemoglobin, 132
of homocyclic systems, 111
of hormonal steroids, 180
of macrocyclic compounds, 131
of macrocyclic ligand complexes, 123
of N-methylated amino-acids, 192
of monocyclic aromatic hydrocarbons, 129
of optically active tertiary phosphines, 115
of organophosphorous compounds, from p-substituted metallophosphines, 115
of organosilicon monomers, 229
of ovarian steroids, 122
of peptides (see Peptide synthesis)
of peptolides, 193
of peresters, metal ion catalyzed, 136
of perfumery materials, 131
of perfumes, with isoprenoid structure, 127
of phospholipids, 107
of polyamides, 227
of polycyclic natural products, 114
of polyphosphinates, 213
of polyphosphine oxides, 213
of proline peptides, 192
of proteins (see Protein synthesis)
of purines, 125
   from imidazoles, 125
of pyrazolidine-3,5-diones, 129
of ribonucleic acids, 122
   cellular sites, 103
of sesquiterpenes, 118
of steroid hormones, 181
of styrchnine, 140
of triglycerides, by the liver, 199
of α,β-unsaturated aldehydes, 130
of α,β-unsaturated ketones, 130
of vitamin B_6 group, 142
structure activity relation, of antibiotics, 108
with phosphene alkenes, 115
with thiophene, 162
Synthetic
   compounds, analysis, 145
   cyclol tripeptide, 193
   fibers, analysis, 143
   hetero-chain polyamides, 227
   methods, 233

Tabadiastase
   ribonucleases in, 103
Tables
   composition, 233
   Craig distribution, 137
   experimental dipole moments, 227
   extraction, 137
   gas chromatography, 226
   identification, 224
   indole alkaloids, 225
   melting point, 234
   of Q-e values, 178
Talanta
   International Journal of Analytical Chemistry, 139
Tannins
   biosynthesis, 155
   vegetable, 104
Tantalum halides
   reaction with pyridine, 150
Tarichatoxin-tetrodotoxin
   potent neurotoxin, 133
Tautomerism, 48, 61
   of heterocyclic compounds, 12
   of hydroxyacylcyclopeptides, 193
Techmetium, 97
Techniques
   laboratory, in organic chemistry, 235
   ω-, LCAO calculations, 141
   modern safety, 226
Teeth, 142
Teichoic acid, 65
Terminology, 224
Terpenes
   biogenesis, 155
   biosynthesis, 230
   chemistry, 137
   mass spectra, 185
   molecular rearrangements, 101
Terpenoids, 67, 222
   structure, 222
Terpolymer, 88
Tertiary amines
   analysis, 139
   for acid extraction, 137
   for water extraction, 137
Tertiary phosphenes
   optically active, properties, 115
      synthesis, 115
Tertiary phosphines
   preparation and properties, 102
Testicular hormones
   protein synthesis, 122
   ribonucleic acid synthesis, 122
Tetracyclic triterpenes, 228
Tetracyclines, 40, 68, 79
Tetrahedral coordination
   in copper (II) chelates, 167
   in nickel (II) chelates, 167
Tetrahedron
   International Journal of Reviews, 140-141
Tetrahydropterins, 206
Tetrahydropyrans, 34
Tetramethylurea, 31
Tetrapyrrole
   biosynthesis, 227
Tetrapyrrole pigments, 159
   coordination chemistry, 159
   physical chemistry, 159
Tetrodotoxin
   structure, 114
Theoretical
   calculations, of isotope effects, 112
   interpretations, Hammett relationship, 105

of structure-reactivity relationships, 105
viewpoints, 230
Theory
  biopolymers, optical properties, 165
  bond, of organometallic compounds, 126
  chemical valence, 135
  collision, of reactions in liquids, 228
  exiton, by second quantization, 120
  free-electron, of conjugated molecules, 229
  infrared, 233
  localized electronic excitations, 119
  molecular, of memory, 133
  molecular orbital, in strained cyclic hydrocarbons, 141
  multiplet, of catalysis, 158
  of adsorption, of gases, 158
  of biochemical systems, 166
  of catalytic hydrogen waves, in polarography, 129
  of complex compounds, 226
  of copolymerization, 177
  of electronic orbitals, in molecules, 128
  of electronic spectra, 228
  of gas chromatography, 156
  of hypochromism, 120
  organophosphorous compounds, bonding, 115
  potential, of adsorption, 158
  quantum, of conjugated molecules, 166
    of reactivity, 187
    of structure, 229
  quasi-equilibrium, of mass spectra, 185
  structural, 223
  transitions, between electronic levels, 115
Thermal
  analysis, of polymers, 26, 93
  diffusion, 154
  methods, of analysis, 234
Thermal degradation
  of polymers, 227
    oxidative mechanism, 129
  of polyolefins, 145
Thermochemistry, 6, 39, 56
  of polymers, 94
  of sulfur compounds, 140
  of sulfur radicals, 140
  plenary lectures, 112
  research, 137
  stepwise complex formation, 112
Thermochromism, 47
Thermodynamics, 112, 173
  activation, 74
  at high temperatures, 131
  data, compilation and review, 138
  equilibria, 154
  free energy, 47
  nonequilibrium, 74
  of irreversible porcesses, 234
  of phase and ordering transitions, 112
  of solutions, 38, 71
  past and future, 112
  plenary lectures, 112
  research, 137
Thermogravimetry, 23, 95

Thermostable fibers, 32
Thiamine, 130
  compounds, 142
Thiatriazoles
  1,2,3,4-, 13
Thietane, 161
Thin-layer chromatography, 1, 33, 50, 61, 64, 125, 128, 222, 230
  coupling, to gas chromatography, 145
  (international symposium), 214
  of amino-acids, 188
  of bile acids, 188
  of polyphenols, 145
  of protein hydrolysates, 188
  of steroids, 188
  of sterol acetates, 145
  of triterpenoids, 188
  on $AgNO_3$-SiO_2 gel, 145
  preparative, of phosphatides, 146
  stationary phases, 145
  with hydroxyapatite, 188
Thin-layer electrophoresis, 214
Thioamine, 130
Thiocarbamides
  analytical chemistry, 143
Thiocarbonyl compounds
  charge distributions, 141
Thiocyanates
  enzymatic conversion, from cyanides, 201
Thioketones, 31
Thiolactones, 131
Thiol groups
  in enzymes, 137
  protection, in cysteine, 191
Thiols
  S-phosphorylated, biochemistry, 136
    chemistry, 136
    synthesis, 136
Thiophenes, 12, 79
  for synthesis, of aliphatic compounds, 162
Thiourea
  qualitative analysis, 143
Thiylation
  of olefins, 128
Thyroid
  gland, in the Fischer rat, 121
  hormone, biosynthesis, 198
    secretion, 198
  interaction, with steroid hormone metabolism, 180
Tilden lecture
  biosynthesis, of alkaloids, 101
  stereoselectivity, in reaction of cyclic compounds, 101
Tin, 59
  hydrides, organic reactions, 16
  organic derivatives, 27
  quadripositive, acceptor properties, 117
Tissue
  culture cells, lipid composition, 114
  distribution, and duration of drug reaction, 201
  electric, of electric eel, 202
Titanium, 30
  in polymerization, 84, 85

Titrations
  acid-base, nomogram for, 139
  EDTA, 223
  radiometric, 139
Titrimetric analysis
  of carbon, 144
  of hydrogen, 144
  organic, 221
Tobacco mosaic virus, 20
Toluene
  competitive deuteration, 113
  -α,α,α-d_3, 113
Topological matrix
  for free-electron theory, 172
Torsion microbalance, 216
Tosyl group
  for basic amino-acid protection, 191
Toxicity, 37, 73
  of aromatic amines, 200
Toxicology
  of boron compounds, 157
  with cyclopeptides of amanita phalloides, 108
Toxins, 30
  tetrodo-, structure, 114
Trace
  analysis, of organic materials, 143
  elements, 111
    determination, in pure substances, 111
  impurities, analysis by gas chromatography, 128
Tracers, 21, 173
Transfer reactions
  biological, 135
Transition metal complexes
  circular dichroism, 102
  magnetic properties, 102
  optical rotatory dispersion in, 123
Transitions
  between electronic levels, 115
  forbidden, 5
Transplantable tumors
  of thyroid, 121
Transplutonium elements, 69
Transport
  ion, and cardiac glycosides, 200
  of drugs, 9
  of electrons, through conjugated systems, 165
  of fatty acids, in blood, 199
  processes, 5
Trapping
  of biradicals, in solution, 190
Trehaloses, 3
Triazine
  1,3,4,6,7,9b-heptazaphenalene, tri-1,3,5-, sym-heptazine, 131
  s-, 69
  2,4,6-trichloro-1,3,5-, cyanuryl chloride, 130
Triazines, 28, 30
Triglycerides
  accumulation, 199
  release, 199
  synthesis, 199
N,N'-bis-(, , -Trinitroethyl)-urea, 138

Trioxane, 89
Tripeptide
  synthetic cyclol, 193
Triple
  bonds, in small rings, 110
  ion equilibria, in acetic acid, 164
Triplet states
  in deoxyribonucleic acid, 165
Triterpenes
  from betuline, nuclear magnetic resonance spectra, 141
  tetracyclic, 228
Triterpenoids
  ion-exchange chromatography, 188
  thin-layer chromatography, 188
Tritiation methods, 21
Tritium, 19, 22, 33
  abnormal isotope effect, 113
  Arrhenius parameter effects, 113
  exchange rate, in diethylmethyl-d_3-malonate, 113
    in diethyl malonate-d, t, 113
  radioactive, 128
N, Trityl-α-amino acid anhydrides
  N=carboxy-, in peptide synthesis, 192
Tropaeolin
  analysis, of pharmaceutical compounds, 146
Tropane alkaloids
  gas chromatography, 146
Tropyl radical
  spectrum, 169
Trypic hydrolysis
  metals in, 166
Trypsin, 76
Tryptamines
  separation, by gas chromatography, 25
Tryptophan, 8
Tubercle bacillus
  lipids, 32
Tuberculosis
  chemiotherapy, 67
Tumors
  thyroid, in Fischer rat, 121
  tranplantable, of thyroid, 121
Tunnelling
  in base-promoted elimination reactions, 113
Turnover rates, 22
"Two spots" method
  for racemization detection, 192
Tyrosine, 8

Ubiquinones, 10, 66, 107
Ultra high vacuum
  instrumentation, 133
  torsion microbalance, 216
Ultramicroanalysis
  of organic materials, 143
Ultraviolet spectra, 6, 196
  of alkylbenzenes, 141
  of chlorine dioxide, 168
  of natural products, 231

of propenal, 169
of steroid hormones, 197
Units
    of measurement, 126
Unpaired spins, 119
Unsaturated bonds
    in polyethylene, infrared analysis, 145
Unsaturated compounds, 121
    acetylenic, 130
    acyclic, 209
    addition, of H atoms to, 113
    aldehydes, α, β, 130
    boron-containing, 131
    chloro-substituted, 129, 130
    disymmetric α, β- ketones, circular dichroism, 168
    fatty acids, 133
    hydrocarbons, 113, 127, 129
    ketones, α, β, 123, 130
    nitro compounds, 116, 229
    polyesters, 222
    polymers, 128
Unstable molecules
    and free radicals, 133
Uranyl
    complexes, with β-diketones, 211
    compounds, photochemistry, 230
        spectroscopy, 230
    salts, 211
Ureas
    alkoxy-, 30
    complexes, of fatty acids, 170
    sulfonyl-, 198
Urinary
    aromatic acids, gas chromatography, 156
    excretion, of drugs, 201
Urine
    analysis, of creatinine, 146
Uterine growth
    estradiol-17β, 181

Vacuum
    Microbalance Techniques, (international symposium), 216
    spectroscopy, development, 143
    ultrahigh, 4
        instrumentation, 133
    ultraviolet, photochemistry, 18
Valence, 23, 72
    isomerization, 27
    theory, 135
Van der Waals
    crystals, Winnier exitons, 119
    dispersion attractions, in DNA, 166
Vapor-phase
    halogenation, of benzene derivatives, 110
    nitration, of saturated hydrocarbons, 189
Vapors
    potential theory, of adsorption, 158

Variations
    in ratio $^{48}Ca/(total\ Ca)$, 113
Vegetable
    oils, glyceride structure, 107
    tannins, 104
Venoms, 66
Vertebrates
    adaptation, to marine environments, 122
    adrenocortical factors, 122
Vibrational spectra, 173
    of carbon subgroup compounds, 129
    of high polymers, 228
Vibronic states
    of dimers, 120
Vinca alkaloids, 7
Vinyl
    acetate, 88
        polymerization, 99
    chloride, 59
        copolymerization, 177
    ethers, 160
    monomers, cationic polymerization, 129
    polymerization, 81, 84
Vinylidene chloride
    copolymerization, 177
Vinylsulfonyl groups
    in dyes, 27
Virial theorem, 171
Viruses
    bacterial, genetic coding, 125
    chemotherapy, 7
    nucleic acids, 6
    plant, genetic coding, 125
    tobacco mosaic, 20
Viscose fibers, 99
Viscosity
    and dielectric relaxation, in alcohols, 183
Viscous flow
    and dipole orientation, 183
Visible spectroscopy, 196
Vitamin A, 16, 142
    biosynthesis, 155
Vitamin B_6, 142
    analysis, in foods, 142
    and hormones, 142
    determination, in biological materials, 142
    group, labelling, 142
        synthesis, 142
Vitamin B_{12}, 10, 31, 36, 48
    coenzyme, 111
    radioactive, analytical applications, 139
Vitamin B, 142
Vitamine
    (collective volume of reviews), 217-218
Vitamin K, 66
Vitamins, 36, 234
    and Hormones, (annual collection of reviews), 142
    biogenesis, 217
    biosynthesis, 224
    water soluble, biosynthesis, 155
Voltage-current characteristics
    of benzene, 164

Wannier exitons
  in Van der Waals crystals, 119
Water
  deuterium isotope effects, 112
  hydrated organic crystals, 125
  two-phase liquid systems, 112
Wave
  functions, of allyl anion, 168
    of allyl cation, 168
    of allyl radical, 168
    of excited states, 168
  mechanics, of carbon compounds, 209
Wet oxidation
  of organic compositions, 139
Wettability
  and adhesion, 151
Willgerodt-Kindler reaction, 30
Wilzbach reaction, 21
Wittig reaction, 100, 115, 117
  for synthesis, 140
  stereochemistry, 140
Wood, 52, 82, 83, 163, 233
  as chemical raw material, 163
  cellulose, 163
  chemical reactions, 163
  composition, 163
  developing, chemistry, 163
  extraneous components, 163
  hemicelluloses, 163
  lignins, 163
  pulp, manufactore, 163
  structure, 163
  supply, 163
  uses, 163
  water relationship, 163

X-ray
  action, on biological systems, 96
  analysis, of natural products, 101, 115
  crystallography, 56, 66, 74
    of liquids, 76
  contrast media, 95
  diffraction, of heterocyclic compounds, 195
    powder method, 126
  diffractometer, 144
  effect, on tissue culture cells, 114
  emission spectroscopy, 79
  fluorescence, 64
  measurements, 94
  microanalysis, 143
  scattering, 203
    DNA structure by, 103
  spectra, 64
  spectroscopy, 42, 63
  structural analysis, 144
  studies, 224
Xylan, 82
Xylene
  oxidation products, analysis, 176

Yeast
  polysaccharides, 111
  protein synthesis, 111
Ylids
  phosphorous, 115
Yohimbe, 35

Zeitschrift
  für Analytische Chemie, 143-146
  für Vitamin-, Hormon-, und Ferment-
    forschung, 146
Zinc, 44
Zone melting, 49, 56, 58, 154

APPENDIX

ADDRESSES OF PUBLISHERS

Academic Press, Inc.
111 Fifth Avenue
New York, N.Y. 10003

Addison-Wesley Publishing Co.
Reading, Massachusetts

Akademie-Verlag G.M.B.H.
Leipziger. Str.
Berlin, W.1 - Germany

Aldine Publishing Company
London University Press
London - England

Allyn and Bacon Inc.
150 Tremont Street
Boston, Massachusetts, 0211

American Association for the
  Advancement of Science
1515 Massachusetts Ave., N.W.
Washington, 5, D.C.

American Book Company
55 Fifth Avenue
New York, N.Y. 10003

American Chemical Society
1155 16th Street, N.W.
Washington, D.C. 20036

American Elsevier Publishing
  Co., Inc.
52 Vanderbilt Avenue
New York, N.Y. 10017

American Petroleum Institute
1155 16th Street, N.W.
Washington, D.C. 20036

Annual Reviews, Inc.
231 Grant Avenue
Palo Alto, California

Appleton-Century-Crofts, Inc.
(Div. Meredith Publishing Co.)
440 Park Avenue, So.
New York, N.Y. 10016

E.J. Arnold & Son, Ltd.
Butterley Street
Leeds 10 - England

Edward Arnold (Publishers) Ltd.
41 Maddox Street
London, W.1 - England

Associated Technical Services
P.O. Box 271
East Orange, New Jersey

The Athlone Press of the
  University of London
2 Gower Street
London, W.C.1 - England

Bailliere, Tindall & Cox, Ltd.
7 & 8 Henrietta Street
London, W.C.2 - England

Barnes & Noble, Inc.
105 Fifth Avenue
New York, N.Y. 10003

J.A. Barth Verlag
Salomonstr. 18B
Leipzig C.1 - Germany

Basic Book Publishers
404 Park Avenue, So.
New York, N.Y. 10016

W.A. Benjamin, Inc.
1 Park Avenue
New York, N.Y. 10016

Birkauser Verlag A.G.
Basel 10 - Switzerland

Blackie & Son Ltd.
16-18 William IV Street
London, W.C.2 - England

Blaisdell Publishing Co.
(Div. of Ginn & Company)
135 West 50th Street
New York, N.Y. 10020

Burgess Publishing Company
426 South 6th Street
Minneapolis, Minn. 55415

Butterworth, Inc.
7235 Wisconsin Avenue
Washington, D.C. 20014

Butterworths Scientific Publications
88 Kingsway
London, W.C.2 - England

Cambridge University Press
Bentley House
200 Euston Road
London, N.W.1 -England

Cambridge University Press
32 East 57th Street
New York, N.Y. 10022

Carnegie Press
Carnegie Institute of
  Technology
Pittsburgh 13, Penna.

The Catholic University of
  America Press
620 Michigan Avenue, N.E.
Washington, D.C. 20017

Chapman & Hall Ltd.
37-39 Essex Street Strand
London, W.C.2 - England

Chemical Education Material
  Study
University of California
Berkeley, California

Chemical Education Publishing
  Company
20th and Northampton Streets
Easton, Pa. 18043

Chemical Publishing Company, Inc.
212 Fifth Avenue
New York, N.Y. 10010

Chemical Society (London)
Burlington House Piccadilly
London, W.1 - England

Chemie-Verlag, G.M.B.H.
Weinheim / Bergstr.
Pappelallee 3 - Germany

J & A Churchill Ltd.
104 Gloucester Place
London, W.1 - England

Clarendon Press
Oxford - England
(see Oxford University Press)

Columbia University Press
2960 Broadway
New York, N.Y. 10027

Consultants Bureau Enterprises, Inc.
227 West 17th Street
New York, N.Y. 10011

Cornell University Press
124 Roberts Place
Ithaca, New York

Thomas Y. Crowell Company
201 Park Avenue, So.
New York, N.Y. 10003

F.A. Davis Company
1914-16 Cherry Street
Philadelphia, Pa. 19103

Marcel Dekker, Inc.
95 Madison Avenue
New York, N.Y. 10016

Directories Publishing Company
510 Madison Avenue
New York, N.Y. 10022

Dover Publications Inc.
180 Varick Street
New York, N.Y. 10014

The Dow Chemical Company
Midland, Michigan

Dunod
92 Rue Bonaparte
Paris 6e - France

ADDRESSES OF PUBLISHERS (CONTINUED)

E.P. Dutton & Co., Inc.
201 Park Avenue, So.
New York, N.Y. 10003

Eclectic Publishers
30 West Washington Street
Chicago, Illinois, 60002

J.W. Edwards (Publishers) Inc.
2500 South State Street
Ann Arbor, Michigan

Elsevier Publishing Company N.V.
Spuistraat 110-2
Amsterdam - The Netherlands

Emerson Books, Inc.
251 West 19th Street
New York, N.Y. 10011

F. Enke Verlag
Hasenbergsteige 3
Stuttgart - Germany

Fachbuch Verlag
Karl-Heine Strasse 16
Leipzig W. 31 - Germany

VEB Gustav Fischer
Villengang 2
Jena - Germany

W.H. Freeman & Company
660 Market Street
San Francisco, Calif. 94104

Gauthier - Villars
55 Quai des Grandes-Augustins
Paris 6^e - France

Ginn & Company
Statler Building
Boston, Massachusetts, 02117

Gordon and Breach
Science Publishers
150 Fifth Avenue
New York, N.Y. 10011

Grune & Stratton, Inc.
381 Park Avenue, So.
New York, N.Y. 10016

23 Bedford Square
London, W.C. 1 - England

Walter de Gruyter & Company
Genthiner Str. 13
Berlin, W. 30 - Germany

Hadrian Press, Inc.
667 Madison Avenue
New York, N.Y. 10021

C. Hanser Verlag
Kolbergerstr. 22
Munchen - Germany

Harcourt, Brace & World, Inc.
750 Third Avenue
New York, N.Y. 10017

Harper & Row
General Division
49 East 33d Street
New York, N.Y. 10016

(For the United Kingdom)
35 Great Russell Street
London, W.C. 1 - England

George G. Harrap & Company Ltd.
1 & 2 High Holborn
London, W.C. 1 - England

Harvard University Press
79 Garden Street
Cambridge 38, Massachusetts

D.C. Heath & Company
285 Columbus, Mass. 02116

W. Heffer & Sons ltd.
Cambridge - England

Heywood & Company Ltd.
Druary House
Russell Street
London, W.C. 2 - England

Hilgert & Watts Ltd.
98 St. Pancras Way
Camden Road
London, N.W. 1 - England

S. Hirzel Verlag in Verw
Schuhmachergasschen 1/3
Leipzig C.1 - Germany

Holden-Day, Inc.
728 Montgomery Street
San Francisco, Calif. 94111

Holt, Rinehart & Winston, Inc.
383 Madison Avenue
New York, N.Y. 10017

Houghton Mifflin Company
110 Tremont Street
Boston, Mass. 02107

Hutchinson Scientific & Technical
  Publications
178-202 Great Portland Street
London, W.1 - England

Iliffe & Sons Ltd.
Dorset House Stamford
London, S.E. 1 - England

Imperial Chemical Industries Ltd.
Imperial Chemical House
Millbank
London, S.W. 1 - England

Information for Industry, Inc.
1000 Connecticutt Ave., N.W.
Washington, D.C. 20006

Information for Scientific
  Information
325 Chestnut Street
Philadelphia, Pa. 19106

Interscience Publishers, Inc.
605 Third Avenue
New York, N.Y. 10016

Iowa State University Press
Press Building
Ames, Iowa

Johnson Reprint Corporation
111 Fifth Avenue
New York, N.Y. 10003

James Kanegis
3907 Madison Street
Hyattsville, Maryland

S. Karger A.G.
Arnold Boecklinstrasse 25
Basel - Switzerland

Lea & Febiger
Washington Square
Philadelphia, Pa. 19106

Paul Lechevalier
87 Bld. Raspail
Paris 6^e - France

H.K. Lewis & Company Ltd.
136 Gower Street
London, W.C. 1 - England

Little, Brown & Company
34 Beacon Street
Boston, Mass. 02006

Livingstone Ltd. (E. & S.)
15, 16 & 17 Teviot Place
Edinburgh - Scotland

Longmans, Green & Company Ltd.
6 & 7 Clifford Street Piccadilly
London, W. 1 - England

Lyons & Carnahan
407 East 25th Street
Chicago, Illinois, 60616

Alex. MacLaren & Sons
268 Argyle Street
Glasgow, C. 2 - Scotland

The Macmillan Company
60 Fifth Avenue
New York, N.Y. 10011

Massachusetts College of

ADDRESSES OF PUBLISHERS (CONTINUED)

Massachusetts College of
   Pharmacy
179 Longwood Avenue
Boston, Mass. 02015

Massachusetts Institute of
   Technology Press
Cambridge 42, Mass.

Masson et Cie
120 Bld. St. Germain
Paris 6e France

Olin Mathieson Chemical
   Corporation
Mathieson Building
Baltimore, Maryland

Merck & Company
126 East Lincoln Avenue
Rahway, New Jersey

Charles E. Merrill Books, Inc.
1300 Alum Creek Drive
Columbus 16, Ohio

Methuen & Company Ltd.
11 New Fetter Lane
London, E.C. 4 - England

Michigan State University Press
Box 752
East Lansing, Michigan

The C.V. Mosby Company
3207 Washington Blvd.
St. Louis, Mo. 63103

Frederick Muller Ltd.
Ludgate House
Fleet Street
London, E.C. 4 - England

Munksgaard Ltd.
Norregade 6
Copenhagen - Denmark

McDowell, Obolensky Inc.
219 East 61st Street
New York, N.Y. 10036

McGraw-Hill Book Company
330 West 42nd Street
New York, N.Y. 10036

National Bureau of Standards
Washington, 25, D.C.

The New American Library of
   World Literature
501 Madison Avenue
New York, N.Y. 10022

George Newnes Ltd.
Tower House
Southampton Street
London, W.C. 2 - England

New York Academy of Sciences
2 East 63rd Street
New York, N.Y. 10021

Noyes Press, Inc.
P.O. Drawer 900
16-18 Railroad Avenue
Pearl River, N.Y. 10965

R. Oldenbourg Verlag
Rosenheimerstr. 145
Munchen 8 - Germany

Oliver & Boyd Ltd.
Tweeddale Court
14 High Street
Edinburgh - Scotland

Oxford University Press
Amen House
London, E.C. 4 - England

417 Fifth Avenue
New York, N.Y. 10022

Paul Parey
Lindstr. 44 / 47
Berlin, S.W. 68 - Germany

Pennsylvania State University
   Press
University Park
Pennsylvania

Pergamon Press, Inc.
122 East 55th Street
New York, N.Y. 10022

Headington Hill Hall
Oxford - England

Pharmaceutical Society of
   Great Britain
17 Bloomsbury Square
London, W.C. 1 - England

Plenum Press, Inc.
227 West 17th Street
New York, N.Y. 10011

Prentice-Hall Inc.
Englewood Cliffs,
New Jersey, 07632
   and
70 Fifth Avenue
New York, N.Y. 10011

Princeton University Press
Princeton, New Jersey

Printing & Publishing Office
National Academy of Science
National Research Council
2101 Constitution Ave., N.W.
Washington, D.C. 20418

Purdue University Press
Lafayette, Indiana

G.P. Putnam's Sons
200 Madison Avenue
New York, N.Y. 10016

Reinhold Publishing Corporation
430 Park Avenue
New York, N.Y. 10022

Editions de la Revue Scientifique
4 Rue Pomereu
Paris 16 - France

The Ronald Press Company
15 East 26th Street
New York, N.Y. 10010

Royal Australian Chemical
   Institute
314 Albert Street
East Melbourne C 2.
Victoria - Australia

Royal Institute of Chemistry
30 Russell Square
London, W.C. 1 - England

Rutgers University Press
30 College Avenue
New Brunswick, New Jersey

W.B. Saunders Company
W. Washington Square
Philadelphia, Pa. 19105

Benno Schwabe & Company
Adolfsallee 49-53
Wiesbaden - Germany

The Sigma Press
2419 M. Street, N.W.
Washington, D.C. 20037

Smith Agricultural Chemical
   Company
618 North Champion
Columbus, Ohio

G.F. Smith Company
867 McKinley
Columbus, Ohio

Societe de Chimie Physique
Ecole de Phisique et de Chimie
10 Rue Vauquelin
Paris V - France

Societe de Productions
   Documentaires
28 Rue St. Dominique
Paris 7e - France

ADDRESSES OF PUBLISHERS (CONTINUED)

Society of Chemical Industry
14 Belgrave Square
London, S.W. 1 - England

E. & F.N. Spon Ltd.
23 Great Smith Street
London, S.W. 1 - England

Springer Verlag
Molkerbastei 5
Wien 1 - Austria

Springer Verlag
Berlin-Gottingen-Heidelberg
Heidelberger Platz 3
Berlin-Wilmersdorf - Germany

Dr. Dietrich Steinkopff
Holzhofallee 35
Darmstadt - Germany

St. Martin's Press Inc.
175 Fifth Avenue
New York, N.Y. 10010

Sugar Research Foundation Inc.
52 Wall Street
New York, N.Y.

Technical Dictionaries Company
Box 144
New York, N.Y. 10031

Verlag Technik, VEB
Oranienburgerstr. 13-14
Berlin C 2 - Germany

Temple Press Books
42 Russell Square
London, W.C. 1 - England

Tennessee Eastman Company
Eastman Road
Kingsport, Tenn.

Texas - U.S. Chemical Company
260 Madison Avenue
New York, N.Y. 10016

G. Thieme Verlag
Herdweg, 63
Stuttgart-Nord - Germany

V.E.B. Georg Thieme
Hainstr. 17-19
Leipzig C.1 - Germany

Charles C. Thomas Publisher
301-327 E. Lawrence Avenue
Springfield, Illinois

United States Department of
    Commerce
Washington, 25, D.C.

U.S. Government Printing Office
Division of Public Documents
Supt. of Documents
Washington, D.C. 20402

University of Chicago Press
5750 Ellis Avenue
Chicago, Ill. 60637

University of London Press Ltd.
Little Paul's House
Warwick Square
London, E.C. 4 - England

University of Michigan Press
Ann Arbor, Michigan

University of North Carolina
    Press
Chapel Hill, North Carolina

University of Notre Dame Press
Notre Dame, Indiana

University of Oklahoma Press
Norman, Oklahoma

University of Pennsylvania
    Press
3436 Walnut Street
Philadelphia 4, Penna.

University of Wisconsin Press
430 Sterling Court
Madison 6, Wisconsin

Urban & Schwarzenberg G.M.B.H.
Frankgasse 4
Wien IX - Austria

D. Van Nostrand Company
120 Alexander Street
Princeton, N.Y. 08540

24 West 40th Street
New York, N.Y. 10018

D. Van Nostrand Company
(Canada) Ltd.
25 Hollinger Road
Toronto 16, Ontario

Varian Associates
611 Hansen Way
Palo Alto, California
(Analytical Instrument Div.)

Verlag Chemie
Postfach 129 / 149
694 Weinheim/Bergstr.
Germany

Verlag Harri Deutsch
6 Frankfurt A.M. - W13
Graefstrasse 47, - Germany

F. Vieweg & Sohn Verlag
Braunschweig - Germany

Wadsworth Publishing Company
Belmont, California

Wayne State University Press
5047 Second Boulevard
Detroit 2, Michigan

John Wiley & Sons, Inc.
605 Third Avenue
New York, N.Y. 10016

The Williams & Wilkins Company
428 E. Preston Street
Baltimore 2, Maryland

Z
5524
O 8
I 5
1963-64

APR 2 1968